Fundamental Genetics

Fundamental Genetics is a concise, nontraditional textbook that explains major topics of modern genetics in 42 mini-chapters. It is designed as a textbook for an introductory general genetics course. *Fundamental Genetics* is also a useful reference or refresher on basic genetics for professionals and students in health sciences and biological sciences. It is organized for ease of learning, beginning with molecular structures and progressing through molecular processes to population genetics and evolution. Students will find the short, focused chapters approachable and more easily digested than the long, more complex chapters of traditional genetics textbooks. Each chapter concentrates on one topic, so that teachers and students can readily tailor the book to their needs by choosing a subset of chapters. The book is extensively illustrated throughout with clear and uncluttered diagrams that are simple enough to be re-drawn by students. This unique textbook provides a compact alternative for introductory genetics courses.

John Ringo is a Professor of Biology at the University of Maine where his research focuses on *Drosophila* genetics, behavior, and evolution. He has carried out laboratory and field studies in the United States and Israel. A member of the Genetics Society of America and the Society for the Study of Evolution, he has published works in various scholarly journals.

Fundamental Genetics

John Ringo
University of Maine

CAMBRIDGE
UNIVERSITY PRESS

PUBLISHED BY THE PRESS SYNDICATE OF THE UNIVERSITY OF CAMBRIDGE
The Pitt Building, Trumpington Street, Cambridge, United Kingdom

CAMBRIDGE UNIVERSITY PRESS
The Edinburgh Building, Cambridge CB2 2RU, UK
40 West 20th Street, New York, NY 10011-4211, USA
477 Williamstown Road, Port Melbourne, VIC 3207, Australia
Ruiz de Alarcón 13, 28014 Madrid, Spain
Dock House, The Waterfront, Cape Town 8001, South Africa

http://www.cambridge.org

First published 2004

Printed in the United States of America

Typeface Swift 10/14 pt. and Gill Sans   *System* LaTeX 2_ε [TB]

A catalog record for this book is available from the British Library.

Library of Congress Cataloging in Publication Data
Ringo, John, 1943-
Fundamental genetics / John Ringo.
      p.   cm.
     Includes bibliographical references and index.
     ISBN 0-521-80934-7 (hb) – ISBN 0-521-00633-3 (pbk.)
     1. Genetics. I. Title.
  QH430.R55   2003
  576.5–dc21   2003048463
ISBN 0 521 80934 7 hardback
ISBN 0 521 00633 3 paperback

This book is dedicated to the memory of my early teachers:

My parents Mary and Gene

My grandmothers Grace and Margaret

My second grade teacher, Myrtle McCullough, who saved me

Contents

Preface

I wrote this book both for the student learning genetics for the first time and for the biologist or health professional looking for background information. *Fundamental Genetics* is a brief account of the basic facts, theories, and experimental approaches of genetics. The book is unconventional. The organizing principle is to progress from structure to function, from simple to complex, and from molecular events to epiphenomena, such as how genes are inherited from parent to child. The history of genetics is scrupulously avoided, mainly to save space but also to avoid the trap of genuflecting to famous geneticists rather than discussing their experiments in the detail necessary for a full understanding – such a discussion, properly done, would fill several large volumes. This is a short book, and each chapter is focused.

A friend of mine says, "It's not how long life is that matters, but how thick." The chapters of *Fundamental Genetics* are short but thick, so if you are a student using the book as a text, it is probably best to read one and only one chapter at a sitting – any less, and you will have trouble getting the whole picture of the chapter; any more, and you will have mental indigestion. If you are using the book as a reference, the glossary may often be a helpful starting point. The teacher who chooses *Fundamental Genetics* as a textbook can skip chapters that are not appropriate for the course; it will be easy to use the chapters in a different order. The chapters are modules.

I have tried hard to be accurate and correct in writing this book, but it will contain mistakes. If you find a mistake, dear reader, or if you disagree with something in the book, or if some point seems opaque to you, contact me. I guarantee that I will answer any legitimate email inquiry.

Acknowledgments

I could not have written the book without the enthusiastic, un-flagging support and encouragement of my wife, Ada Zohar, who kept me on track through the writing. Also, her many valuable comments on the early stages of the manuscript made for substantive improvements, especially in the introductions to each chapter.

Many colleagues helped me by explaining things and giving me source material; these include Danny Segal, Dusty Dowse, Bill Glanz, Irv Kornfield, Becky Van Beneden, Rondi Butler, Keith Hutchison, Malcolm Shick, Sara Orgad, Sarit Cohen, and Knud Nierhaus. Larry Beauregard, head of the Genetics Department, Affiliated Laboratories, Bangor, Maine, generously supplied micrographs of human chromosomes. Andrew Trevor, Tamar Schlick, Juergen Suehnel of IMB Jena Image Library, and the Protein Data Bank graciously consented to my using their images. I thank Alexei Khodjakov, and Nikon Corporation, for allowing me to use his beautiful micrograph for the cover.

I thank anonymous reviewers for their comments at an early stage in the writing. I am very grateful for the helpful, scholarly comments of Jeff Hall and Irv Kornfield, made late in the development of the book. Jeff's comments, insightful and spirited, were voluminous enough to fill a small book.

Katrina Halliday is my excellent editor. I thank her for her wisdom and understanding as much as for her help, as she shepherded the book along its way. Eleanor Umali, senior project manager at Techbooks, has been nothing short of astounding. She is a wizard of organization and efficiency, and I was lucky indeed to have had her on the team.

Chapter 1

Life Forms and Their Origins

Overview

This chapter introduces several basic genetic concepts, without going into detail about any of them. These genetic concepts are as follows:

- life form
- nucleic acid
- gene
- chromosome
- organism
- virus
- semiautonomous organelle

The origin of life and the evolution of the three domains of life are described briefly.

Life Forms Are Genetic Systems

Two essential components of every life form are proteins and nucleic acids. Nucleic acids (DNA and RNA) are thread-like coding molecules, the building material of genes and chromosomes. Genetics is about genes and chromosomes – their structure and function, their behavior and misbehavior, their evolution, and methods of studying them. Because genes are the coding molecules of life, they are complicated and varied. It is difficult to pin down the term "gene" in a simple definition, but, to a first approximation, a gene is a segment of nucleic acid whose immediate function is to encode a piece of RNA (Figure 1.1). The key concepts here are **replication** (copying) of genes and coding. The replication of genes and their coding properties are described in detail in later chapters.

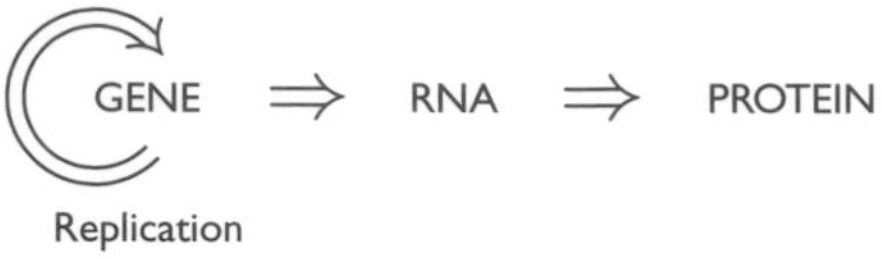

Fig 1.1 Genes replicate and code for RNA. For most genes, the final gene product is protein.

GENE ⟹ RNA ⟹ PROTEIN

Replication

From a genetic point of view, a life form is an assemblage of large molecules capable of reproducing itself and including at least one chromosome. A chromosome is a long, thin thread made of DNA or, in some cases, RNA and may also contain proteins. To qualify as a chromosome, a nucleic acid molecule must contain one or more genes, be replicated faithfully in a regulated manner, and be transmitted from a life form to its descendants in a reproductive cycle. Not every molecule of nucleic acid is a chromosome, even if it contains genes. The nucleic acid part of a life form's set of chromosomes is its **genome**.

All life forms arise from preexisting life forms via a reproductive cycle during which chromosomes are copied and the copies are passed on from parent to progeny (Figure 1.2). According to this broad, genetically based definition, life forms include organisms (cellular forms), viruses, mitochondria, and chloroplasts. This book concentrates on the genetics of organisms.

Organisms

Organisms are made of cells, membrane-bound structures capable of reproduction, growth, and metabolism. The genome of a cell encodes all the proteins required for that cell's survival. Every cell has at least one chromosome, which is made of DNA and proteins. Cells also have many ribosomes, micromachines for synthesizing proteins. A membrane surrounds every cell. In some organisms, the cell envelope includes a cell wall and one or more additional membranes. An organism can be a single cell or many cells joined together.

Organisms comprise three major divisions or domains: Bacteria, Archaea, and Eukarya (Figure 1.3). The compelling genetic evidence for this broad taxonomic division comes from DNA sequences of slowly evolving genes. Despite their genetic and biochemical differences, bacterial and archaeal cells are morphologically similar: they lack nuclei, and they reproduce asexually, by simple cell division. Little is known about the genetics of archaea.

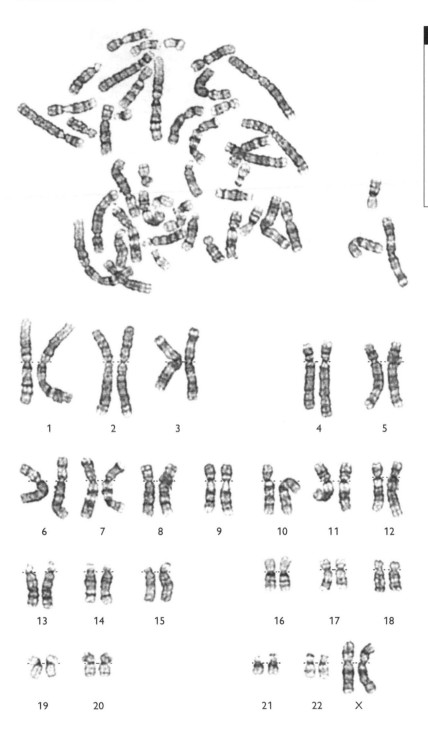

Fig 1.2 Above: Stained chromosomes of a normal female human, from a cell nearing division. Below: The same 46 chromosomes rearranged into numbered pairs, the karyotype. Photograph by Dr. Laurent Beauregard, Genetics Department, Affiliated Laboratory, Inc., Bangor, Maine.

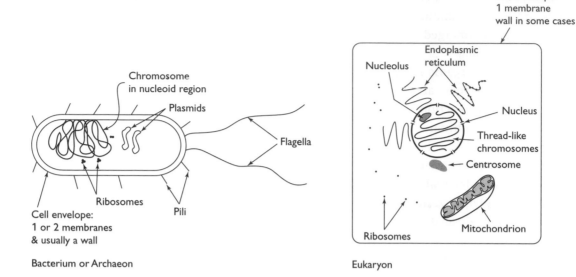

Bacterium or Archaeon

Eukaryon

Fig 1.3 A quick look at two kinds of cell.

In contrast to bacteria and archaea, eukaryal cells possess membrane-bound nuclei, an internal system of membranes, and a cytoskeleton made of microtubules.

The Origin of Life

From the time the earth began to form, 4.6×10^9 years (4600 Ma = megaanum, or million years) ago, until it cooled sufficiently for liquid water to exist on its surface, 4400 to 4200 Ma ago, the temperature was too high for life to exist. Meteorites bombarded early earth, and some geophysicists believe these ocean-vaporizing impacts likely did not abate sufficiently for life to emerge until 4200 to 4000 Ma ago (Figure 1.4). The $^{12}C:^{13}C$ ratio of organic carbon is higher than that of inorganic carbon. This isotopic ratio in fossils of the most ancient sediments known suggests that life was abundant 3900 Ma ago or a bit earlier; sedimentary apatite

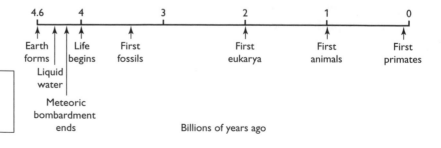

Fig 1.4 Time line of earth's history. The times given for biological "firsts" are very approximate.

(a mineral consisting of calcium phosphates) is a biomarker first appearing in large amounts 3800 Ma ago. The implications are remarkable: life emerged from non-life during a period lasting only 100 to 300 Ma.

Fundamental organic molecules required for life (e.g., amino acids and nucleotides) are thought to have originated through natural chemical reactions starting with simple molecules such as methane, ammonium, phosphate, and water, with the energy for the reactions being heat and electrical discharges in the atmosphere. Modern experiments have shown these reactions to be feasible. Also, under realistic conditions not involving enzymes, amino acids polymerize into polypeptides and nucleotides polymerize into nucleic acids.

The First Organisms: RNA-Based?

All living organisms have genes made of DNA, which code for RNA. RNA molecules are intermediate coding molecules in the synthesis of proteins, which make important structures of the cell and carry out virtually all the metabolic functions (Figure 1.5). According to one theory, the original life forms used RNA for coding and for metabolic functions. Some RNAs act as enzymes; these are ribozymes. Biochemists are finding many chemical reactions that are catalyzed by RNAs. If ancient proto-organisms possessed RNAs capable of directing the synthesis of more copies of RNA molecules, then both genes and enzymes could have been made of RNA in those ancient times, perhaps between 4000 and 3500 Ma ago. In this "RNA world" RNA served double duty: genes and enzymes.

The ancestral cells or protocells had evolved into bacteria-like cells by 3500 Ma ago; fossil cells that resemble bacteria were very abundant by then. The split between archaea and bacteria occurred between 3500 and 1900 Ma ago, and the eukarya-archaea split probably occurred between 1900 and 1500 Ma ago

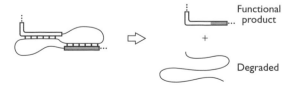

Functional
product

+

Degraded

Fig 1.5 Some RNAs can both cut themselves and ligate the pieces.

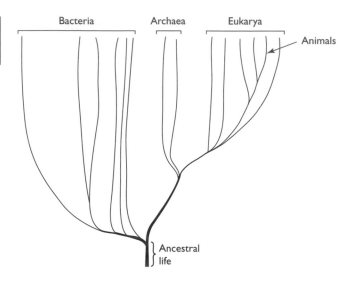

Fig 1.6 Evolutionary tree of three domains of life. Each line is a major taxonomic lineage.

(Figure 1.6). The eukarya are enormously diverse; taxonomists classify them into a stupefyingly detailed and complex hierarchy of taxa.

Mitochondria and Chloroplasts: Semiautonomous Organelles

Eukaryal cells contain several organelles or "mini-organs" inside the cell. Two important membrane-bound organelles found in many eukarya are the mitochondrion [pl., mitochondria] and the chloroplast. Their main functions are oxidative metabolism (mitochondria) and photosynthesis (chloroplasts). They evolved from purple bacteria and cyanobacteria, respectively (Figure 1.7). The ancestral bacteria became mutualistic endosymbionts in eukaryal

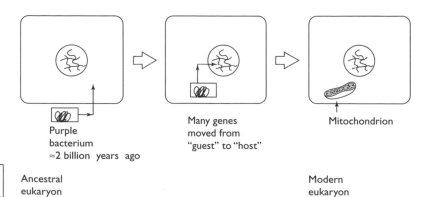

Fig 1.7 Evolution of mitochondria.

cells, meaning that both the host cell and the life form that living inside it benefited. Mutualistic endosymbiosis is not rare. However, in these cases, the now-organelles have clearly lost their status as bacterial cells, for genes in the "host" eukaryon's nucleus encode many proteins of these organelles. Mitochondria and chloroplasts are therefore genetically parasitic. On the other hand, every mitochondrion and every chloroplast has its own chromosome and its own protein-coding machinery. Furthermore, mitochondria and plastids (chloroplasts and related organelles) are unlike any other membrane-bound organelle, such as the nucleus: mitochondria and chloroplasts are never disassembled and reassembled; instead, they reproduce by division, as did their ancestral bacteria.

Viruses, the Completely Acellular Life Forms

Viruses are acellular life forms: obligate intracellular parasites possessing one or more chromosomes. During the infectious stage of a virus's life cycle, the virus is a virion the viral genome encapsulated in a structure made of protein. Sometimes the virion includes a membrane envelope stolen from the host's cytoplasmic membrane (Figure 1.8). Most viruses are genetically parasitic, relying on its host for enzymes used in genetic processes. Only viruses with relatively large genomes code for many of the proteins required for their own reproduction. Viruses infect all three domains of life and may be classified according to host, genetic material, or phylogeny. An overall phylogeny of viruses is not appropriate, for viruses appear to have evolved independently many times. Any phylogeny of viruses, except for closely related ones, is difficult to establish, owing to rapid evolution and the tendency of viruses to acquire cellular genes. The going theory is

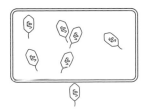

Bacterial viruses

HIV budding off host
cell membrane

Fig 1.8 Some infectious virus particles.

that virus genomes evolved from bits and pieces of their hosts' genomes.

A Few Odd Forms

Plasmids are small, nonessential, "extra" chromosomes of cells. Plasmids code for proteins, including proteins useful to the cell (e.g., genes conferring antibiotic resistance). Some plasmids move between cells and may have genes encoding the machinery for intercellular movement. Are plasmids life forms? Some say yes, because plasmids are parasite-like, but I opt for the idea that plasmids are merely small, inessential chromosomes. Another category of DNA molecule with some of the basic properties of a life form is the **transposon**. Transposons are sequences of DNA that can move about the genome, within or between chromosomes; transposons code for proteins that help them to move.

There are nucleic acid molecules that do not qualify as life forms by the definition offered here but that some biologists do consider living, or at least lifelike. These are **viroids** and **virusoids**, small circular RNA molecules that do not code for protein. Viroids are parasites of plants and cause significant economic damage. Virusoids are parasites of viruses.

The strangest of all life-oid things is the **prion**, an infectious protein that can cause the modification of similar proteins in a cell, ultimately leading to the cell's death. Prions cause certain slow, infectious, neurological diseases, including bovine spongiform encephalopathy ("mad cow disease").

Most of Genetics Is Based on a Restricted Sample of Organisms

There are over 10 million species in the three domains of life. Much of what is known about the genetics of cellular organisms has been learned from intensive study of a limited sample of species clustered in a few branches of the evolutionary tree of organisms, most prominently two bacteria (*Escherichia coli* and *Bacillus subtilis*), a few fungi (notably the mold *Neurospora crassa* and the bread yeast *Saccharomyces cervisiae*), two flowering plants (*Zea mays* and *Arabidopsis thaliana*), and four animals (the fruit fly *Drosophila melanogaster*, the nematode *Caenorhabditis elegans*, the mouse *Mus musculus*, and *Homo sapiens*). Important genetic phenomena have been substantially investigated in hundreds of other species of bacteria, fungi, plants, animals, and ciliates, as well as in many viruses – only viral species, though, that infect bacteria or multicellular/multinucleate eukarya. The most conspicuous and troubling gaps in knowledge of basic genetics occur in the archaea and in the early-branching taxa of eukarya – troubling, because we have no idea how large those gaps may be. Fortunately, straightforward analysis of DNA sequences is beginning to allow us to infer a great deal about the genetics of these organisms.

Further Reading

Gesteland RF, Cech T, Atkins JF. 1999. *The RNA World*, 2nd ed., Cold Spring Harbor Laboratory Press, Cold Spring Harbor, NY.

Gray MW. 1999. Evolution of organellar genomes. *Curr. Opin. Genet. Dev.* 9:678–687.

Holland HD. 1997. Evidence for life on earth more than 3850 million years ago. *Science* 275:38–39.

Lazcano A, Miller SL. 1996. The origin and early evolution of life. *Cell* 85:793–798.

Levine A. 1991. *Viruses*. WH Freeman, New York.

Woese CR, Kandler O, Wheelis ML. 1990. Towards a natural system of organisms: Proposal for the domains Archaea, Bacteria, and Eucarya. *Proc. Natl. Acad. Sci. USA* 87:4576–4579.

Chapter 2

Nucleic Acids

Overview

To understand genes, one must first consider nucleic acid, for nucleic acid is the stuff that genes are made of. Inasmuch as function follows structure, a clear picture of nucleic acid will illuminate all genetic processes.

The genetic material of all life forms is nucleic acid, either deoxyribonucleic acid (**DNA**) or ribonucleic acid (**RNA**). DNA and RNA are linear chains made of subunits called **nucleotides**. Two strands of DNA typically associate in a two-stranded helix. This chapter describes the structure of nucleotides, the way nucleotides are connected to make chains, and a bit about the shapes of RNA and DNA, including some properties of double helices.

Polymers

Nucleic acids, the coding molecules of life, are linear polymers. A linear polymer is an unbranched chain made of many subunits connected by covalent chemical bonds; short polymers are known as oligomers (Figure 2.1). The subunit of a nucleic acid polymer is the nucleotide. Energy [typically about 400 kJ/mol] is required to make or to break a covalent bond between subunits of a polymer.

In this simplified representation of an RNA segment, the circles and triangles stand for nucleotides and each line stands for a covalent bond joining adjacent nucleotides. A = adenosine, G = guanidine, U = uridine, and C = cytidine.

Nucleotides

Every nucleotide has three parts: sugar, nitrogenous base, and one to three phosphates; the subunits of nucleic acid are

monophosphates. A **nucleoside** is a nucleotide minus the phosphate (nucleoside = sugar + nitrogenous base) (Figure 2.2). Ribose is the sugar of RNA (ribonucleic acid) and 2′-deoxyribose is the sugar of DNA (deoxyribonucleic acid).

In ribose, the sugar of ribonucleotides, and 2′-deoxyribose, the sugar of deoxynucleotides, carbon atoms are numbered 1′ ("one-prime") through 5′ (Figure 2.3). Numbers without primes designate the carbon and nitrogen atoms in the nitrogenous bases.

The nitrogenous bases are derivatives of **purine** and **pyrimidine** and are called by those names, even though RNA and DNA do not contain the parent molecules purine and pyrimidine (Figure 2.4). Purine and its derivatives have a double ring structure; pyrimidine and its derivatives have a single ring structure. The most common purine derivatives are guanine and adenine; the most common pyrimidine derivatives are cytosine, uracil, and thymine. Only four nucleotides are used for the synthesis of RNA in organisms — those containing the bases adenine, guanine, cytosine, and uracil. DNA synthesis in organisms is similar, except that thymine is used instead of uracil. In RNA molecules, there are in addition about 50 less common bases, produced by enzymatic modification of bases within an RNA chain. The DNA of some viruses contains uncommon bases. In the DNA of plants and animals, roughly 5% of the cytosines have a methyl group added, at position 5.

The deoxynucleotides are sometimes abbreviated with an initial "d"; thus, deoxyguanidine is sometimes abbreviated dG. Adding a "d" for deoxy is done only when RNA and DNA sequences are given together, to avoid confusion. A single nucleotide may be a monophosphate, diphosphate, or triphosphate. Thus, AMP, ADP, and ATP are the mono-, di-, and triphosphates of adenosine (Figure 2.5).

The names of common bases and corresponding nucleotides are given in Figure 2.6.

Fig 2.1 An RNA polymer.

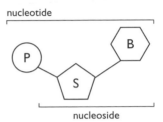

Fig 2.2 Nucleotide and nucleoside. S = sugar, B = base, P = phosphate.

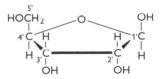

ribose

2′-deoxyribose

Fig 2.3 Ribose and 2′-deoxyribose

The Phosphodiester Bond between Nucleotides

In all known life forms, nucleotides in a polymer are joined covalently by a phosphodiester bond between the 3′ carbon of one sugar and the 5′ carbon of the next sugar. The string of sugars

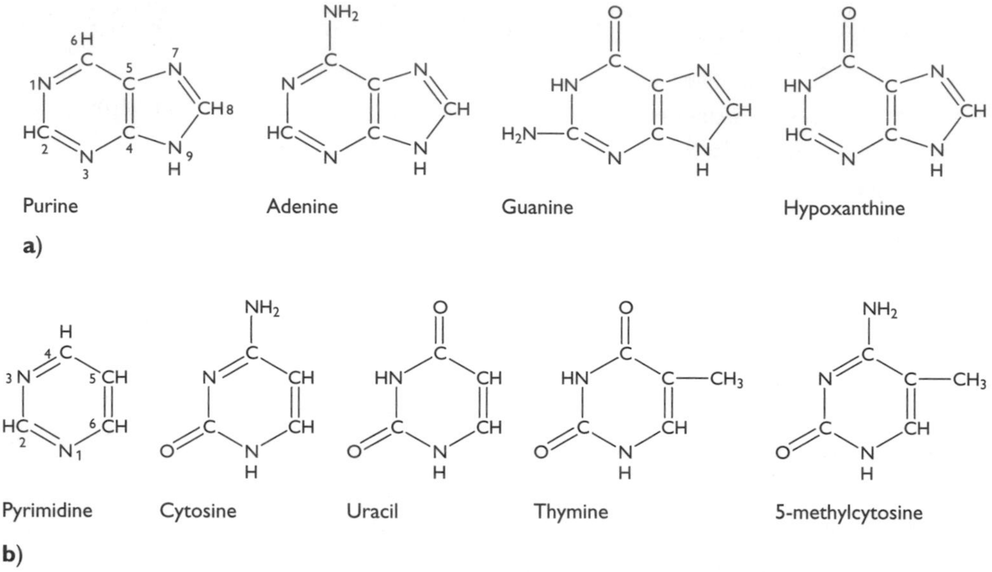

Purine   Adenine   Guanine   Hypoxanthine

a)

Pyrimidine   Cytosine   Uracil   Thymine   5-methylcytosine

b)

Fig 2.4 Examples of bases found in DNA and RNA.
(**a**) Purine and its derivatives. Adenine and guanine are the commonest purine bases. Hypoxanthine, less common, is found in many tRNAs (small RNAs that carry amino acids into the ribosome during protein synthesis).
(**b**) Pyrimidine and its derivatives. Cytosine, uracil, and thymine are the commonest pyrimidine bases.

Fig 2.5 Nucleotide structure: two examples.

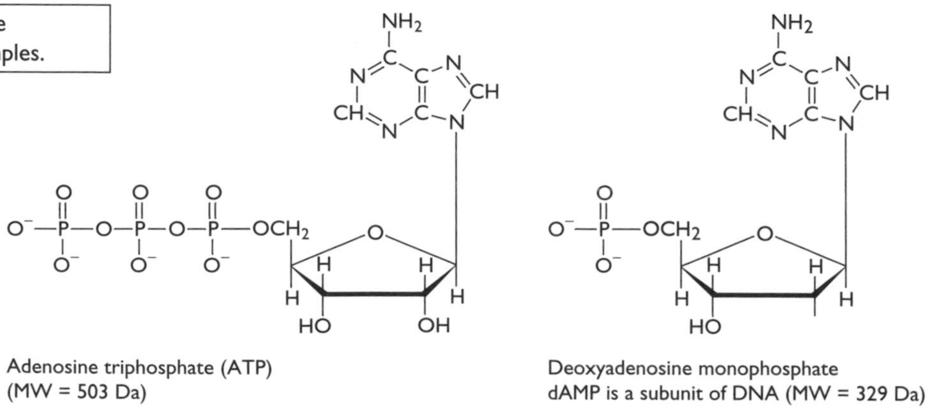

Adenosine triphosphate (ATP)
(MW = 503 Da)

Deoxyadenosine monophosphate
dAMP is a subunit of DNA (MW = 329 Da)

Purine base	Ribonucleotide	Pyrimidine base	Ribonucleotide
adenine	adenosine (A)	cytosine	cytidine (C)
guanine	guanidine (G)	thymine	thymidine (T)
hypoxanthine	inosine (I)*	uracil	uridine (U)

*An exception to the pattern of a nucleotide's name being derived from the name of its base.
unspecified nucleotide (N) unspecified purine nucleotide (Pu) unspecified pyrimidine nucleotide (Py)

Fig 2.6 Names and abbreviations of common bases and the corresponding nucleotides.

and phosphates makes the "backbone" of the polymer, and bases stick out from the sugars (Figure 2.7).

Nucleotides Pair via Hydrogen Bonds

Nucleotides tend to pair via hydrogen bonds between bases. A hydrogen bond is weak relative to a covalent bond and takes little energy to make or break [roughly 5 kJ/mol]. Hydrogen bonds form between adenine and thymine (AT), between adenine and uracil (AU), and between guanine and cytosine (GC). AT and AU pairs have two hydrogen bonds, and GC pairs have three hydrogen bonds (Figure 2.8). Other nucleotide pairings are possible between RNAs in special contexts. Pairing may occur between two strands of DNA, two strands of RNA, or strands of DNA and RNA.

Nucleic Acids Make Double Helices by Base-Pairing

The pairing of nucleic acids promotes formation of a helical duplex. A single strand of nucleic acid may bend around and pair with itself, in which case a short, single-stranded duplex is

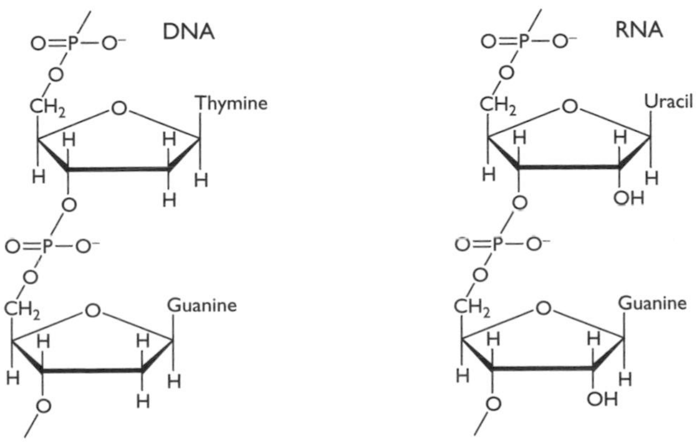

Fig 2.7 Sugar-phosphate backbone of nucleotides.

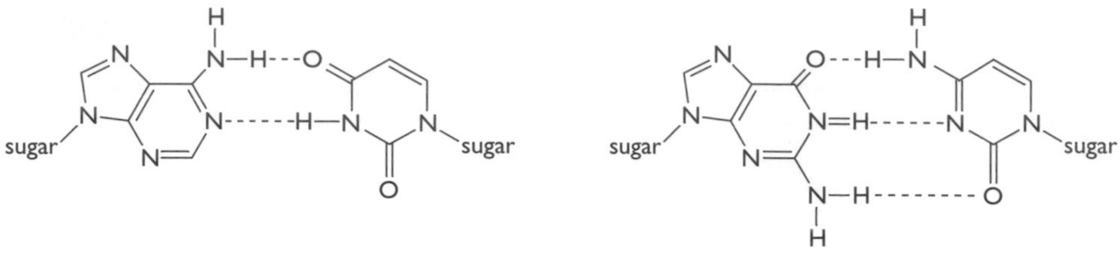

A:T base pair (MW = 615 Da)   G:C base pair (MW = 616 Da)

Fig 2.8 Normal base-pairing between nucleotides of DNA.

produced. When two separate strands pair they make a long, helical, double-stranded duplex. The chromosomes of all organisms are made of double-stranded DNA. A region of pairing may contain as few as 2 base pairs, or as many as $\sim 10^8$ base pairs — an entire, large chromosome (Figure 2.9).

When two strands of nucleic acid pair and intertwine they usually make a right-handed double helix. Grasp a pen in your right hand and point your thumb up; your four fingers turn in a right-handed direction, like the strands of DNA. The bases are on the inside of the double helix, and the sugar-phosphate backbones are on the outside. As shown in the figure on the previous page, the two strands of the double helix are oriented in opposite directions: looking down the length of a double helix, the linkages between nucleotides run from 5′ carbon to 3′ carbon in one strand and from 3′ carbon to 5′ carbon in the other strand. The overall appearance is that of a twisted ladder whose rungs are the base pairs. There are grooves in the helix between the pair of backbones. The helix is asymmetrical, with major and minor grooves.

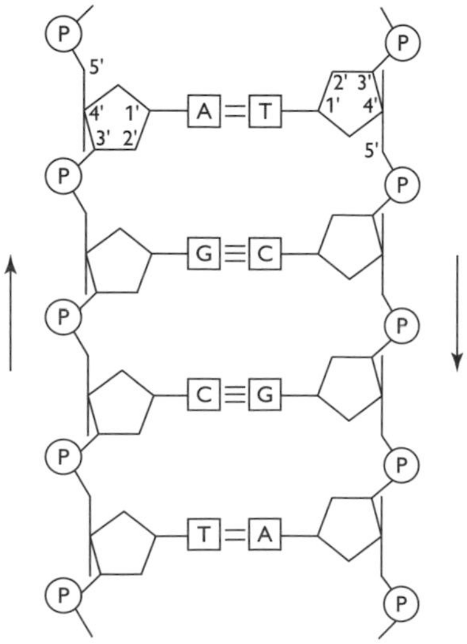

Fig 2.9 Double-stranded nucleic acid. This picture is flattened to simplify the view; the helical shape is shown on the next page. The two strands point in opposite directions.

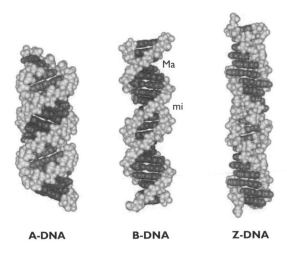

A-DNA   **B-DNA**   **Z-DNA**

Fig 2.10 Nucleic acid helices. IMB Jena Image Library of Biological Macromolecules, 50 years of double DNA helix structure.

In life, two main forms of nucleic acid duplex are known, A-form and B-form, both right-handed helices. The form of a helix heavily depends on the kind of sugar (ribose or deoxyribose). Ribose contains a bulky OH group on carbon 2', which influences bond angles in the sugars; these bond angles, in turn, influence the shape of the polymer. RNA and RNA-DNA hybrids are A-form, and DNA is mainly B-form. A rare form of duplex DNA is called Z-form, a skinny, left-handed, zigzag helix that occurs in short regions (Figure 2.10). Short triplex regions very rarely form when one strand loops out of the double helix at one point and adds to a nearby stretch; this is called H-form DNA.

A- and B-form helices differ in several ways (Figs. 2.10 and 2.11). In the B-form helix, the base pairs are centered deep inside and lie flat in a plane perpendicular to the axis of the helix; the phosphates are on the outside; and the sugars lie between bases and phosphates. In the thicker A-form helix, the base pairs sit halfway between the center and the outside of the helix and are tilted 20° from the horizontal; both the phosphates and the sugars are on the outside. Minor, local variations in helix structure are caused by the sequence of nucleotides (base content) and interactions between the nucleic acid and proteins.

Fig 2.11 Some approximate average characteristics of A-, B-, and Z-form nucleic acid. RNA and RNA-DNA hybrids are A-form. Most double-stranded DNA is B-form. The rare Z-form DNA occurs in short sequences.

Form	Mean thickness	Mean number of bases/twist	Length of twist	Direction of twist
A	2.2 nm	11	2.5 nm	righthanded
B	2.0 nm	10	3.3 nm	righthanded
Z	1.8 nm	9	4.6 nm	lefthanded

Other Characteristics of Double-Stranded DNA

The shape of the B-form DNA is neither invariant nor constant. The dimensions of the helix vary slightly from place to place. In addition, base pairs come apart and then re-pair — the double helix "breathes."

Water associates with DNA and influences its structure. In life, the sugar-phosphate backbone of B-form DNA is fully hydrated, and water occupies both the major groove and the minor groove. Water in the minor groove stabilizes the helix.

Double-stranded DNA as described above is called "relaxed DNA" because it has no extra twists. Relaxed DNA is rather stiff; the smallest circle made of it would be at least 170 nucleotide pairs long. DNA-binding proteins bend DNA, causing it to make tighter turns than it could by itself (Figure 2.12). DNA-bending proteins are commonplace in all organisms. In addition to bending, other variations in the shape of B-form DNA, mostly local variations, occur (Figure 2.13): single-stranded regions, loops and hairpins, and supercoiling.

All DNA chromosomes contain supercoiled regions. Everyday objects become supercoiled – string, rubber band, telephone cord, garden hose. Take a piece of multistrand string or rope in hand

Fig 2.12   Proteins can bend DNA.

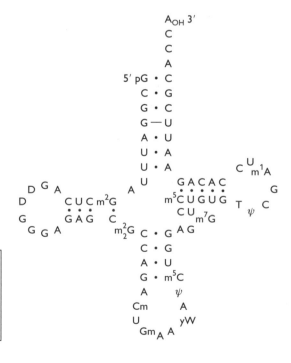

Fig 2.13   Hairpins result from intrastrand base-pairing. In this flattened picture of tRNA, 13 of the 76 nucleotides contain modified bases.

Relaxed chromosome

Supercoiled chromosome

Fig 2.14 The appearance of a small, circular chromosome, relaxed and supercoiled, based on electron micrographs.

and twist it in the same direction as the strands twist; the result is overwinding, or positive supercoiling. Take the same piece of multistrand string or rope in hand again and twist it in the direction opposite to the strand twisting; the result this time is underwinding, or negative supercoiling.

DNA in chromosomes can easily become supercoiled. In cells, DNA is slightly underwound through the agency of DNA-binding proteins, so that the mean number of nucleotides per twist is increased over that of fully relaxed DNA by about 6%. Moreover, when duplex DNA is unwound, which must happen every time RNA is synthesized and every time DNA is replicated, positive supercoiling rapidly develops ahead of the unwinding point, which requires the action of special enzymes to relieve this supercoiling (Figure 2.14).

Some Commonly Encountered Abbreviations

Double-stranded DNA is often abbreviated **dsDNA**; single-stranded DNA is often abbreviated **ssDNA**. The size of a nucleic acid polymer is usually given as the number of nucleotides or nucleotide pairs. A standard unit of size is the **kilobase (kb)**. This is applied to DNA and RNA and to both double-stranded and single-stranded nucleic acid. For example, a piece of ssRNA that consists of 3200 nucleotides is 3.2 kb long, and a piece of dsDNA that consists of 3200 nucleotide pairs is 3.2 kb long. One thousand kilobases is a **megabase (Mb)**.

Further Reading

Adams RLP, *et al.* 1992. *The Biochemistry of the Nucleic Acids,* 11th ed. Chapman and Hall, London.

Dickerson RE, *et al.* 1982. The anatomy of A-, B-, and Z-DNA. *Science* 216: 475–485.

Travers A. 1993. *DNA-Protein Interactions.* Chapman and Hall, London.

Chapter 3

Proteins

Overview

Proteins are linear polymers of **amino acids**. Genes encode all proteins, and proteins perform essential roles in all genetic processes, including the synthesis of DNA, RNA, and proteins. Some proteins bind to DNA and RNA, and some proteins are enzymes that act on nucleic acids. Some proteins bind to specific nucleotide sequences, but others bind equally well to any nucleotide sequence. This chapter describes the main points of protein size and structure.

Amino Acids

Protein is a generic term for a linear polymer made of amino acids as well as an aggregate of these polymers. An amino acid is a small carboxylic acid with an amino group and a side group that defines it. There are hundreds of different amino acids, but only 22 that are known to be genetically encoded, and two of these – selenocysteine and pyrrolysine – are found in only a handful of proteins. The molecular masses of the 20 common, genetically encoded amino acids range between 75 and 204 Da (Figure 3.1). Notice that amino acids are smaller than nucleotides.

Peptides

A peptide is a short polymer of amino acids, usually 30 amino acids or fewer. Adjacent amino acids in peptides are held together by a **peptide bond** (Figure 3.2) – a covalent bond between the carboxyl carbon atom of one amino acid and the core amino nitrogen atom of the other. A polypeptide is a large polymer of amino acids, usually 100 amino acids or more. There is no standard

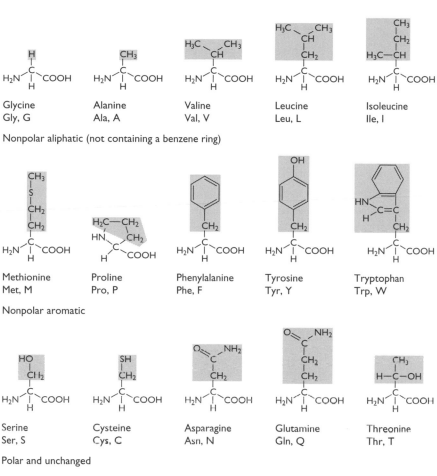

Glycine
Gly, G

Alanine
Ala, A

Valine
Val, V

Leucine
Leu, L

Isoleucine
Ile, I

Nonpolar aliphatic (not containing a benzene ring)

Methionine
Met, M

Proline
Pro, P

Phenylalanine
Phe, F

Tyrosine
Tyr, Y

Tryptophan
Trp, W

Nonpolar aromatic

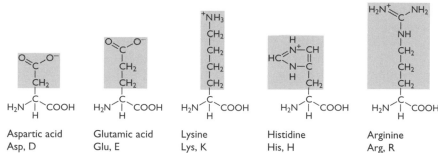

Serine
Ser, S

Cysteine
Cys, C

Asparagine
Asn, N

Glutamine
Gln, Q

Threonine
Thr, T

Polar and unchanged

Aspartic acid
Asp, D

Glutamic acid
Glu, E

Lysine
Lys, K

Histidine
His, H

Arginine
Arg, R

Polar and changed

Fig 3.1 The 20 common, genetically encoded amino acids, with three-letter and one-letter abbreviations. They are arranged according to attributes of the side group. User-friendly three-letter abbreviations of amino acids are being supplanted by less easily remembered one-letter abbreviations.

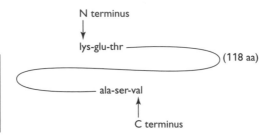

Glycine   Alanine

                → H₂O

Fig 3.2 | The peptide bond.

convention to delineate peptides and polypeptides. A protein may be a single polypeptide, or it may comprise many polypeptides held together by weak bonds such as hydrogen bonds.

Protein structure has several levels of organization: **primary**, **secondary**, **tertiary**, and **quaternary**.

Primary Structure

Primary structure is the sequence of amino acids. The thread of amino acids is considerably thinner than a single strand of RNA or DNA, because amino acids are smaller than nucleotides. The end of the polypeptide with a free amino group is called the N terminus, and the end with a free carboxylic acid group is called the C terminus (Figure 3.3).

Fig 3.3 | Primary structure of bovine ribonuclease A, a 124-amino acid. N terminus at top right 14 kDa protein; C terminus at left center.

N terminus

lys-glu-thr ———————— (118 aa)

ala-ser-val

C terminus

Secondary Structure

Secondary structure is the shape assumed spontaneously by each segment of a polypeptide; the most common ones are alpha (α) and beta (β) (Figure 3.4). The α-helix is a right-handed helix about 0.3 nm thick, averaging 3.6 amino acids per turn; the length of one turn is about 0.6 nm. The β-ribbon is a flat section of amino acids; two or more β-ribbons often associate into a β-sheet, formed by hydrogen bonds between two or more of the ribbons. β-sheets impart strength to a protein and are found in proteins with structural roles.

Fig 3.4 (Left) Three α-helices in part of a DNA-binding protein. (Right) β-sheet (arrows) and 3 α-helices in part of a DNA-binding protein. Travers 1993, *DNA-Protein Interactions*, Chapman & Hall.

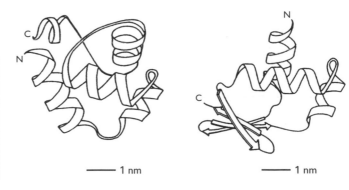

——— 1 nm   ——— 1 nm

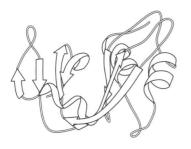

Fig 3.5 Bovine Ribonuclease A, a 124-amino acid, 14-kDa protein; this protein is a dimer, but for simplicity only one of the two identical subunits is shown. Protein Data Bank, Sadasivan et al. 1998.

Tertiary Structure

Tertiary structure is the folding pattern of secondary structures such as α-helices into a three-dimensional conformation (Figure 3.5). Proteins called chaperonins help in the formation of tertiary structure. In diagrams of tertiary structure, the strands of α-helices and β-ribbons are shown as flattened sections, while the connecting regions are depicted as thin threads.

Quaternary Structure

Quaternary structure is the association of two or more polypeptides via weak chemical bonds. Each polypeptide is then a subunit of the whole protein, and subunits are sometimes given Greek letters (Figure 3.6). For example, the fully active RNA polymerase enzyme in the bacterium *Escherichia coli* is a pentamer = $\alpha_2\beta\beta'\sigma$, and the adult human hemoglobin molecule is a tetramer = $\alpha_2\beta_2$.

The number of subunits in a protein or other molecule is designated by a prefix. The following table lists the most-used of these prefixes. Examples: (1) Bacterial RNA polymerase (Chapter 7) is a pentamer (it is an association of 5 polypeptides). (2) A typical

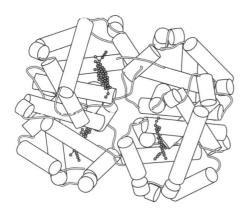

Fig 3.6 Adult human hemoglobin, a 66-kDa tetramer; each α chain has 141 amino acids, and each β chain has 146 amino acids. A heme molecule sits at the heart of each polypeptide. In this figure a cylinder represents an α helix. Protein Data Bank (primary citation not available).

Number of subunits	Name	Number of subunits	Name
1	monomer	11	undecamer
2	dimer	12	dodecamer
3	trimer	13	tridecamer
4	tetramer	14	tetradecamer
5	pentamer	15	pentadecamer
6	hexamer	20	eicomer
7	heptamer	21	uneicomer
8	octamer	22	doeicomer
9	nonamer	30	tricomer
10	decamer		

polymerase chain reaction primer (Chapter 28) is an eicomer (it is a chain of 20 nucleotides).

If the subunits of a protein are identical, use the prefix "homo"; if not, use the prefix "hetero" (Figure 3.7).

Every polypeptide is encoded by an RNA molecule. After a polypeptide has been synthesized, it may be trimmed or even spliced, and some of its amino acids may be modified enzymatically – for example, by the addition of an acetyl group or a sugar. Peptides are made by cutting a polypeptide into pieces.

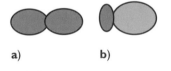

a)   b)

Fig 3.7 Homodimer (**a**) and heterodimer (**b**).

Protein Domains

A protein, even a single polypeptide, may have several structurally and functionally distinct regions called **domains** (Figure 3.8). A domain contains several secondary structures (α-helix, β-ribbon). Domains are connected in such a way that they can move separately from each other. A function may be carried out by a single domain, or by two domains working together. On the order of

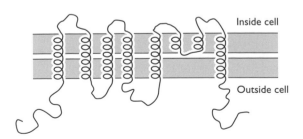

Fig 3.8 Receptor with transmembrane domain. A receptor binds to a small molecule (e.g., estrogen) that acts as signal to the cell.

Inside cell

Outside cell

10^3 domain motifs associated with specific functions have been found.

Proteins Assist Genetic Processes

Neither RNA nor DNA works by itself. Proteins participate in every genetic process – for example, synthesis and degradation of DNA, mutation, repair of mutation, recombination, synthesis and degradation of RNA, transport of nucleic acids, protein synthesis, positioning and moving chromosomes, nuclear division, and cell division. Proteins also make up part of the chromosomes of organisms. While carrying out these functions, proteins must bind specifically to other proteins, bind to nucleic acids, and, in some cases, catalyze chemical reactions.

Protein–Protein Binding

Proteins tend to stick to each other. Much of a protein's stickiness is highly specific, promoting the marriage of some proteins but not others, and ensuring that each copy of a multimeric protein is put together in the same way. Proteins are held together primarily by weak chemical bonds (e.g., hydrogen bonds). The association of proteins through many weak bonds is strong enough to withstand random, thermal movement, yet weak enough to allow for disaggregation in response to molecular signals.

Non-Sequence-Specific DNA Binding

Histones are important proteins that bind to DNA in a non-sequence-specific way. DNA wraps around histones to make the nucleosome, an essential molecular structure in eukaryal chromosomes (Chapter 5). Some enzymes that act on DNA bind equally well to all nucleotide sequences (Figure 3.9). How do proteins bind to DNA without regard to sequence? Such proteins recognize DNA by its shape, binding mainly to the DNA's sugar-phosphate backbone.

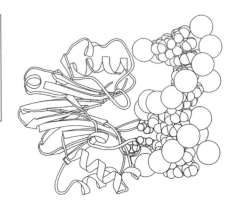

Fig 3.9 DNase I, an enzyme that cuts DNA, binds to the minor groove of DNA. Ma = major groove; protein contacts minor groove. Travers 1993, *DNA-Protein Interactions,* Chapman and Hall.

Sequence-Specific DNA Binding

Proteins that bind to specific sequences of double-stranded DNA include enzymes that synthesize RNA, enzymes that cut DNA, and regulatory proteins (Figure 3.10). Proteins that regulate genetic processes – RNA synthesis, DNA synthesis, recombination, DNA repair – may bind to DNA at specific sequences to do their work. Proteins that bind to specific sequences fit into major or minor grooves, and binding is based on hydrogen bonds between amino acids and specific purine and pyrimidine bases.

RNA-Binding Proteins

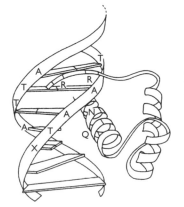

Fig 3.10 A regulatory protein inserts into DNA grooves and binds to specific nucleotides. Travers 1993, *DNA-Protein Interactions,* Chapman and Hall.

Many proteins bind to RNA, including enzymes, ribosomal proteins, and, in eukarya, proteins that carry RNAs from the nucleus to the cytoplasm. RNAs are usually single stranded but form secondary structures such as short A-form duplex regions (usually less than a single twist) and open loops. RNA-binding proteins seem to be of two broad types: groove binders and single-strand binders. Groove binders recognize features of RNA duplex regions and either sit in the broad, shallow minor groove or else insert into the major groove from the ends. Single-strand binders recognize specific bases, often in segments of RNA that lack secondary structure (single-stranded, nonduplex regions). The binding domains of these proteins are often either α-helices or β-ribbons.

Further Reading

Dickerson RE, Geis I. 1983. *Hemoglobin.* Benjamin/Cummings, Menlo Park, CA.

Doolittle RF. 1995. The multiplicity of domains in proteins. *Annu. Rev. Biochem.* 64:287–314.

Draper DE. 1999. Themes in RNA protein recognition. *J. Mol. Biol.* 293:255–270.

Sadasivan C *et al.* 1998. Plasticity, hydration, and accessibility in ribonuclease A. *Acta Cryst. D. Biol. Chryst.* 54 pp. 1343.

Travers A. 1993. *DNA-Protein Interactions.* Chapman and Hall, London.

Chapter 4

Simple Chromosomes

Overview

Every life form's genetic material is packaged in one or more chromosomes. Its **genome** is the nucleic acid in one complete set of chromosomes, excluding nonessential ones. This chapter is about the composition, size, shape, and number of chromosomes in bacteria, archaea, mitochondria, chloroplasts, and viruses – all but the eukarya. The nucleic acid of chromosomes is double-stranded DNA, except for a few single-stranded DNA plasmids and except for some viruses.

Bacteria and Archaea

Bacteria and archaea have two kinds of chromosomes, **essential chromosomes**, which are required for the survival and normal functioning of the cell, and **plasmids**, which are not absolutely necessary for survival and reproduction. Most bacteria and archaea whose genomes have been analyzed have only one essential chromosome, and in nondividing (nonreproducing) cells there is one copy of that chromosome. The number of different types of plasmids per cell varies from zero to several, and the number of copies of each plasmid ranges from 1 to $\sim 10^2$, depending on the plasmid. Usually, essential chromosome is simply referred to as chromosome.

In most bacteria the genome is located in an amorphous region, the **nucleoid** (Figure 4.1), which takes up from a quarter to half the cell's volume. It is not known whether plasmids are also restricted to the nucleoid. The parts of the chromosome where DNA replication begins and ends are attached to the cell's membrane. Unlike the nucleus of a eukaryal cell, a membrane does not bound the nucleoid region. An exception is the bacterium *Mycoplasma gallisepticum*, whose chromosome sits inside a membrane-bound organelle; perhaps other species of bacteria have this nucleus-like structure.

Essential Chromosomes

The total genome size of bacteria and archaea ranges from 0.5 to 9.2 megabases (Mb). Most species have one essential chromosome, but some species have two or more (Figure 4.2). A chromosome in bacteria and archaea is made of one molecule of double-stranded DNA plus associated proteins. In most taxa chromosomes are closed circles, but in a few cases chromosomes are linear. Bacteria undergoing cell division cycles have multiple, identical copies of each chromosome.

E. coli has a single essential chromosome, fairly typical in size for bacteria: a 4.6-Mb circle of DNA with a mass of $\approx 3 \times 10^9$ Da, roughly 1% of the cell's total mass. The DNA of E. coli's chromosome, if it were relaxed, would be 1.6 mm long and 2 nm thick – 800,000 times as long as it is thick and about 500 times as long as the 3-μm cell, making chromosome packaging a serious "problem." Proteins bind to DNA to condense it, so that it fits into the nucleoid region. The E. coli chromosome makes about 50 loops, about 100 kilobases (kb) per loop, and each loop is

Fig 4.1 Nucleoid region.

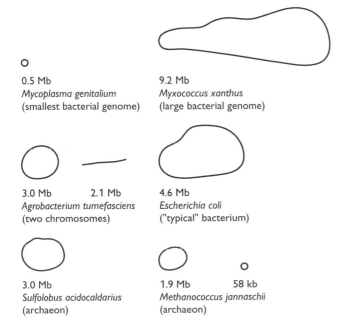

0.5 Mb
Mycoplasma genitalium
(smallest bacterial genome)

9.2 Mb
Myxococcus xanthus
(large bacterial genome)

3.0 Mb   2.1 Mb
Agrobacterium tumefasciens
(two chromosomes)

4.6 Mb
Escherichia coli
("typical" bacterium)

3.0 Mb
Sulfolobus acidocaldarius
(archaeon)

1.9 Mb   58 kb
Methanococcus jannaschii
(archaeon)

Fig 4.2 Diverse chromosomes in selected bacteria and archaea. These examples were chosen to emphasize diversity in the size and number of essential chromosomes. Chromosomes, approximately to scale, are cartooned as simple circles or rods. The 58-kb chromosome of *M. jannaschii* was reported as an "extrachromosomal element," but it contains genes that code for proteins that are essential to this cell, including two histones and two restriction enzymes.

Proteins bound
to DNA (HU, IHF)

a)

b)

Fig 4.3 Looped chromosome (a) and chromosome loop (b).

compacted about 10-fold compared with free, relaxed, B-form DNA (Figure 4.3).

In *E. coli*, the chromosome is 80% to 90% DNA and 10% to 20% protein by mass. Most of its DNA is completely free of protein. A dozen different chromosomal proteins make up the bulk of the protein in the nucleoid region, bound to the B-form DNA with weak chemical bonds. Some of the chromosomal proteins, such as HU, bind with no sequence specificity, but others, such as IHF, bind to specific DNA sequences. Only a minority of the DNA is bound to chromosomal proteins. Overall, the DNA is slightly negatively supercoiled (underwound).

HU is a small, basic heterodimeric protein ($\alpha\beta$, 19 kDa) that binds to DNA with any sequence of nucleotides and bends it. HU is more concentrated at the edges of the nucleoid, where transcription takes place; HU participates in transcription, replication, and recombination. HU and DNA form bead-like structures *in vitro*. IHF, which is similar to HU in amino acid sequences, also makes small heterodimers ($\alpha\beta$, 22 kDa) and also bends DNA. In contrast to HU, IHF binds preferentially to regulatory DNA sequences. Like HU, IHF serves in many genetic processes.

The DNA of archaeal chromosomes is wrapped around small, basic proteins called archaeal histones, homologous to eukaryal histones (Figure 4.4). The archaeon *Methanothermus fervidus* has two histones, HmfA and HmfB, each about 70 amino acids, which make tetramers. Positively supercoiled DNA (90 to 150 base pairs (bp) wraps around a tetramer of HmfA and HmfB with no sequence specificity, to make a particle similar to the nucleosome of eukarya. Native archaeal chromosomes vary from place to

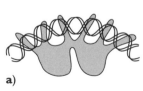

a)

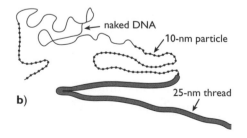

naked DNA

10-nm particle

25-nm thread

b)

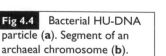

Fig 4.4 Bacterial HU-DNA particle (a). Segment of an archaeal chromosome (b).

place, including areas of naked DNA, regions of DNA and histones arranged like ≈10-nm beads on a string, and regions of rough ≈25-nm threads.

Plasmids

In addition to its essential chromosome(s), a cell may have one or more plasmids – small, extra chromosomes that are not needed by the cell in a natural environment. The term plasmid is in wide use today, although one sometimes sees the ill-conceived synonym extrachromosomal element. Plasmids are usually double-stranded DNA, rarely single-stranded DNA. They are usually circular but may be linear. Most are in the size range 3 to 100 kb. Plasmids contain functional genes and are replicated, although their replication is independent of replication of the main chromosome(s). Most small chromosomes of bacteria and archaea whose DNA sequences have been analyzed appear not to have essential genes. For this reason, they are likely plasmids.

Some plasmids are **conjugative** – capable of being transferred from a donor cell to a recipient during conjugation, when two cells make physical contact and a passage forms between them. Conjugative plasmids have genes that encode proteins required for conjugation, and nonconjugative plasmids lack them.

Plasmids conferring resistance to an antibiotic are often designated R, followed by letters or numbers; R plasmids in pathogenic bacteria are medically important. Some plasmids encode a bactericidal protein (colicin) as well as an "antitoxin"; these are called Col plasmids. Plasmids appear to be as commonplace in archaea as they are in bacteria.

In abundance, any given plasmid falls into one of three categories: low copy number, about 1 per cell; medium copy number, about 10 per cell; and high copy number, 30 to 100 per cell.

Origin of Replication

All bacterial and archaeal chromosomes, including plasmids, contain a short sequence where chromosome replication begins, termed the origin of replication, or simply **origin**. Proteins of the replication machinery bind to the origin before chromosome replication begins. In *E. coli* the origin, *oriC*, is about 200 bp long.

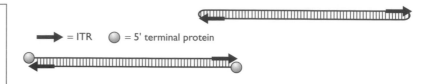

Fig 4.5 Telomeres of linear plasmids contain inverted terminal repeats (ITR). Telomeres of plasmid in *Streptomyces* (left). Hairpin telomeres of plasmid in *Borelia* (right).

⟶ = ITR   ◯ = 5' terminal protein

Telomeres

An obvious difference between a circular chromosome and a linear chromosome is that the former is endless, whereas the latter has two ends. Constitutively linear chromosomes of bacteria and archaea, including linear plasmids, contain structures at their ends called **telomeres**. These are short terminal sequences of DNA; in some cases protein, part of the telomere, is attached covalently to a terminal nucleotide. Telomeres facilitate replication of the chromosome's ends and may also help protect the linear chromosome from degradation. The main features of bacterial telomeres are (1) inverted terminal repeat (ITR) sequences, and (2) either hairpin loops or 5′-terminal proteins (Figure 4.5). ITRs get their name from the fact that the nucleotide sequence at one terminus is identical to that at the other terminus, except for pointing in the opposite direction ($5′{\rightarrow}3′$ vs. $3′{\rightarrow}5′$).

Mitochondria

Mitochondria, as bacteria, have essential chromosomes and may also have plasmids, localized in the nucleoid region. The mitochondrial chromosomes are double-stranded (ds) DNA, and plasmids are usually double-stranded DNA but may be single-stranded (ss) DNA. Mitochondrial genes encode the protein-synthesizing machinery (rRNA, tRNA, ribosomal proteins; Chapter 11) and some enzymes of aerobic metabolism. Many proteins in mitochondria are encoded by nuclear genes and synthesized in the cell's cytoplasm – on the endoplasmic reticulum – and then imported into mitochondria.

Most mitochondria have a single essential chromosome, usually present in several copies; one or more plasmids, also in multiple copy number, may be present. Mitochondrial chromosomes are usually circular, but linear mitochondrial chromosomes occur in ciliates, algae, animals, and fungi. Mitochondrial genomes vary in size from 14 to 2500 kb. Linear mitochondrial plasmids of

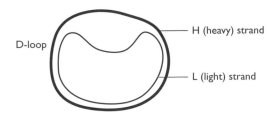

D-loop

H (heavy) strand

L (light) strand

Fig 4.6 Mammalian mitochondrial chromosome.

the inverted repeat type are known: they have terminal inverted repeats and proteins linked covalently to the 5′ ends.

The chromosome of mammalian mitochondria is circular and consists of a purine-rich heavy (H) and pyrimidine-rich light (L) strand (Figure 4.6). During the normal development of a mammal's mitochondrion a structure called the D-loop (displacement loop) forms in the mitochondrial chromosome, where part of a new H strand is synthesized. Each strand has a separate origin of replication, O_H on the heavy strand and O_L on the light strand. Each strand has a single promoter (site where RNA synthesis is initiated). O_H and the two promoters are located in the D-loop. Plant mitochondrial chromosomes have many promoters, and there are large regions of noncoding DNA sequences, interspersed between genes.

Kinetoplasts

The mitochondria of the Protozoan order Kinetoplastida (*Trypanosoma* and *Leishmania*) are bizarre. In addition to a main chromosome in the nucleoid region, the mitochondrion of these protozoans contains a disk-like body, the **kinetoplast**, which contains 40 to 50 large chromosomes (maxicircles) and 5000 to 10,000 small ones (minicircles), all interconnected in a giant network. Maxicircles contain 25 to 35 kb and minicircles contain 0.65 to 2.50 kb of DNA. The maxicircles, which appear to be identical copies of the main chromosome, contain genes typical of mitochondrial genomes. Hundreds of different minicircles are present, each in multiple copies.

Plastids

Plastids (chloroplasts and their relatives) are organelles found in green algae and plants. Chloroplasts descended from cyanobacteria, which became endosymbionts of plants and algae. Other plastids evolved from chloroplasts; these include chromoplasts (pigment-containing plastids of flowers) and amyloplasts (starch-storing plastids). The evolution of chloroplasts is complex compared with that of mitochondria: there were at least two separate endosymbiotic captures of cyanobacteria, and some chloroplasts evolved by secondary capture – a chloroplast-containing eukaryon became the chloroplast of a new host (e.g., Cryptomonads).

Chloroplasts have one circular essential chromosome, usually present in multiple copies. The chromosome (120 to 200 kb of dsDNA) encodes four rRNAs (RNAs that are part of the ribosome), about 30 tRNAs, and about 100 proteins. Noncoding sequences are interspersed between genes. A chloroplast may also have one or more multiple-copy plasmids.

Fig 4.7 Diverse viral genomes. Host in parentheses: B = bacterium, An = animal, P = plant. Genome size in kb, number of chromosomes, number of genes, and chromosome shape are given for each virus; ss = single strand, ds = double strand. For single-strand viruses, (+) = chromosome is equivalent to encoded RNA; (−) = chromosome is complementary to encoded RNA.

Viruses

Viral genomes may be RNA or DNA and may be double stranded or single stranded (Figure 4.7). The genome may consist of a single chromosome or several chromosomes. Total genome size ranges from 2 to 670 kb, so that the largest viral genome (of virus G, infecting *Bacillus subtilis*) is bigger than the smallest bacterial genome. Chromosomes may be circular or linear. Linear viral chromosomes lack telomeres. The genomes of some viruses reproduce independently of the host genome. In other viruses, one or more copies of the genome may integrate into the host genome, where they reside for part of the virus's life cycle.

Type	Name	Host	Genome	Chromosomes per genome	(+) or (−)
ssRNA	MS2	B	3.6 kb	1 linear	(+)
	polio	An	7 kb	1 linear	(+)
	rabies	An	12 kb	1 linear	(−)
	HIV-1	An	8.5 kb	1 linear	(+)
	influenza-A	An	13.6 kb	8 linear	(−)
dsRNA	reovirus	An	30 kb	10 linear	
ssDNA	Ff	B	6.4 kb	1 circular	
	parvovirus	An	2 kb	1 linear	(+) or (−)
ss/dsDNA	hepatitis-B	An	3.3 kb	1 circular	
dsDNA	λ	B	48 kb	1 linear	
	T4	B	170 kb	1 circular	
	reovirus	An	30 kb	10 linear	
	HSV-1	An	153 kb	1 linear	
	vaccinia	An	200 kb	1 linear	
	G	B	670 kb	1 circular	

Further Reading

Casjens S. 1998. The diverse and dynamic structure of bacterial genomes. *Annu. Rev. Genet.* 32:339–377.

Gray MW. 1989. Origin and evolution of mitochondrial DNA. *Annu. Rev. Cell Biol.* 5:25–50.

Lang BF *et al.* 1997. An ancestral mitochondrial DNA resembling a eubacterial genome in miniature. *Nature* 387:493–497.

Neidhardt FC (ed.). 1996. Cellular and molecular biology of *Escherichia coli* and *Salmonella,* 2nd ed. ASM Press, Washington, DC.

Sato S *et al.* 1999. Complete structure of the chloroplast genome of *Arabidopsis thaliana. DNA Res.* 6:283–290.

Takayanagi S *et al.* 1992. Chromosomal structure of the halophilic archaebacterium *Halobacterium salinarium. J. Bacteriol.* 174:7207–7216.

Chapter 5

Chromosomes of Eukarya

Overview

The eukaryal genome is located in the cell's nucleus, packaged in structurally complex chromosomes. Through the cell cycle, chromosomes condense and decondense, undergoing dramatic changes in thickness and length. With few exceptions, eukaryal chromosomes consist of thread-like **chromatin**, composed of particles of histones wrapped by DNA, called **nucleosomes**. Nearly all eukaryal chromosomes are linear and possess a pair of telomeres. In most species, each chromosome has a single **centromere**, which functions in chromosome separation and movement.

Condensed versus Uncondensed Nuclear Chromosomes

In eukarya, chromosomes reside in the nucleus, a large spherical organelle bounded by a double-membraned envelope. In most cases chromosomes are visible by light microscopy only during **mitosis**, the brief nuclear division phase. Mitotic chromosomes are short and thick (d $\sim1^2$ μm, "packed to travel") and take up colorful stains used to detect their presence. They consist of twin copies called **sister chromatids** and are attached to spindle fibers (microtubules that move chromosomes). Genes are not transcribed into RNA during mitosis. Mitotic chromosomes are useful for the study of chromosome structure, even though they bear little resemblance to "chromosomes at work," and even though mitosis takes up a small fraction of the life span of a typical cell (Figure 5.1).

Chromosomes perform most of their genetic functions during **interphase**, the cell's lengthy, nondividing period. In contrast

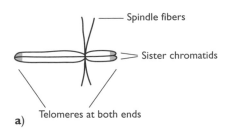

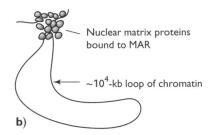

to mitotic chromosomes, interphase chromosomes are long and thin (less than a tenth of the wavelength of visible light); they are dispersed widely in the nucleus; during much or most of interphase they exist in a single copy, not as sister chromatids; and their structure is difficult to discern even in the best electron micrographs.

In Figure 5.1, the mitotic chromosome consists of two identical sister chromatids, each having a centromere attached to spindle fibers. The centromere divides the chromosome into two arms. Telomeres cannot be seen microscopically but are diagrammed in Figure 5.1 as dark spots at the ends. On the right-hand side of Fig. 5.1, a tiny part of an interphase chromosome is diagrammed as a loop of chromatin, the basic fiber making up eukaryal chromosomes, attached to the nuclear matrix, a network of protein fibers that crisscross the nucleus. Matrix attachment regions (MARs) are DNA sequences that bind to the nuclear matrix.

Fig 5.1 Mitotic chromosomes (**a**) are short and compact, consist of two identical copies called sister chromatids, stain in patterns of dark and light bands, and are attached to spindle fibers at centromeres. Interphase chromosomes (**b**) are long and thin, are dispersed throughout the nucleus, exist in single copies during G_1, are not visible with light microscopy, are attached to the nuclear matrix at the matrix attachment region (MAR).

The Number of Chromosomes per Nucleus

Species of eukarya differ in the number of chromosomes per nucleus. Taking two extreme examples, the nematode *Ascaris megalocephala* has two chromosomes per somatic (nonreproductive) cell and one per gamete (egg or sperm); the fern *Ophioglossum reticulatum* has 1260 chromosomes per somatic cell. Ciliates have a nonreproductive **macronucleus** containing ~10^4 tiny (1 to 10 kb) chromosomes.

Chromatin

Chromatin is the fiber that makes up the eukaryal chromosome at all stages of the cell cycle except when it is being replicated or transcribed. Chromatin is a lumpy thread 20 to 40 nm thick. Except for dinoflagellates, chromatin is built from nucleosomes,

≈10-nm particles made of histones and DNA. During interphase, chromatin fibers are spread out through the nucleus and attached to the nuclear matrix. During mitosis, chromatin fibers are gathered together into relatively thick, visible chromosomes.

The Nucleosome

Nuclear chromosomes contain two kinds of protein, histones and nonhistone proteins. When interphase chromosomes are treated with chemicals to remove all the proteins not tightly bound to DNA and then viewed with an electron microscope, nucleosomes can be observed; these are particles linked by a thread (DNA) (Figure 5.2).

A nucleosome (sometimes called **core nucleosome**) is DNA wound around an octamer of histones – two copies each of four different histones: H2A, H2B, H3, and H4 (Figure 5.3). The eight histones are arranged in three parts: a tetramer, $(H3-H4)_2$, sandwiched between two dimers, (H2A-H2B); the octamer is roughly disk-shaped, 6.5 nm in diameter, and 6 nm thick. B-form DNA (≈146 bp) wraps around the histone disk 1.65 times in a left-handed superhelix, making the core nucleosome's dimensions about 10.5×6 nm. The DNA fits snugly around the histone octamer, which strongly bends the DNA. A linker, DNA associated with histone H1 or with nonhistone proteins, separates adjacent core nucleosomes. Linker size varies among species, 10 to 100 bp of DNA.

The 11- to 15-kDa core histones, H2A, H2B, H3, and H4, are 100 to 150 amino acids long and rich in lysine and arginine, while the 17- to 28-kDa linker histone, H1, is rich in lysine. Core histones have been highly conserved in evolution, while H1 has been moderately conserved. In mammals, a core nucleosome's mass is ≈200 kDa (DNA plus protein).

Histones are positively charged and bind to B-form DNA with hydrogen bonds in a non-sequence-specific way. Despite the lack of sequence specificity, there is some evidence of phasing – i.e.,

Native chromatin: a lumpy, 30-nm-thick thread

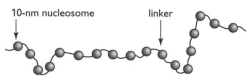

Chromatin minus nonhistones

Fig 5.2 Native chromatin: a lumpy, 30-nm-thick thread (left). Chromatin minus nonhistones (right).

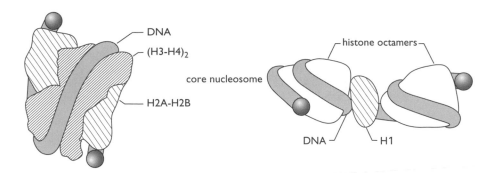

the association of specific sequences of DNA with nucleosomes. Phasing is likely caused by interactions between linker DNA and nonhistone, sequence-specific DNA-binding proteins.

Fig 5.3 Core nucleosome.

Chromatin Architecture

Chromatin is three times thicker than the nucleosome, implying that nucleosomes are packed together, side by side. Several models of nucleosome packing have been proposed (Figure 5.4). In the solenoid model, nucleosomes are packed in a spiral array, producing a thread of uniform thickness, interrupted by regions where nonhistone proteins bind. In the zigzag model, pairs of nucleosomes hang together and the spacer region alternates direction in a zigzag fashion. This model explains the lumpiness of chromatin fibers and better conforms to details of electron micrographs of nucleosomes than does the solenoid model.

Chromatin contains a large and diverse set of nonhistone proteins. The structure of chromatin is heterogeneous and complex, and higher-order structure is variable within and among genomes.

Euchromatin and Heterochromatin

Euchromatin and heterochromatin are major categories of chromatin based on light microscopy. Euchromatin is dispersed through the nucleus in interphase and is condensed into compact chromosomes during mitosis. Heterochromatin is highly condensed and darkly staining during most of interphase and is located near the nuclear envelope (Figure 5.5). By and large, genes being actively transcribed into RNA are located in euchromatin, not in heterochromatin, most of whose DNA is nontranscribable. Heterochromatin is rich in highly repeated, noncoding sequences. It is replicated late in the cell's DNA-synthesizing phase.

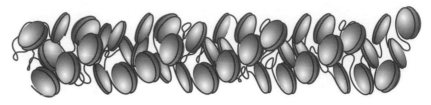

a)

b)

Interphase Euchromatin Arranged on the Nuclear Matrix

Interphase euchromatin is connected to the nuclear matrix (also called the nuclear scaffold), a network of 10-nm protein fibers. DNA sequences that bind to the nuclear scaffold are free of histones. These scaffold-binding sequences, known as MARs, are 0.7-kb AT-rich sequences occurring on average about every 70 kb within each chromosome. This arrangement produces loops of chromatin, each containing 30 to 300 kb of DNA (Figure 5.6). Thus, the nucleus of a diploid human somatic cell has ~10^4 chromatin loops. Chromatin loops appear to be functional units that

Heterochromatin
near centromere & telomeres

Mitotic chromosome

— Nucleolus

Heterochromatin near the nuclear envelope

Interphase nucleus

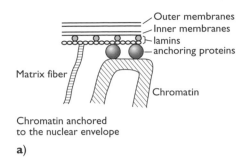

Chromatin anchored
to the nuclear envelope

a)

Many chromatin loops originate
from each anchor point

b)

Fig 5.6 Chromatin anchored to the nuclear envelope (**a**). Many chromatin loops originate from each anchor point (**b**).

facilitate transcription of DNA into RNA. Chromatin loops are anchored to the inside of the nuclear envelope by proteins.

Telomeres

Telomeres cap the ends of every eukaryal chromosome. Unlike the telomeres of linear bacterial chromosomes, eukaryal telomeres are complex structures. Telomeres typically have two kinds of DNA sequences – terminal and proterminal; various telomeric proteins bind to these sequences. The terminal region is usually 100 bp to 100 kb long. In protozoa, animals, and plants, terminal sequences are 2- to 6-bp tandem repeats; the sequence is G-rich on the 3'-terminating strand (Figure 5.7). The very tip of the telomere of ciliates and yeast is single-stranded and forms a hairpin loop by intrastrand base pairing. The proterminal region is usually 10 to 1000 kb long, and it often contains repeated sequences.

Fig 5.7 General layout of telomeres (left). Hairpin loop of telomere in ciliates and yeast (right).

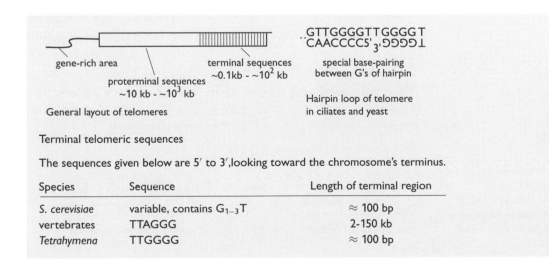

gene-rich area

proterminal sequences
~10 kb - ~10^3 kb

terminal sequences
~0.1kb - ~10^2 kb

General layout of telomeres

GTTGGGGTTGGGGT
CAACCCC5'₃'GGGGT

special base-pairing
between G's of hairpin

Hairpin loop of telomere
in ciliates and yeast

Terminal telomeric sequences

The sequences given below are 5' to 3', looking toward the chromosome's terminus.

Species	Sequence	Length of terminal region
S. cerevisiae	variable, contains $G_{1-3}T$	$\approx$ 100 bp
vertebrates	TTAGGG	2-150 kb
Tetrahymena	TTGGGG	$\approx$ 100 bp

In some species, telomeres are adjacent to the nuclear envelope, presumably bound to it by proteins, while in other species, telomeres are located in the interior of the nucleus; in some taxa, some of the telomeres are near the nucleoli. Telomeres are commonly found in clusters or pairs.

Telomeres protect the chromosome from degradation and from fusion with other DNA and provide a way to complete replication of the terminal sequences. Telomeres also may play a role in chromosome pairing during meiosis, a process central to sexual reproduction.

Mitotic Chromosomes

During nuclear division chromosomes are highly condensed to help the orderly movement of one copy of each chromosome into each daughter nucleus. Microtubules called spindle fibers attach to and move each chromosome during division, causing sister chromatids to segregate (move apart).

Centromeres

A chromosome's centromere is a DNA sequence with several functions. It has essential roles in the attachment of spindle fibers to the chromosome, in chromosome segregation during mitosis, and in the control of the cell cycle. The structure of centromeres varies markedly among taxa and may or may not be localized to one place in the chromosome (Figure 5.8).

Monocentric chromosomes have single, localized centromeres. **Kinetochores** – large structures made of protein – bind

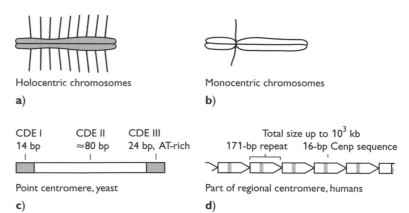

Holocentric chromosomes
a)

Monocentric chromosomes
b)

Fig 5.8 Monocentric chromosomes (**a**). Holocentric chromosomes (**b**). Point centrome, yeast (**c**). Part of regional centromere, humans (**d**).

CDE I   CDE II   CDE III
14 bp   ≈80 bp   24 bp, AT-rich

Point centromere, yeast
c)

Total size up to 10^3 kb
171-bp repeat   16-bp Cenp sequence

Part of regional centromere, humans
d)

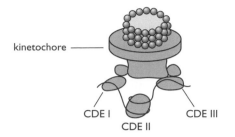

kinetochore ——

CDE I   CDE III
        CDE II

Fig 5.9 Possible structure of centromere, kinetochore, and spindle fiber in yeast spindle fiber.

to localized centromeres; during nuclear division spindle fibers attach to the kinetochores (Figure 5.9). Localized centromeres are of two kinds: **point centromeres** are short DNA sequences ($\sim 10^2$ bp), as in *Saccharomyces cerevisiae*, and **regional centromeres** are long DNA sequences (10^1 to 10^3 kb), as in humans.

Spindle fibers attach to many points in **holocentric** chromosomes during mitosis. It is not known whether holocentric chromosomes possess centromeres, and, if they do, what the DNA sequences are. *Caenorhabditis elegans* has holocentric chromosomes, as do many plants and arthropods.

Chromosome Morphology

In monocentric chromosomes, the regions flanking the centromere are called **chromosome arms**. In species whose chromosomes have been studied, the shorter arm is designated **p** (petit) and the longer arm is called **q**; if the arms appear to be equal in length, then they are arbitrarily designated p and q. Chromosomes whose arms are equal in length are **metacentric**, chromosomes whose arms are unequal in length are **submetacentric**, and chromosomes whose p arm is very short or even impossible to see under the light microscope are **acrocentric** (Figure 5.10).

Condensed chromosomes in mitosis are double, consisting of sister chromatids – identical, connected copies of the chromosome, which are products of chromosome replication in the cell's synthesis phase. The term chromatid only applies to members of a pair; after mitosis each chromosome is a single structure, and it is wrong to say that an unreplicated chromosome has a single chromatid.

q   p   q   p   q   p

Metacentric   Submetacentric   Acrocentric

Fig 5.10 Mitotic chromosomes classified by position of centromere.

In vertebrate genomes the GC content varies from region to region. In condensed chromosomes these two kinds of region stain differently. Each chromosome can be identified by its characteristic staining pattern. For example, Giemsa preferentially stains the GC-poor regions, producing **G bands**; GC-rich, Giemsa-light regions produce **R bands**. Each Giemsa-stained chromosome has a characteristic set of alternating G and R bands, giving each chromosome a unique banding pattern.

In polyploid tissues of some insects and some flowering plants, all the copies of each chromosome synapse (pair), making multistrand **polytene** chromosomes. In *Drosophila*, notably the larval salivary gland cells, each polytene chromosome contains thousands of copies, and the synapsis is precise, rendering a fine, distinct staining pattern of light and dark bands, unique to each chromosome. Pairing is imprecise in plant polytene chromosomes, yielding indistinct bands.

Further Reading

Arents G *et al.* 1991. The nucleosomal core histone octamer at 3.1. Åresolution: A tripartite protein assembly and a left-handed superhelix. *Proc. Nat. Acad. Sci. USA* 88:10148–10152.

Beard D, Schlick T. 2001. Computational modeling predicts the structure and dynamics of chromatin fiber. *Structure.* 9:105–114.

Daban JR, Bermudez A. 1998. Interdigitated solenoid model for compact fibers. *Biochemistry* 37:4299–4304.

Gerace L, Foisner R. 1994. Integral membrane proteins and dynamic organization of the nuclear envelope. *Trends Cell Biol.* 4:127–131.

Pluta AF *et al.* 1995. The centromere: hub of chromosomal activities. *Science* 270:1591–1594.

Rye NJ. 1995. The proterminal regions and telomeres of human chromosomes. *Adv. Genet.* 32:273–307.

Woodcock CL, Horowitz RA. 1995. Chromatin organization re-viewed. *Trends Cell Biol.* 5:272–277.

Chapter 6

Genome Content

Overview

This chapter inventories classes of DNA sequences in the genomes of organisms, mitochondria, chloroplasts, and viruses.

Part of an organism's genome consists of functional sequences – genes, regulatory sequences, telomeres, centromeres, and origins of replication. However, genomes also include DNA sequences that either are clearly nonfunctional or appear to be nonfunctional. The genomes of bacteria and archaea are mostly free of nonfunctional sequences, but, in many eukaryal genomes, the junkyard is larger than the portion that encodes RNA.

Some genes are present in multiple, identical copies. There are also sets of functionally related, homologous genes, called **gene families**. Nonfunctional DNA sequences may also be present in multiple copies. Repeated DNA sequences are often arranged in tandem.

What Is a Gene?

Dear reader, I have bad news for you: geneticists cannot agree on what a gene is, even though we do agree that genes are fundamental biological objects. Worse still, gene can change its meaning with context. Though this situation wants a strong remedy, none is available. The best I can offer you is a simple, natural concept of gene, contrasted with widely used, alternative concepts.

Consistently in this book, a gene is defined as a chromosomal segment of nucleic acid that encodes an RNA transcript (a freshly synthesized piece of RNA), along with nearby regulatory sequences needed to initiate RNA synthesis. This applies even when the transcript codes for several peptides or polypeptides.

Alternatively, some geneticists define gene as a segment of nucleic acid that encodes one final gene product, such as a polypeptide or an rRNA molecule. Often, gene is defined operationally, based on the results of a kind of experiment called the complementation test. Each of these ideas has strengths and weaknesses. In this book, gene concepts will be revisited as the need arises.

Bacteria and Archaea

More than 95% of the DNA in bacterial and archaeal chromosomes codes for RNA or else regulates genetic processes – transcription, replication, or recombination. Some sequences – transposons – can move from place to place within a genome. Much of the nonfunctional DNA consists of repeated sequences.

Genes

The three main kinds of RNA are ribosomal RNA (rRNA, which is part of the ribosome), transfer RNA (tRNA, which carries amino acids into ribosomes), and messenger RNA (mRNA, which codes for polypeptides). A minority of genes encode rRNA or tRNA, or both; most genes encode mRNA. A typical rRNA gene is ≈ 6 kb; it codes for three different rRNAs (called 5S, 16S, and 23S, names that indicate size) and in some instances it also encodes a few tRNAs. Most tRNA genes are ≈ 100 bp and encode single tRNAs. Protein-coding genes vary in size from $\sim 10^2$ to $\sim 10^4$ bp; an mRNA may encode one or more polypeptides. There are typically several copies of each rRNA and tRNA gene but only one copy of each mRNA gene. The rRNA and tRNA genes are scattered throughout the genome. In archaea and bacteria, genes are arranged contiguously, or nearly so.

Transposable Elements

Bacteria and archaea have DNA sequences capable of moving from one chromosomal **locus** (place on a chromosome) to another by recombination. Moveable sequences are **transposable elements**. When a transposable element moves to a new locus, it is said to transpose (i.e., jump to a new location). Transposable elements range in size from ≈ 1 to ≈ 10 kb, and each element exists in 1 to ≈ 50 copies per chromosome. The bacterium *Escherichia coli*

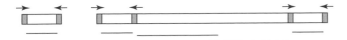

has two kinds of transposable element: **insertion sequences (ISs)** and **transposons** (Figure 6.1). Every IS (1 to 3 kb) contains 31-bp inverted repeat sequences at its ends and codes for transposase, an enzyme that promotes transposition to a new chromosomal locus. Transposons (5 to 10 kb) contain one or more genes sandwiched between two ISs; transposons may contain genes conferring resistance to antibiotics. There are several **gene families** – multiple, related copies – of ISs and of transposons.

Noncoding DNA

A small part of a bacterium's or archaeon's DNA does not code for RNA, ≈5% in *E. coli*. *E. coli's* genome contains six families of repeated, noncoding DNA sequences, with >30 copies of each type. These DNA sequences have no known function. The most abundant of these is the BIME sequence. BIMEs are 40 to 500 bp long, and there are ≈500 copies of BIME per chromosome. Another repeated sequence is the palindromic unit, a family of repetitive ≈40-bp sequences that make stem-and-loop structures via internal base pairing (Figure 6.2). A DNA palindrome is a sequence that is identical when read forward in one strand of DNA and backwards in the other strand (e.g., ATAT).

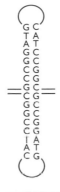

Fig 6.2 Typical palindromic unit (PU) sequence.

Eukarya

Genome Size

Eukaryal genomes span a huge range in size, from ≈10 to ≈ 96,000 Mb per haploid nucleus (one copy of the genome). The human genome is about average for mammals, ≈3000 Mb per haploid nucleus. The smallest genomes are found in simple unicellular species, such as bread yeast, and the largest are found in amphibians and flowering plants; e.g., the salamander *Necturus lewisi* (5.9×10^4 Mb) and the lily *Fritillaria davisii* (9.6×10^4 Mb). Most of the size variation is due to noncoding sequences, which make up over 99% of the largest genomes.

Fig 6.3 The proportions of three categories of DNA sequences, in three organisms.

Degree of repetition	Fruit fly	Human	Tobacco
Unique	70	60	33
Moderately	12	20	65
Highly	17	20	7

Sequence Copy Number

It is useful to classify DNA sequences into three broad groups according to copy number: unique sequences, moderately repeated sequences, and highly repeated sequences (Figure 6.3):

- **Unique sequences.** One copy (sometimes a few copies or similar copies). This includes most functional, protein-coding genes. However, only a small part of a typical gene encodes the final gene product.
- **Moderately repeated sequences.** 10 to 10^4 copies. Moderately repeated sequences vary in length between $\sim 10^2$ and $\sim 10^3$ nucleotides (nt), and are generally interspersed between unique sequences. Some code for RNA, but most do not.
- **Highly repeated sequences.** $>10^4$ copies. Highly repeated sequences are arranged tandemly, vary in length between 2 and $\sim 10^2$ nt and have no known function.

Genes

The RNA of eukarya can conveniently be classified into four categories: rRNA, tRNA, mRNA, and miscellaneous small RNAs. As in bacteria and archaea, a tiny fraction of the genome encodes the small RNAs, and their genes are present in multiple copies.

Genes Encoding rRNA. Eukarya have four kinds of rRNA: 5S, 5.8S, 18S, and 28S. In a few eukarya, one multicopy gene encodes a large precursor of all four rRNAs. In animals, plants, and some fungi, there are two size classes of rRNA genes, small (5S) and large. The large size class ($\approx$7 to 13 kb) encodes 45S RNA, which is cut into 18S, 5.8S, and 28S pieces. Multiple copies of the large class are arranged in tandem at one or more sites called **nucleolus organizer regions (NORs)**; there are 150 to 600 of these genes per genome, depending on the species. The human genome has 150 big rRNA genes in five NORs, which reside in the nucleolus, inside the nucleus. The small size class ($\approx$200 bp), encoding 5S rRNA, exists in hundreds or thousands of copies arranged in tandem clusters; the clusters are not part of the NORs. The human genome has ≈2000 5S rRNA genes.

Genes Encoding tRNA and Miscellaneous Small RNAs.   Genes for tRNA (≈100 bp) are scattered throughout the genome. Any given species has about 40 different kinds of tRNA, 10 to 180 copies of the gene encoding each one, and a total of 400 to 7000 tRNA genes per genome.

Miscellaneous small RNAs include small nuclear RNAs (snRNAs) and small cytoplasmic RNAs (scRNAs). snRNAs include seven kinds of U RNA, U1 to U7, which participate in RNA processing. One abundant scRNA is 7S RNA, which helps translocate newly synthesized proteins across membranes. Genes encoding small RNAs are $\sim10^2$ bp, and typically they are present in 10 to 500 copies per genome.

The smallest RNAs are **microRNAs** (miRNAs, 20–40 nt) and **small interfering RNAs** (siRNAs, 20–22 nt). There are many kinds of miRNAs and siRNAs, which regulate gene expression, principally by stopping or repressing protein synthesis (Chapter 11). The genes that encode these very small RNAs are under intensive investigation.

Genes Encoding Polypeptides.   The vast majority of genes encode mRNA. Such genes range in size from <1 to ≈1000 kb. Eukaryal genomes contain $\sim10^4$ mRNA-coding genes. Most polypeptides are encoded by a single, unique gene, although there are many families of genes with homologous sequences (i.e., similar by virtue of common evolutionary origins). For example in humans, eight globin genes encode seven different hemoglobins (there are two copies of the α-globin gene). Genes of one family often cluster together, with short or long spacers between genes. Most mRNA-coding genes are surrounded by noncoding sequences. Exceptional cases, in which many copies of a gene exist, include the histone genes; for each histone gene, yeast has 2 copies and humans have ≈40 copies.

Noncoding Sequences

A small fraction of noncoding sequences have known functions: centromeres, telomeres, and DNA regulatory sequences (mainly binding sites for proteins that comprise various genetic machines). Typically, a eukaryal chromosome has more than 1000 origins of replication, which are short, AT-rich sequences, spaced at intervals of 40 to 200 kb. The regulatory sequences controlling RNA synthesis are very short and are located in and around genes.

Much of the genome is transcribed into mRNA but does not serve as a template for protein synthesis. Some of these sequences are regulatory. For many genes, the precursor of mRNA is shortened by the removal of **intervening sequences** (**introns**). Although introns do not encode parts of the final gene product, they may nonetheless serve important regulatory functions (Chapter 10).

In plants and animals, the noncoding part is a heterogeneous collection of unique and repeated sequences, with the repeated sequences predominating. Noncoding sequences vary enormously among taxa in overall proportion of the genome, in length, and in nucleotide sequence. Three categories of nonfunctional sequences are (1) pseudogenes, (2) transposon-derived sequences, and (3) simple sequence repeats.

Pseudogenes. A pseudogene is a nonfunctional copy of a gene; many gene families include pseudogenes. Abbreviations of the names of pseudogenes often begin with the Greek letter Ψ. No final gene product is made from a pseudogene, either because it cannot be transcribed or because the transcript lacks some essential feature required for protein synthesis. Pseudogenes abound in the genomes of vertebrates. The human hemoglobin gene family has five pseudogenes.

Transposon-Derived Sequences. Most eukaryal genomes are littered with transposons and sequences derived from them. For example, short interspersed nuclear elements (SINEs) and long interspersed nuclear elements (LINEs) are derived from retrotransposons and make up a large fraction of the human genome. A retrotransposon is a sequence of DNA that proliferates in evolution via RNA intermediates. An enzyme, reverse transcriptase, uses the RNA intermediate as a template for synthesizing a new DNA copy, which inserts into another chromosomal location. SINEs and LINEs appear in diverse eukarya but are best characterized in mammals (Figure 6.4). They exist in many families, with sequence variation among members of a family. SINEs and LINEs are scattered throughout the genome. Proterminal regions of telomeres are rich in SINEs and LINEs.

pseudogene   reverse transcriptase gene

genes for
7S snRNA

AT-rich regions

Fig 6.4 *L1*, a LINE in the human genome (left). *Alu*, a SINE common in primate genomes (right).

SINEs are short (0.1 to 0.5 kb), interspersed, repeated sequences (10^2 to 10^6 copies). *Alu* is a large family of SINEs in Old World primates, which gets its name from the restriction enzyme that cuts its ends, *Alu* I. *Alu* appears to be homologous to 7S RNA, a part of the signal recognition particle that assists entry of nascent polypeptides into the lumen of the endoplasmic reticulum. There are nearly a million copies of *Alu* in the human genome ($\approx$9% of the genome), in which an *Alu* sequence occurs on average every 5 kb. Some *Alu* elements are still capable of retrotransposition.

LINEs are long (up to $\approx$7 kb), interspersed, repeated ($\sim 10^3$ copies) sequences. *L1* is a family of LINEs common to mammals. Sequence length varies within and between species, the full, ancestral-like sequence being $\approx$6.5 kb and truncated versions being 0.5 to 5.0 kb. The human genome contains $\approx$4000 full-size, 6.5-kb *L1* sequences and $\approx$5000 truncated versions of *L1*.

Simple Sequence Repeats. Tandem arrays of short, repeated sequences are another common kind of multicopy sequence. This category includes telomeres, satellites, microsatellites, and minisatellites. Telomeres, described in the previous chapter, are 2 to 6 nt repeated in blocks of 0.1 to 20 kb found at the ends of chromosomes.

Satellites are tandemly arranged, very highly repeated (10^3 to 10^6 copies), short sequences (mostly 5 to 200 bp) (Figure 6.5). Satellites are concentrated in centromeric and telomeric heterochromatin. Regional centromeres, as in mammals, consist of satellite sequences. Satellite DNA was first identified from minor fractions of genomic DNA separated by physical methods, owing to high or low GC content.

Fig 6.5 Satellite sequences from three organisms.

Species	Name	Length	Copy number	Other information
Fly*	I	7	$\sim 10^7$	ACAAACT
Cow	1.706	2350	$\approx$5000	
Human	α	340	$\sim 10^6$	double 170-bp sequence; common to primates

Drosophila virilis

Minisatellites, also known as variable number tandem repeats, are tandemly arranged, repeated (10 to 10^3 copies), short sequences (usually 10 to 70 bp), in a block of 0.1 to 20 kb. Minisatellites are scattered throughout animal and plant euchromatin. Each minisatellite family is hypervariable and is found at 2 to 50 chromosome loci. A minisatellite can influence the expression of a nearby gene. In many species, the proterminal regions of telomeres contain minisatellites.

Microsatellites are repeated (10 to 100 copies), 1- to 5-bp sequences, most commonly di-, tri-, or tetranucleotide sequences in a block of 10 to 400 nt. About 10^5 microsatellite loci are scattered throughout animal and plant euchromatin.

Minisatellite (human example)   Microsatellite (human example)
$(AGAGGTGGGCAGGTGG)_n$ $n \approx 30$ $(CA)_n$ $15 < n < 100$

Mitochondria and Chloroplasts

Fig 6.6 Genomes in the mitochondria of various taxa.

Genomes of diverse mitochondria

Type of genes	human	yeast	plant[1]	flagellate[2]
23S rRNA	1	1	1	1
16S rRNA	1	1	1	1
5S rRNA	0	0	1	1
tRNA	22	24	22	26
RNase P[3]	0	1	0	1
Ribosomal proteins	0	1	7	27
Elongation factor	0	0	0	1
RNA polymerase	0	0	0	4
Respiration	13	16	16	24
Protein import/ maturation	0	0	3	6
Chromosome (kb)	17	75	367	69

[1]The mustard plant *Arabidopsis thaliana*   [2]The flagellate *Reclinomonas americana*
[3]An enzyme that cuts dsRNA

Further Reading

Casjens S. 1998. The diverse and dynamic structure of bacterial genomes. *Annu. Rev. Genet.* 32:339–377.

Cerutti H. 2003. RNA interference. *Trends Genet.* 19:39–46.

Elder JF, Turner BJ. 1995. Concerted evolution of repetitive DNA sequences in eukaryotes. *Z. Rev. Biol.* 70:297–320.

Kidner CA, Martienssen RA 2003. Macro effects of microRNAs in plants. *Trends Genet.* 19:13–826.

Liebert CA *et al.* 1999. Transposon Tn21, flagship of the floating genome. *Microbiol. Molec. Biol. Rev.* 63:507–522.

Sato S *et al.* 1999. Complete structure of the chloroplast genome of *Arabidopsis thaliana*. *DNA Res.* 6:283–290.

Chapter 7

RNA Synthesis 1: Transcription

Overview

Every gene has a single role: to encode RNA via **transcription**. Transcription is the synthesis of a single strand of RNA – a **transcript** – whose nucleotide sequence is complementary to a portion of a gene. As molecular processes go, transcription is highly regulated, moderately accurate, and, once started, fast-paced. Transcription is performed by a small troop of large proteins, chief among which is **RNA polymerase**. RNA polymerase uses the **template strand** of DNA to synthesize a complementary strand of RNA in a **5′ to 3′** direction.

Transcription is a play in four acts: (1) **binding** of RNA polymerase to a gene's **promoter**, (2) **initiation** of the RNA chain, (3) **elongation** of the RNA chain, and (4) **termination** of transcription.

In many cases, RNA synthesis does not end with transcription, because some transcripts are processed into mature RNAs by enzymatic modifications. The next chapter describes posttranscriptional processing.

Polymerization Reaction

The key chemical reaction in RNA polymerization is an esterification, in which the α phosphate of ribonucleotide triphosphate (NTP) is added to the 3′ oxygen of the growing polynucleotide chain, yielding a lengthened chain plus diphosphate. The substrates for the reaction are the four NTPs: ATP, GTP, CTP, and UTP. Synthesis of an RNA molecule begins at its 5′ end, proceeds by adding one nucleotide at a time, and stops at its 3′-OH end. RNA polymerase catalyzes the polymerization reaction (Figure 7.1).

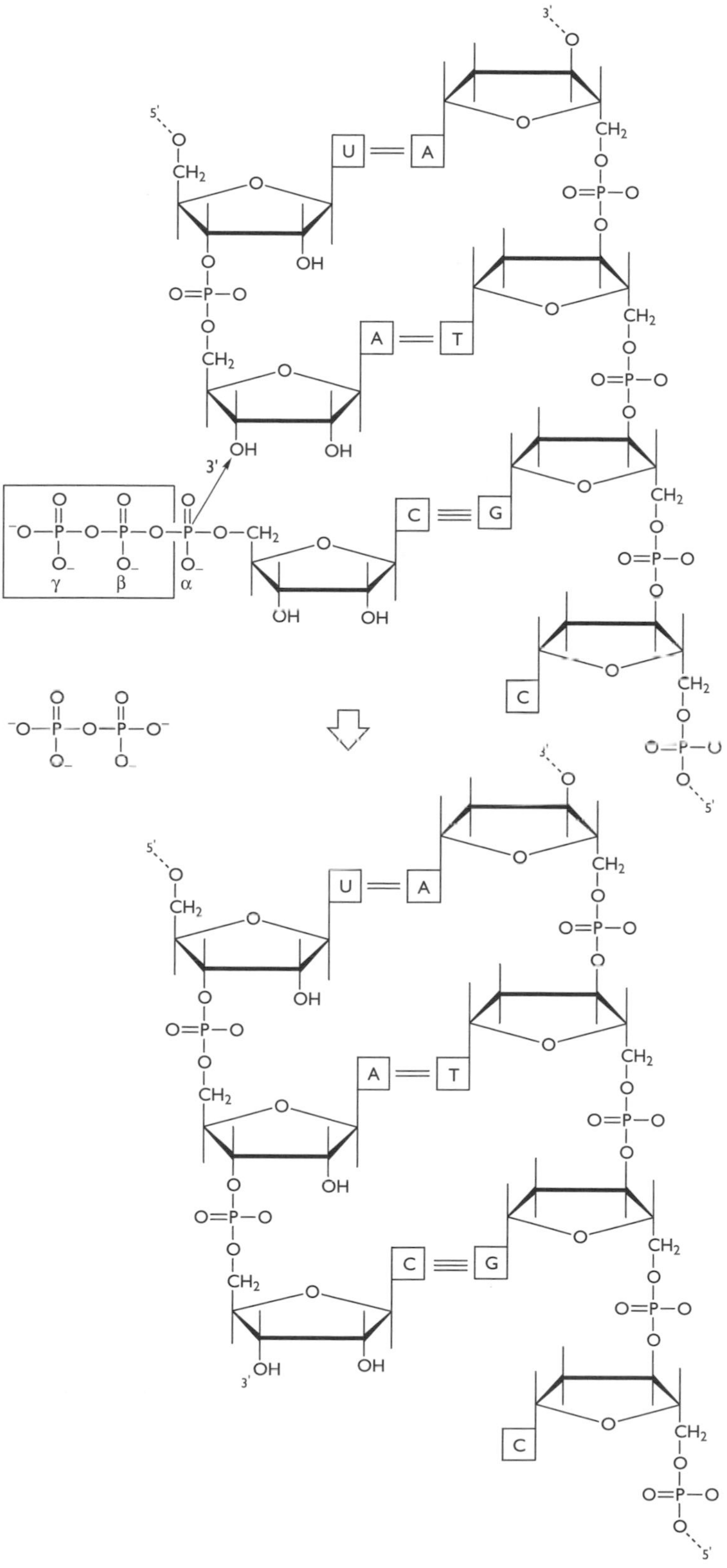

Fig 7.1 Polymerization of RNA.

RNA Polymerase

RNA polymerase is the main actor in each step of transcription, including the polymerization reaction. Every life form must use a version of this enzyme. Bacterial RNA polymerase is described here, and the corresponding enzymes of other life forms are described at the end of the chapter.

Bacterial RNA polymerase exists in two main forms, a holoenzyme (holo = complete), which is required for the binding and initiation steps, and a core enzyme, which can catalyze the polymerization of RNA but cannot initiate its synthesis. In bacteria the holoenzyme is a large pentamer, $\alpha_2\beta\beta'\sigma$, and the core enzyme is a tetramer, $\alpha_2\beta\beta'$. There are $\sim 10^4$ copies of RNA polymerase holoenzyme per cell of *Escherichia coli* (Figure 7.2). A single gene encodes each subunit of the core enzyme, α, β, and β', but *E. coli* has at least six different holoenzymes, as six different σs have been identified. *Bacillus subtilis* is similar, except that it has at least nine different σs. Each σ protein binds specifically to a subset of regulatory sequences characteristic of genes. The RNA polymerase of *E. coli* has other auxiliary subunits, including ρ and NusA. Both ρ and NusA are termination factors that help RNA polymerase stop transcribing at the appropriate place.

Transcript's Template

The sequence of nucleotides in the transcript is precisely complementary to the sequence of nucleotides in its gene. Except for RNA viruses, the template for transcription is a segment of one strand of DNA.

A gene is a single, continuous segment of a chromosome, containing both coding and regulatory sequences necessary for transcription to occur. The strand of DNA that acts as a template for RNA synthesis is the **template strand**; its complementary strand is

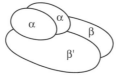

Fig 7.2 RNA polymerase in *E. coli*.

Core enzyme   Holoenzyme

β - 151 kDa
β' - 156 kDa
α - 36 kDa
σ - 70 kDa, predominantly
Total ≈ 450 kDa

16 nm, as long as 48 bp of dsDNA

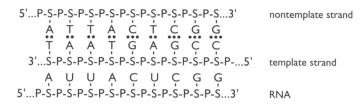

Fig 7.3 The template strand and the nontemplate strand of DNA.

the **nontemplate strand** (Figure 7.3). Common synonyms for non-template strand are complementary strand and coding strand. Each gene is said to have a 5′ and a 3′ end, corresponding directly to the 5′ and 3′ ends of the transcript; by this convention, the gene's nucleotide sequence is the nontemplate sequence rather than the template sequence. Upstream and downstream refer to the 5′ and 3′ directions of the nontemplate strand of DNA.

RNA polymerase begins transcription consistently at one particular starting nucleotide – not one nucleotide to either side of it – and transcription stops precisely at a signaled position in the gene, as well. How does RNA polymerase work with such precision? Before initiating transcription, RNA polymerase first binds to a regulatory sequence of DNA known as the **promoter**. A gene's promoter (some genes have more than one promoter) is close to the starting site of RNA synthesis and, in most genes, upstream of the start site. RNA polymerase binds to the promoter before synthesis is initiated. Auxiliary binding sites for regulatory proteins are typically close to the promoter and farther upstream; if they are distant, they are called **enhancers**. The sequence at the 3′ end of a gene that facilitates the termination of transcription is the **terminator**. In Figure 7.4, R is an auxiliary binding site (regulatory sequence), P is the promoter, and T is the terminator. Nucleotides of a gene whose DNA sequence is known are numbered: the starting site of transcription is +1 and the residues upstream receive negative numbers. A "zero" position is defined not to exist, just as there was no "year zero" in the most commonly used Western calendar.

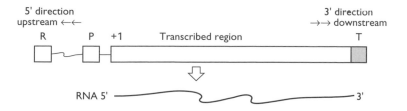

Fig 7.4 Anatomy of a gene.

Transcription in Bacteria

Binding of RNA Polymerase to the Promoter

Every journey begins with a single step. The first step of transcription is the binding of RNA polymerase to a promoter. A gene's rate of transcription, including whether it happens at all, depends on the rate of binding. The maximal speed of binding varies greatly among promoters – from once per second to once per many minutes in active genes; at the other end of the spectrum, binding may be blocked for a cell's lifetime. In bacterial protein-coding genes and rRNA-coding genes, promoters have two common elements 35 and 10 bp upstream of the starting site of transcription. The **spacer region** is between the -35 and -10 sites, and the **discriminator region** is between the -10 and $+1$ sites. The holoenzyme binds to dsDNA at the -35 site and helps melt double-stranded DNA (dsDNA) at the -10 site. The -35 and -10 sequences vary from promoter to promoter; rarely are both sequences identical to the consensus sequences. (Every nucleotide of a consensus sequence is the most common one at that position.) Each of several σ subunits recognizes different sets of promoter sequences. Promoters vary in efficiency and strength. The stronger the promoter, the faster RNA polymerase binds to it. Most enzymes encounter substrates by random collisions in three-dimensional space. Not bacterial RNA polymerase. It locates the promoter much faster, by zipping along the DNA molecule – one-dimensional diffusion.

Many bacterial genes have upstream sites, typically $\sim 10^2$ bp upstream of the -35 site, where regulatory proteins bind (Figure 7.5). After binding to the upstream site, these regulatory proteins bend the DNA nearby; RNA polymerase binds preferentially to bent DNA. After RNA polymerase binds to a promoter, the enzyme's downstream domains melt one turn of the DNA double helix at the transcription starting point ($+1$); the enzyme then binds tightly to dsDNA, downstream. The RNA polymerase bound to the ≈ 11 bp of melted DNA in the region around the $+1$ nucleotide is called the **open complex**, which is ready for the next step, chain initiation (Figure 7.6).

Fig 7.5 The σ-binding promoter of *E. coli*, nontemplate consensus sequence.

| TTGACA | | TATAAT | | A |
| -35 | 16 bp
spacer | -10 | 8 bp
discriminator | $+1$ |

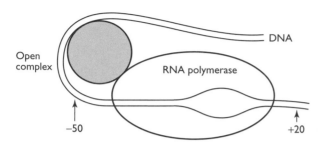

Fig 7.6 RNA polymerase bound to partially melted DNA; auxiliary protein (hatched) bound to upstream regulatory sequences.

Chain Initiation

After RNA polymerase binds to the promoter and makes an open complex, transcription begins by initiating synthesis of an RNA chain. To start, the first nucleotide, usually ATP or GTP, base-pairs with the first nucleotide of the DNA template strand and also binds to an initiation site in RNA polymerase. Following this, an RNA oligonucleotide of ≈10 nt is synthesized. More frequently than not, chain initiation is aborted, yielding a 2- to 9-nt fragment of RNA. In the event of abortive initiation, RNA polymerase holoenzyme remains bound to DNA and the initiation step restarts (Figure 7.7). If chain growth continues, then σ dissociates from RNA polymerase, and chain elongation begins. This step is usually short, ≈1 second.

Chain Elongation

After chain initiation, transcription suddenly gains speed. RNA polymerase core enzyme translocates downstream, and the transcript grows one nucleotide at a time. Weak chemical bonds hold together the complex of DNA, RNA, and RNA polymerase. RNA polymerase has two DNA-binding sites: site I binds downstream DNA and site II binds upstream DNA. Similarly, the enzyme has two RNA-binding sites: site I binds downstream RNA and site II binds upstream RNA.

As RNA polymerase unwinds dsDNA, the DNA becomes positively supercoiled immediately downstream of the unwinding point, for as DNA unwinds it spins one rotation every 10 bp. DNA

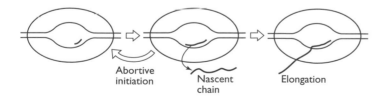

Fig 7.7 Chain initiation aborts or else leads to chain elongation.

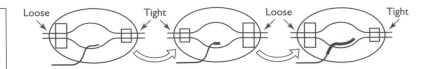

| Fig 7.8 | Chain elongation step. Upstream and downstream DNA-binding sites (boxes) alternatively hold DNA in place; thick line at 3′ end of transcript represents newly synthesized sequence. |

gyrase relaxes this positively supercoiled region. In *E. coli*, DNA gyrase is a 374-kDa heterodimer; it requires ATP and cleaves both strands of DNA, thus preventing the entire DNA molecule from rotating and writhing.

It is widely held that RNA polymerase works like an inchworm, moving discontinuously and changing shape as the RNA chain lengthens (Figure 7.8). In this model:

- DNA binding site I and RNA binding site I (the sites that bind downstream DNA and RNA) are locked into place as a short stretch of RNA is synthesized. RNA polymerase's upstream binding sites slide forward 1 nt at a time.
- DNA binding site II (upstream) locks the DNA and the RNA binding site II locks the RNA as DNA binding site I and RNA binding site I relax and the polymerase molecule moves downstream. RNA polymerase melts dsDNA as the enzyme moves down the helix.
- A stable RNA-DNA hybrid does not form; instead, newly synthesized RNA binds to polymerase.
- The cycle repeats.

RNA polymerase may pause for seconds or minutes when it encounters certain template sequences or hairpins in the RNA. NusA, which binds to RNA polymerase early in the elongation phase, can facilitate and prolong pausing.

Chain Termination

Transcription ends precisely, by one of two distinct mechanisms in *E. coli*, **intrinsic termination** and **ρ-dependent termination** (Figure 7.9); some genes have both intrinsic and ρ-dependent terminators. Intrinsic terminators are DNA sequences at the 3′ ends of genes, which code for RNA hairpins that weaken binding to RNA polymerase. The intrinsic terminator sequence encodes RNA that self-base-pairs to make a hairpin structure and, immediately downstream, a U-rich sequence. The hairpin and the U-rich sequence displace RNA bound to the RNA binding site of RNA polymerase, thereby causing the transcript to

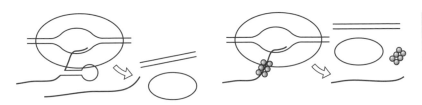

Fig 7.9 Intrinsic termination (left). ρ-dependent termination (right).

dissociate from the complex. ρ-dependent termination depends on ρ, a 276-kDa homohexamer that binds to sequences at the 3′ end of a transcript, causing RNA polymerase to dissociate from both template and RNA; ρ binds to a ribosome-free zone of RNA downstream of sequences that signal the end of protein synthesis (translation).

Transcription in Archaea

Archaea possess a single RNA polymerase ($\approx$500 kDa), but unlike bacterial RNA polymerase, it has many subunits (12 subunits in *Methanococcus jannaschii*) homologous to those of eukaryal RNA polymerases. The consensus promoter of archaeal genes contains TATAAAA, 25 bp upstream of the +1 starting site; this 7-bp sequence is called a **TATA box**. Most archaeal regulatory sites lie upstream of the TATA box. Regulatory proteins called **transcription factors** (TFs) bind to regulatory sites and to the TATA box. RNA polymerase then binds to these TFs during the binding step of transcription (Figure 7.10). Archaea also possess terminator proteins and antiterminator proteins.

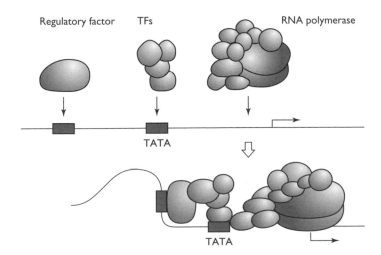

Fig 7.10 The binding step in archaea.

RNA polymerase	Type of RNA transcribed
I	large ribosomal RNA, the precursor of 5.8S, 18S, and 28S rRNA
II	pre-mRNA and most snRNAs (small nuclear RNAs, which help splice pre-mRNA)
III	various small RNAs: tRNA, 5S rRNA, one snRNA, snoRNA (snoRNAs help process pre-rRNA), and scRNAs (small cytoplasmic RNAs = regulators of protein synthesis)

Fig 7.11 RNA polymerase and types of RNA transcribed in eukarya.

Transcription in Eukarya

In stark contrast to bacteria and archaea, eukarya have nuclei, where the genome is housed and where all cellular RNA synthesis happens. Only after RNA synthesis is complete – i.e., only after posttranscriptional processing – do RNA molecules move from nucleus to cytoplasm. Eukarya are also unique in possessing three distinct RNA polymerases, each transcribing a distinct class or type of gene (Figure 7.11).

Eukaryal core RNA polymerases are large (400 to 700 kDa) and have ≈10 subunits, many of which are homologous to archaeal RNA polymerase subunits. Each RNA polymerase has two large subunits (>100 kDa) homologous to β and β' subunits of *E. coli*, and 4 to 12 smaller subunits. If a holoenzyme is defined to include TFs, then it can be very large indeed. Many eukaryal TFs are named with letters and numbers indicating the class of gene whose transcription they stimulate; e.g., TFIIA assists RNA polymerase II.

Eukaryal RNA polymerases locate promoters (Figure 7.12) by first binding to regulatory proteins, including TFs, which bind to the promoter or to nearby sequences. Selectivity of RNA polymerases is ensured by the joint effect of many TFs.

Histones and some nonhistone proteins of chromatin repress transcription by blocking the binding step. Proteins must be removed from DNA for a gene to be transcribed, because RNA polymerase surrounds DNA during the initiation and

Fig 7.12 Three classes of genes in eukarya are transcribed by different RNA polymerases.

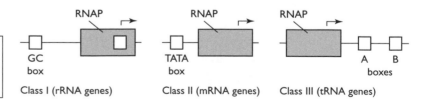

Class I (rRNA genes)	Class II (mRNA genes)	Class III (tRNA genes)

Fig 7.13 Binding step, RNA polymerase I.

elongation phases. Nucleosomes may re-form immediately after transcription.

RNA Polymerase I

For RNA polymerase I, which transcribes 45S rRNA in the nucleolus, the binding step requires a few DNA-binding proteins, one to bind to the promoter and sometimes another to bind upstream regulatory sequences (Figure 7.13). Vertebrates use two regulatory proteins, SL1 and UBF. UBF binds to an upstream sequence; then and only then does SL1 bind to the promoter, after which RNA polymerase I binds to both SL1 and to DNA, completing the binding step. Transcription is terminated by DNA sequences $\sim 10^2$ nt downstream of the 3' end of the coding portion. Proteins bind to these terminators and help RNA polymerase dissociate from DNA and RNA.

RNA Polymerase III

For RNA polymerase III, which transcribes the precursors of various small RNAs, the binding step requires TFIIIB, A, and C (Figure 7.14). Terminators for RNA polymerase III are short downstream sequences.

RNA Polymerase II

RNA polymerase II uses far more TFs than the other RNA polymerases (Figure 7.15). This makes sense, for the number of class II genes in a genome is huge compared with the numbers of class I or class III genes. The promoters of class II genes contain a TATA box, to which TFIID binds with the help of TFIIA. Subsequently, RNA polymerase II and several other TFs bind to DNA to complete the binding step. Many TFs are themselves multimers; for example, TFIID consists of **TATA-binding protein (TBP)** and several other proteins known as **TBP-associated factors (TAFs)**. As in bacteria, abortive initiation and extensive pausing during elongation are common.

Fig 7.14 Binding step, RNA polymerase III. A-C = TFIIIA-C.

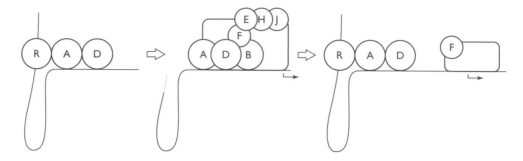

Fig 7.15 Binding step, RNA polymerase II R = regulatory protein bound to enhancer; A-J = TFIIA-J.

Termination of transcription is signaled, in many cases, by A-rich sequences on the template strand, $\sim 10^2$ nt downstream from the 3′ end of the coding portion.

Transcription in Mitochondria and Chloroplasts

The RNA polymerase of mitochondria is small (140 kDa) and consists of a single polypeptide encoded by a nuclear gene. Mitochondrial RNA polymerase is homologous to that of bacteriophage T7, not to any known cellular RNA polymerase subunit. Human mitochondrial chromosomes have two promoters. Mitochondrial transcription requires at least one transcription factor. Three RNAs are transcribed in the nucleus and imported into mitochondria: some tRNAs, RNase P, and RNase MRP. The latter two RNAs are required for posttranscriptional RNA processing.

Chloroplasts have two RNA polymerases. The plastid chromosome encodes one, and a nuclear gene encodes the other RNA polymerase, a 110-kDa monomer resembling RNA polymerase of bacteriophage T7. The two RNA polymerases recognize different promoters and transcribe different genes.

Transcription in Viruses

Viruses, which are a heterogeneous collection of acellular parasites, many with bizarre genetic systems, have evolved many times from cellular genomes. Accordingly, their transcriptional mechanisms are highly varied. Most viruses use the host's RNA polymerase to transcribe viral genes, but some large DNA viruses and some RNA viruses encode their own RNA polymerases.

Bacteriophage RNA polymerases are small monomers. In bacteriophage T7, it is a 110-kb monomer that transcribes ≈80% of the T7 genome (the host RNA polymerase transcribes the other 20%). T7 RNA polymerase recognizes only its own promoters, all of which contain a 16-bp sequence. T7 polymerase also recognizes a single terminator. Coliphage N4 encodes two RNA polymerases. Vaccinia, an animal virus, encodes its own RNA polymerase.

Some bacteriophages, such as T4 and λ, modify host RNA polymerase. Phage T4 modifies the α subunits by ADP-ribosylation, phosphorylates the β and β' subunits, and makes other chemical modifications. T4 transcription factors activate transcription of viral genes. Phage λ has two proteins, N (14 kDa) and Q (23 kDa), which bind to host RNA polymerase and cause it to read through terminators in the λ phage chromosome.

Most RNA viruses have single-stranded RNA chromosomes that code directly for protein [(+) strands]. Regardless of whether an RNA virus genome is single-stranded or double-stranded, the genome must be replicated. If RNA is used as a template for RNA synthesis, then the RNA-dependent RNA polymerase is called a **replicase** or **transcriptase**, depending on its exact role in the viral life cycle. These RNA polymerases transcribe the entire viral genome.

Retroviruses replicate their genomes via a DNA intermediate, synthesized by **reverse transcriptase** (RNA-dependent DNA polymerase). Retroviruses use host RNA polymerase II to transcribe the DNA copy of their genome.

Speed and Accuracy of Transcription

In *E. coli*, the average rate of RNA chain growth during chain elongation is approximately 45 nt s^{-1} (nucleotides per second), for genes transcribed with little pausing. That rate of synthesis – 2700 nucleotides per minute – requires that DNA be unwound at the stately speed of about 270 rpm. The *in vivo* rate of elongation for class II genes of eukarya is $\approx$25 nt s^{-1}. What makes eukarya such slowpokes? For one thing, DNA in eukaryal chromosomes is wrapped around histones as well as being bound to other proteins that need removing before transcription can proceed. The speed of elongation by T7 RNA polymerase is remarkably high, up to 400 nt s^{-1}.

In organisms, transcription has an error rate of $\sim$10^{-4}, which is to say that 1 in every 10,000 or so nucleotides is a wrong one. Given the large number of essential cellular proteins, this error rate seems extraordinarily high. Viral replicases and transcriptases are also error-prone, so that RNA viruses have a high rate of mutation (stably inherited changes in the genome).

Further Reading

Baumann P *et al.* 1995. Transcription: new insights from studies on Archaea. *Trends Genet.* 11:279283.

Bell SD, Jackson SP. 1998. Transcription and translation in archaea: a mosaic of eukaryal and bacterial features. *Trends Microbiol.* 6:222–228.

Latchman DS. 1998. *Eukaryotic Transcription Factors,* 3rd ed. Academic Press, London.

Nudler E *et al.* 1994. Discontinuous mechanism of transcription elongation. *Science* 265:793–796.

Roberts JW. 1996. Transcription termination and its control. In Lin ECC, Lynch AS (eds.) *Regulation of Gene Expression in Escherichia coli.* R.G. Landes, New York.

Uptain SM *et al.* 1997. Basic mechanisms of transcript elongation and its regulation. *Annu. Rev. Biochem.* 66:117–172.

Chapter 8

RNA Synthesis 2: Processing

Overview

In many cases, newly transcribed RNA is not ready for duty; instead, it is a precursor of mature RNA and must undergo posttranscriptional **RNA processing** to become functional. Also known as **RNA maturation**, posttranscriptional processing falls into seven categories:

- cleaving and trimming
- 3′ nucleotide addition
- base modification
- splicing
- capping (modifying the 3′ end of pre-mRNA)
- polyadenylation (adding a sequence of adenosines to the 5′ end of pre-mRNA)
- editing (altering the coding sequence of pre-mRNA)

The exact nature of every maturation process depends on the life form and the type of RNA (rRNA, tRNA, mRNA). This chapter describes the seven types of processing in turn, noting differences among life forms. At chapter's end, some processing pathways are summarized.

Cleaving and Trimming

Some fresh transcripts contain extraneous sequences, and some contain two or more sequences that need to be separated. Endonucleases may cleave transcripts into smaller pieces to remove end sequences or internal spacers. In some instances, exonucleases trim the ends, removing nucleotides one by one.

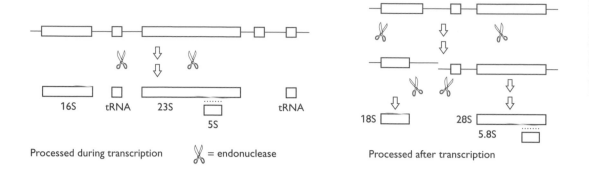

Processed during transcription   ✂ = endonuclease   Processed after transcription

Fig 8.1 Pre-rRNA cleaving in bacteria (left). Pre-rRNA cleaving in mammals (right).

Large rRNAs are encoded by genes whose transcripts are cleaved in multiple steps to yield smaller RNAs. In bacteria, each of the six copies of the rRNA gene contains 16S, 23S, and 5S rRNA sequences and one or two tRNAs; the functional and nonfunctional sequences are interspersed (Figure 8.1). Processing by ribonucleases (RNases) begins before transcription is complete. In eukarya, the entire pre-rRNA molecule is transcribed before it is processed; both steps are carried out in nucleoli (Figure 8.1). In mammals, a 45S primary transcript matures into 18S, 5.8S, and 28S rRNAs. rRNA is cleaved in a similar way in mitochondria and chloroplasts.

Nearly all tRNA is cleaved and trimmed during maturation (Figure 8.2). In bacteria, cotranscription of multiple tRNAs is commonplace. An endonuclease cleaves the transcript, and exonucleases trim the cleaved products. In *Escherichia coli* six distinct 3' exonucleases participate in tRNA trimming. In archaea and eukarya, each tRNA is transcribed individually. The pre-tRNAs of archaea, eukarya, mitochondria, and chloroplasts are cleaved and, in some instances, trimmed as well.

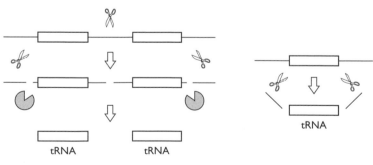

Fig 8.2 Pre-tRNA cleaving and trimming in bacteria (left). Pre-tRNA cleaving in eukarya (right).

Processed during transcription   ◖ = exonuclease   Processed after transcription

3′ Nucleotide Addition (tRNA)

All tRNAs terminate with -CCA on the 3′ end, although in many cases the transcript lacks one to three of these nucleotides. Most bacterial tRNA genes include the full sequence; if not, the nucleotides are added enzymatically to complete the tRNA sequence (Figure 8.3). It is not known whether archaeal tRNAs undergo nucleotide addition. Most eukaryal pre-tRNAs lack -CCA, so the three nucleotides are added during posttranscriptional maturation by a nucleotidyl transferase after the 5′ and 3′ ends are trimmed.

Base Modification

Both rRNAs and tRNAs contain dozens of bases other than A, G, U, and C. Unusual bases cannot be incorporated regularly during transcription, so rRNAs and tRNAs undergo extensive base modification. Each modification is catalyzed by one or more enzymes; E. coli has about 50 of them. Little is known about the enzymology of base modification. Nucleoli contain more than 100 kinds of small nucleolar RNAs, which pair with pre-rRNA near methylation sites and bind to 2′-O-methylases, guiding these enzymes to methylate the 2′ oxygen of the appropriate ribose. For historical reasons, the modification of bases in pre-mRNA is called editing, described near the end of this chapter.

Splicing

For some genes, one or more interior sequences (**intervening sequences** or **introns**) are present in a transcript but absent from the mature RNA molecule. An intron is flanked on both sides by **exons**, which include the protein-coding parts. Intron and exon refer both to the transcribed RNA sequences and to the DNA

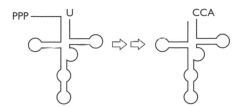

Fig 8.3 Trimming eukaryal pre-tRNA and adding-CCA to the 3′-OH end.

sequences that encode them. Splicing is the process of removing introns and ligating exons. Introns are common in eukarya and are found in the genomes of some representatives of all the other life forms. Some pre-tRNAs and some pre-mRNAs are spliced.

In plants and animals, multi-intron genes are more the rule than the exception, and multiple introns may be spliced in alternative patterns. Alternative splicing (Chapter 12, 25) expands the number of final gene products encoded by a single gene.

Four main mechanisms of splicing are known; each is associated with a distinct type of intron: **tRNA, spliceosomic, group I,** and **group II introns.**

Pre-tRNA Introns

In eukarya and archaea, but not in bacteria, some tRNA genes have introns. Yeast pre-tRNA introns range in size from 14 to about 60 nt, commencing 1 nt downstream of the anticodon. The splicing endonuclease, which in yeast is a 124-kDa trimer, cleaves pre-tRNA at both ends of the intron, generating a 2′,3′-cyclic phosphoryl terminus on the upstream exon and a 5′ OH terminus on the downstream exon, then ligates the ends, restoring a normal 5′-3′ linkage (Figure 8.4).

Spliceosomic Introns

Protein-coding genes of eukarya often contain introns, although there are plenty of intronless eukaryal genes. Most protein-coding genes of *Saccharomyces cerevisiae*, for example, lack introns. In plants and mammals genes have on average ≈ 4 introns, $\sim 10^2$ nucleotides each.

Introns are removed from pre-mRNA in nuclei by enzyme complexes known as spliceosomes, big aggregates of protein and RNA located in speckled bodies of the nucleus. The RNAs of the spliceosome include the small nuclear RNAs (snRNAs) U1, U2, U4,

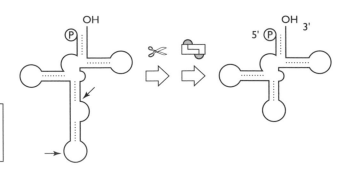

Fig 8.4 Removing introns from pre-tRNA and splicing. = ligase.

Exon	Intron	Exon
... sAG<u>GU</u>...	ACAUCAU...	<u>AG</u>NNNN

Fig 8.5 Consensus sequence, vertebrate splice junctions.

U6, and U5. These snRNAs were given U designations because they are rich in uridine. snRNAs associate with proteins to make particles called small nuclear ribonucleoproteins (snRNPs) ("snurps"), usually one snRNA and ≈10 proteins per snRNP. A spliceosome, made of many snRNPs, is a huge molecular machine, ≈25 × 25 × 50 nm and massing 3 to 5 MDa – a bit larger than a ribosome.

The splice junctions contain conserved sequences, and the U-RNAs base-pair with nucleotides at the splice junctions (Figure 8.5). Most commonly in vertebrates, introns have GU on the 5′ side, AG on the 3′ side, and A near the splice junction.

Splicing has two main steps: (1) attack of the 3′-most G in the upstream exon by the 2′ OH of an A in the intron, cutting the exon intron bond and making a lariat structure, and (2) attack of the 5′ nucleotide in the downstream exon by the 3′ OH of the G in the upstream exon, completing the splice (Figure 8.6).

Group I and Group II Introns

Group I and group II introns are small transposons, and many of them are completely or partly self-splicing; that is, the RNA of

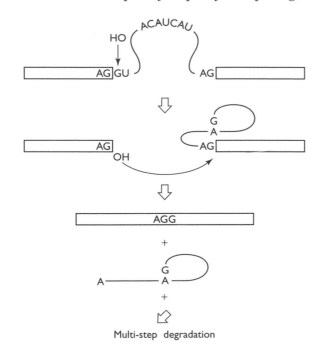

Multi-step degradation

Fig 8.6 Pre-mRNA in eukarya spliced by spliceosomes.

the intron itself is a ribozyme that performs or assists splicing. Transposition of one of these introns may create a novel intron in another gene.

Group I introns, ranging from 0.2 to >1 kb, have been found (1) in the rRNA genes of some early-evolving eukarya (e.g., *Tetrahymena* and *Physarum*) and of some mitochondria and chloroplasts, and (2) in some protein-coding genes of bacteria, mitochondria, chloroplasts, and a few bacteriophages.

Group II introns, typically ≈2 kb, have been found in protein-coding genes of several life forms: (1) purple bacteria, (2) cyanobacteria, (3) mitochondria of plants and fungi, (4) chloroplasts, and (5) algae.

During splicing, group I and II introns form lariat structures, but the chemical reactions are different for each and different in turn from those of spliceosomic splicing.

Capping

Caps, unique to eukarya and some of their viruses, are unexpected structures at the front ends of most class II transcripts (mRNAs and all the U-RNAs except U6): an alkylated guanosine is connected at the 5' end by a 5'-5' triphosphate bridge. Capped mRNA

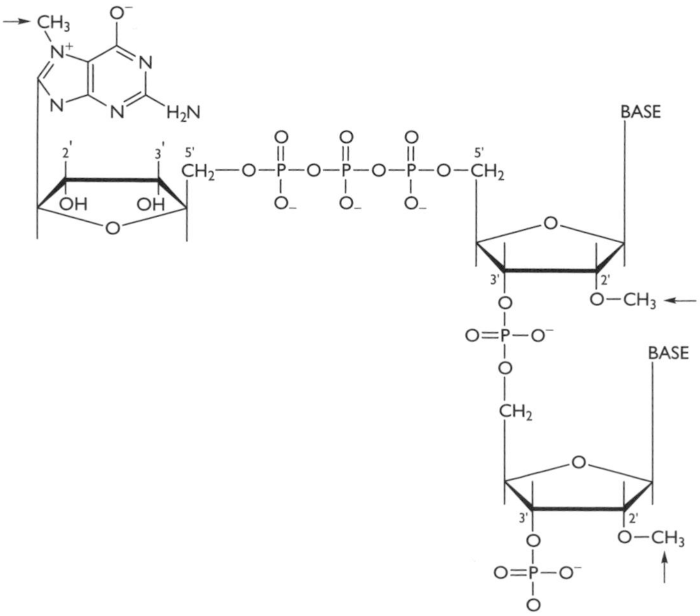

Fig 8.7 Structure of the cap on mRNA; arrows point to added methyl groups.

is resistant to digestion by RNases that attack the 5′ ends of RNA molecules. Therefore capping appears to be a means of extending the half-life of mRNAs in eukarya.

The capping reaction takes place immediately after the 5′ end of the RNA molecule has been synthesized (Figure 8.7). In pre-mRNA, first guanosine triphosphate (GTP) reacts with the 5′nucleotide, forming a 5′-5′ triphosphate bridge. Next, the terminal guanosine is methylated, and subsequently nucleotides +1 or +1 and +2 are methylated. In the cap of U-RNA, the terminal guanosine is trimethylated

Polyadenylation

Freshly transcribed mRNA is vulnerable to attack by RNases at the 3′ end. Addition of a short poly(A) tail to the 3′ end of a transcript protects mRNA and extends its half-life in the cytoplasm.

A template-independent RNA polymerase adds a poly(A) tail ~200 bases long to the 3′ end of pre-mRNA, at a site 10 to 30 bp downstream of a polyadenylation signal, AAUAAA (Figure 8.8). The RNA is first cleaved by a complex including snRNPs and a protein that binds specifically to the polyadenylation signal. In some cases, the region downstream of the cleavage site is extensive, with as many as thousands of nucleotides.

Editing

In nearly all mRNAs, the protein-coding part is encoded directly by DNA. Editing is an exception to this rule. The editing of mRNA involves the insertion or deletion of single nucleotides or oxidation of single bases.

Most of the mRNA of mitochondria in *Trypanosoma* and related unicellular eukarya is edited by the precise insertion or

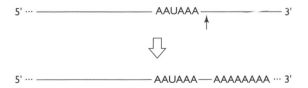

Fig 8.8 Polyadenylation of pre-mRNA; arrow marks site where endonuclease cuts RNA.

Fig 8.9 Editing of
mitochondrial pre-mRNA in
Trypanosoma.

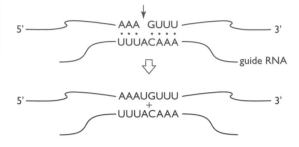

deletion of single uridines (U) in pre-mRNA, consequently affecting profoundly the sequence of amino acids encoded by the mature mRNA; 95% of editing events are insertions, and 5% are deletions (Figure 8.9). As many as 100 or so uridines are inserted into some transcripts. Guide RNAs facilitate the insertion of uridines; guide RNAs are short RNAs that are complementary to edited sequences. During insertion, a uridine derived from free uridine triphosphate (UTP) is added enzymatically to the 3′ end of an upstream pre-mRNA cleavage product, while during deletion a uridine is taken away from the 3′ end of an upstream pre-mRNA cleavage product to yield uridine monophosphate (UMP).

Some of the mRNA of animals (*Caenorhabditis elegans, Drosophila melanogaster,* and *Bombyx mori*), and perhaps of other eukarya, is modified by the enzyme double-stranded RNA adenosine deaminase (abbreviated dsRAD or DRADA). dsRAD, whose only substrate is duplex RNA greater than about 35 bp, catalyzes the conversion of adenine to inosine, which is read as guanosine during protein synthesis. Most pre-mRNAs can form regions of duplex RNA. Another deaminating enzyme, apolipoprotein B (ApoB), converts C to U. The consequence of editing is to alter the amino acid sequence of a polypeptide translated from the mRNA (Figure 8.10).

Fig 8.10 mRNA editing and changes in coding sequence.

Change	Enzyme	In protein synthesis
A → I	dsRAD, works on dsRNA	I can pair with C, U, or A
C → U	ApoB, depends on nearby pre-mRNA sequences	U pairs with A more easily than with G

Maturation Pathways

Figure 8.11 shows RNA maturation pathways in eukarya.

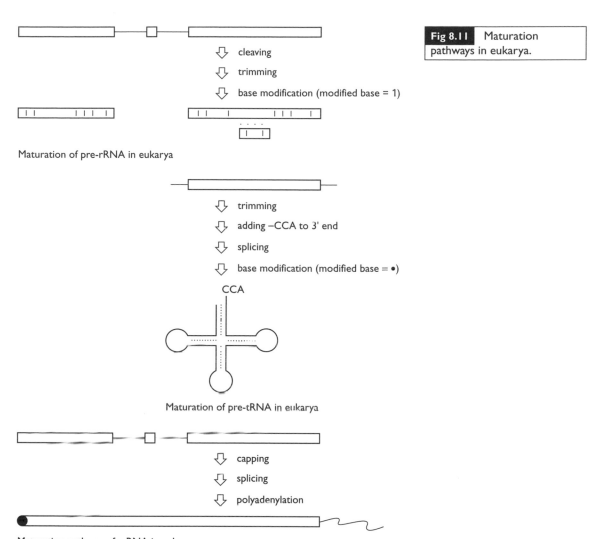

Fig 8.11 Maturation pathways in eukarya.

Maturation of pre-rRNA in eukarya

Maturation of pre-tRNA in eukarya

Maturation pathway of mRNA in eukarya

Further Reading

Eichler DC, Craig N. 1994. Processing of eukaryotic ribosomal RNA. *Prog. Nucleic Acid Res. Mol. Biol.* 49:197–239.

Krainer AR (ed.). 1997. *Eukaryotic mRNA Processing.* IRL Press, Oxford.

Söll D, RajBhandary (eds.). 1995. *tRNA Structure, Biosynthesis, and Function.* ASM Press, Washington, DC.

Srivastava AK, Schlessinger D. 1990. Mechanism and regulation of bacterial ribosomal RNA processing. *Annu. Rev. Microbiol.* 44:105–129.

Chapter 9

Abundance of RNAs in Bacteria

Overview

All the gene-encoded molecules of a cell – proteins and RNAs – are needed in amounts that differ from gene product to gene product and that change over the cell's lifetime. For example, if a cell has too little ribosomal RNA (rRNA), then protein synthesis is retarded; similarly, if molecule X becomes the sole carbon source for a cell and it fails to respond by synthesizing enzymes to catabolize X, then it will be starved for carbon and energy. Because RNAs encode proteins, one way to control the quantity of all gene products is to control the abundance of each kind of RNA.

Opposing forces determine the abundance of RNA: synthesizing RNA increases its cellular concentration, while enzymatic degradation of RNA, cell growth, and cell division reduce its concentration. This chapter is about the molecular machinery that controls rates of RNA synthesis and degradation. Other ways of controlling the amount of gene products are presented elsewhere.

Some of the key concepts of RNA regulation are (1) **RNA stability**, (2) positive and negative regulatory proteins (**activators** and **repressors**) that bind to (3) regulatory DNA sequences (**promoters**, **upstream activating sequences**, and **enhancers**), and (4) a regulatory system, the **operon**. This chapter focuses on bacteria; other simple forms are considered very briefly.

Abundance of Stable RNAs – rRNA and tRNA

In *Escherichia coli* undergoing exponential growth, about 97% of RNA is rRNA and tRNA. There are two reasons for this large excess: rRNAs and tRNAs are much more stable than mRNA (less susceptible to degradation), and the rates of synthesis of rRNA and tRNA genes are very high.

The cell is occupied by squadrons of ribonucleases (RNases), which together degrade RNA into single nucleotides. In the absence of new synthesis, every species of RNA is degraded at a constant, characteristic rate (first-order kinetics). Both rRNAs and tRNAs are resistant to RNases and have long half-lives, which greatly exceed the time between cell divisions. By contrast, mRNA is short-lived, with a half-life of 30 seconds to 20 minutes.

Bacterial genomes have several rRNA genes, each of which encodes the three rRNAs (5S, 16S, and 23S). In exponentially growing cells, transcription of each rRNA gene is initiated as fast as once per second. Similarly, the rates of transcription initiation are very high for tRNA genes, which like rRNA genes exist in multiple copies. About half of the active RNA polymerase molecules in growing cells are transcribing stable RNAs. The genes encoding stable RNAs are co-regulated. The rate of transcription of rRNA and tRNA genes depends on three components: (1) regulatory DNA sequences of these genes, (2) proteins that bind to the regulatory sequences, and (3) small molecules that inhibit RNA polymerase. Only rRNA genes are considered here, as more is known about their regulation.

Regulatory Sequences of rRNA Genes and Proteins That Bind to Them

E. coli has seven rRNA genes, each with two strong promoters, P1 and P2, and three binding sites for Fis (factor for inversion stimulation) just upstream of P1. Furthermore, the 5′ part of each promoter is an UP (upstream) element, an additional binding site for RNA polymerase. Together, the Fis sites and UP elements stimulate transcription of rRNA genes ~10^2-fold compared with mutant genes lacking these sequences (Figure 9.1).

Fig 9.1 An rRNA gene (*rrnB*): P1, P2 = promoters; T1, T2 = terminators; transcription starts at arrows.

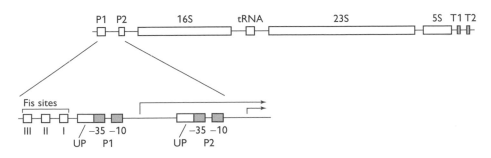

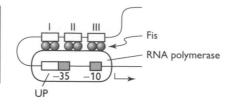

Fig 9.2 Fis stimulates the binding of RNA polymerase to promoters of genes encoding stable RNAs; upstream promoter (P1) shown here.

Fis (encoded by the *fis* gene) is a 22-kDa dimer that binds to one of three 15-bp sites, bends the DNA, and also binds to RNA polymerase, thereby stimulating P1-RNA polymerase binding (Figure 9.2). The concentration of Fis in the cell peaks at the onset of exponential growth, when demand for ribosomes is high. An influx of nutrients from the cell's medium stimulates a burst of transcription of *fis*, followed by synthesis of the Fis protein. Transcription is initiated at P1 much more than at P2 during periods of rapid growth, partly because of the stimulating effects of Fis.

Regulation of rRNA Genes by (p)ppGpp

In addition to the Fis protein, two unusual guanosine nucleotides regulate rRNA gene transcription: guanosine 5′-diphosphate 3′-diphosphate (ppGpp) and guanosine 5′-triphosphate 3′-diphosphate (pppGpp), together designated (p)ppGpp. (p)ppGpp binds to RNA polymerase, thereby blocking the ability of RNA polymerase to bind to the promoters, P1 and P2 (Figure 9.3).

The regulation of rRNA genes by (p)ppGpp is an adaptation to changeable conditions. Bacteria live in a world of feast or famine. In nutrient-rich media, bacteria must have a large supply of ribosomes and rapidly manufacture new ones to support exponential growth. In nutrient-poor media, growth slows or stops, and there is less need for ribosomes. The cellular concentration of (p)ppGpp is inversely related to the availability of energy-containing nutrients. The regulation of rRNA gene expression to match the cell's carbon input is **growth rate control**.

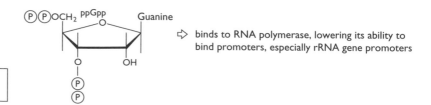

Fig 9.3 Regulation of gene transcription by ppGpp.

A second and stronger kind of rRNA gene regulation accompanies the **stringent response**, a coordinated physiological response to nitrogen starvation, which causes deficiencies of amino acids. Low levels of amino acids trigger the synthesis of an enzyme that makes more (p)ppGpp, which, in turn, shuts down transcription of stable RNAs by binding to RNA polymerase and blocking its binding to rRNA gene promoters.

Regulation of mRNA Synthesis

Operons and Regulation at the Binding Step

Because the binding step usually limits the rate of mRNA synthesis, the control of binding is discussed first. Two kinds of regulatory protein, activators and repressors, exert powerful effects on the rate that RNA polymerase binds to promoters. In addition, the binding of RNA polymerase to a promoter depends on the affinity of σ (usually σ^{70}) for the promoter; σ has a high affinity for strong promoters and a low affinity for weak promoters. The best way to approach this subject is to consider a fundamental kind of genetic system that controls bacterial transcription, the operon.

An operon is a group of genes that encode a functionally related set of final gene products, which are synthesized in a coordinated way. Every operon has at least one regulatory gene and at least one structural gene. Regulatory genes encode proteins that regulate transcription of other parts of the operon, and structural genes encode at least one but usually several final gene products. The gene products (e.g., enzymes) have related physiological, reproductive, or structural roles in the cell. A bacterial genome typically has hundreds of operons.

The **lactose (lac) operon** of *E. coli* is an excellent example; its structural portion encodes three enzymes (Figure 9.4). Geneticists who define genes in terms of final gene products rather than in terms of transcript refer to the enzyme-coding part of the *lac* operon as three separate, structural genes, one gene per enzyme; the three enzymes are encoded by a single mRNA.

In the diagram below, *lacZ*, *lacY*, and *lacA* code, respectively, for β-galactosidase, permease, and transacetylase, which are necessarily co-regulated, because one mRNA encodes them all.

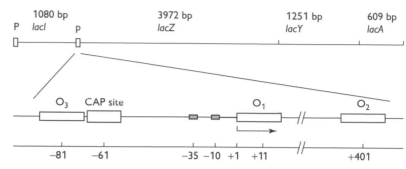

Fig 9.4 The *lac* operon of *E. coli*.

P = promoter; O_1, O_2, and O_3 = binding sites for *lac* repressor; CAP is a regulatory protein.

The basic facts are simple.

(1) If glucose is the sole carbon source for the bacterium, then *lacZ* is not transcribed and the cell lacks β-galactosidase.

(2) If glucose is switched to lactose, then the cell quickly responds by transcribing *lacZ* and synthesizing β-galactosidase; this enzyme is said to be induced.

(3) A high concentration of glucose in the medium inhibits the ability of lactose to induce β-galactosidase; this is the **glucose effect** (Figure 9.5).

The regulatory gene in the operon is **lacI**, which encodes **lac repressor**, a 152-kDa homotetramer. *lacI* has a weak promoter but no other regulatory sequences; therefore, it is transcribed constitutively at a low rate, resulting in a low concentration of this protein all the time.

In the absence of lactose, *lac* repressor binds to three 21-bp **operators** – O_1, O_2, and O_3. When *lac* repressor binds to these sequences, RNA polymerase is blocked from binding to the promoter. Binding of repressor is cooperative: when it binds to one operator, it also binds more easily to a second operator (Figure 9.6).

When lactose comes into the cell, some of it is converted to allolactose, the natural **inducer** in this system. Allolactose binds to an allosteric site of *lac* repressor, drastically reducing its

Fig 9.5 Regulation of the *lac* operon by lactose and glucose.

Carbon source	transcription of *lacZ*	β-galactosidase
Glucose	No	Absent
Lactose	Yes	Present
Glucose + lactose	Very little	Little present

O_3 O_1 O_2 O_3 O_1 O_2

Fig 9.6 Two of several possible repressor-binding configurations.

affinity for the operators (Figure 9.7). With *lac* repressor absent from the operators, RNA polymerase binds to the promoter, makes an open complex, and enters the chain initiation step of transcription.

Meanwhile, cyclic AMP (cAMP), plentiful in the cell when glucose is scarce but scarce in the cell when glucose is plentiful, mediates the glucose effect. cAMP binds to an allosteric site of **catabolite gene activator protein** (**CAP**), a 45-kDa homodimer. CAP-cAMP binds to a 14-bp upstream site centered at −61, and probably also to RNA polymerase. CAP-cAMP helps RNA polymerase to bind to the promoter — this is cooperative binding (Figure 9.8). The *lac* operon has a weak promoter, in that RNA polymerase cannot by itself efficiently bind and form an open complex with a melted region; CAP-cAMP compensates for the "weakness." The glucose effect occurs when glucose concentration rises, causing cAMP concentration to fall and decreasing the availability of CAP-cAMP.

Genetic elements of the operon that affect only sequences on the same DNA molecule are called **cis-acting**, and genetic elements that can affect sequences on other DNA molecules (e.g., another copy of the operon located on a plasmid) are called **trans-acting**. Thus, *lacI* is trans-acting, and the three operators are cis-acting.

Operons can be **inducible**, like the *lac* operon, or **repressible**, like the **his operon**, which codes for enzymes used in the synthesis of the amino acid histidine. In this operon, histidine is a **co-repressor**, which activates the repressor by binding to an allosteric site. Inducers and co-repressors are small molecules that control regulatory proteins by binding to allosteric sites of these proteins. The inducer is a small molecule (e.g., sugar) related to

Repressor = tetramer (circles); ▲ = inducer

Fig 9.7 When inducer molecules bind to repressor, it no longer binds to operators.

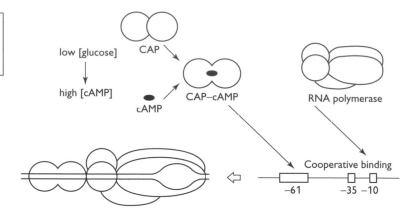

Fig 9.8 cAMP activates CAP. CAP-cAMP and RNA polymerase bind cooperatively.

the substrate of the enzymes being induced, and the co-repressor is a small molecule (e.g., amino acid) related to the product of a reaction catalyzed by the enzyme being repressed.

Operon control is either **positive** or **negative**, depending on the nature of the main regulatory protein (Figure 9.9). Control is positive if an activator turns on transcription and negative if a repressor turns off transcription. Any given operon may be controlled both positively and negatively, but inducible and repressible control are mutually exclusive.

Coordinately Expressed Groups of Operons

In some cases, operons with related functions are expressed together. Two kinds of higher-order sets of operons are **regulons** and **modulons**. A regulon is a small system of co-regulated operons and simple genes encoding many proteins with a common physiological or developmental role. For example, the phosphate (*Pho*) regulon of E. coli comprises eight operons scattered around the chromosome, encoding over 30 proteins participating in the uptake and metabolism of inorganic phosphate. Inorganic phosphate concentration controls the expression of all eight operons. When the cell is starved of inorganic phosphate the *Pho* regulon is

	Positive	Negative

Inducible	Activator control
	Inducer makes
	    activator work
		Repressor control
		Inducer inactivates
		    repressor

Repressible	Activator control
	Co-repressor
	    inactivates activator
		Repressor control
		Co-repressor makes
		    repressor work

Fig 9.9 Types of control of operons.

switched on, but when the concentration of inorganic phosphate is high, the *Pho* regulon is switched off. A modulon is bigger than a regulon: it is a large system of scattered operons and simple genes encoding proteins that control a major physiological process or metabolic mode. For instance, in *E. coli*, the *FNR* modulon operates the master switch between aerobic and anaerobic metabolism. In aerobic conditions the *FNR* modulon is switched off, and in anaerobic conditions it is switched on, synthesizing 70 proteins used in anaerobic metabolism.

Regulation at the Elongation Step

Pausing and Early Termination. Pausing may regulate the elongation phase of transcription. The duration of a pause ranges from seconds to minutes and depends on the DNA sequence that signals pausing as well as regulatory proteins. NusA is a 55-kDa protein that enhances pausing, while NusG is a 21-kDa protein that suppresses pausing. NusA and NusG act independently. When RNA polymerase pauses immediately downstream of an RNA hairpin (stem-loop structure), NusA binds to RNA polymerase and to RNA and increases the length of pauses, thus slowing down transcription. NusG also binds to RNA polymerase; it decreases the length of pauses, thus speeding up transcription.

Attenuation, a Mechanism of Early Termination. Attenuation is the early termination of transcription, at a site downstream of the promoter and well upstream of the normal terminators. Numerous bacterial genes are partly regulated by attenuation. Attenuation can work in several ways: (1) in some biosynthetic operons, by coupling transcription and translation; (2) in some catabolic operons, through the binding of an antiterminator protein to an RNA hairpin structure near the 5′ end of the nascent transcript; and (3) in aminoacyl synthetase (a kind of enzyme that adds an amino acid to a tRNA) genes, through tRNA-mRNA binding.

Attenuation was discovered in *E. coli*'s *trp* operon, whose gene products catalyze steps in the synthesis of tryptophan. The primary regulation of the *trp* operon is negative and repressible. The *trp* operon is negative because a repressor protein controls its expression; it is repressible because the repressor protein works only if the co-repressor tryptophan binds to it; repressor without co-repressor is ineffective. Secondarily here means in a less-important way and by a mechanism that follows (in time) the main (=primary) mechanism of regulation. The *trp* operon is

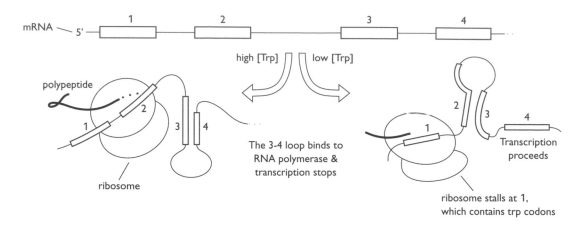

Fig 9.10 Attenuation of transcription of the *trp* operon, via tryptophan concentration.

also regulated by attenuation, precipitated by a coupling of transcription and translation. Translation begins when a 5′ end of mRNA becomes available, long before transcription is complete. When tryptophan is plentiful, transcription of the *trp* operon is halted by attenuation, but when tryptophan is scarce, there is no attenuation of transcription (Figure 9.10).

The attenuator is a region located between +50 and +140 in the mRNA. Within this region are four sequences with the potential for forming duplexes. Pairing of regions 3 and 4 makes a stem-loop structure that binds to RNA polymerase, causing it to dissociate from DNA. Pairing of 2 and 3 merely stalls translation without effect on transcription.

Whether the 3–4 hairpin or the 2–3 hairpin forms depends on the progress of the ribosome through the attenuator region, which in turn depends on tryptophan concentration, as sequence 1 contains two tryptophan codons.

When tryptophan is abundant, the ribosome is not stalled at sequence 1 and it quickly proceeds through sequence 2, allowing 3 and 4 to pair. The 3–4 duplex binds to RNA polymerase, causing early termination of transcription.

When tryptophan is scarce, the ribosome stalls at sequence 1 until rare tryptophan-charged tRNAs enter the ribosome. This allows pairing of sequences 2 and 3, which precludes formation of the 3–4 termination signal. Pairing is short-lived for all the duplexes, and the ribosome eventually progresses through the attenuator region and completes translation of the first coding sequence. Meanwhile, transcription continues; once RNA polymerase has "escaped" this region, attenuation cannot occur.

Degradation of RNA

The extensive duplex regions of rRNA and tRNA may contribute to their stability. mRNAs are destroyed by multiprotein complexes called degradosomes, composed of endonucleases, RNA helicases, and exonuclease-like enzymes (PNPases). Exonucleases typically digest RNA from the 3′ end, but many mRNAs form hairpin structures at their 3′ ends, which protect against digestion (Figure 9.11). Hairpin loops are attacked in two ways. They may be unwound by RNA helicases, and, after unwinding, exonucleases degrade the single-stranded region 3′ to the loop. Alternatively, an enzyme [poly(A) polymerase] adds a poly(A) tail to a hairpin, after which endonucleases attack the single-stranded part.

Other Simple Life Forms: Archaea, Mitochondria, and Viruses

In archaea, much of the DNA is complexed with histones, and the RNA polymerase is like those of eukarya. Archaea have operons, activators, and repressors.

The coding part of mitochondrial genomes is tiny. In many cases, there are few transcription units (genes, by this book's definition). In mammals, two RNAs are transcribed and then processed. Transcription is not elaborately regulated – e.g., transcription is not inducible or repressible.

Many viruses, even those with genomes smaller than 50 kb, have exquisitely regulated gene expression. In λ phage, a genetic

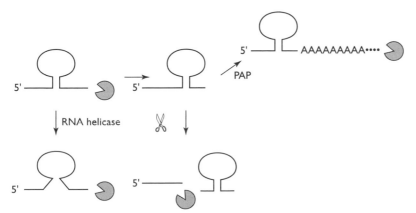

Fig 9.11 3′ to 5′ degradation of mRNA by exonuclease; PAP = poly(A) polymerase.

= exonuclease.

switch controls transcription of genes for the two main phases of the life cycle, lysogeny (a quiescent state) vs. an infectious, host-killing phase. A protein, λ repressor, exerts negative control over *cro*, a master gene that initiates the infectious phase. λ repressor exerts positive control over its own gene, *cI*. If the lysogenic cell is exposed to ultraviolet light, a bacterial protein that helps repair DNA cleaves λ repressor, thereby releasing the repression of *cro*, which leads to synthesis of viral DNA, RNA, and proteins, and, within a few minutes, death of the cell and the release of hundreds of virus particles. Gene regulation in some viruses has been studied intensively.

Further Reading

Adhya S. 1996. The *lac* and *gal* operons today. In: Lin ECC, Lynch AS (eds.). *Regulation of Gene Expression in Escherichia coli.* Chapman and Hall, New York, pp. 181–200.

Carpusis AJ, Vanzo NF, Raynal LC. 1999. mRNA degradation: a tale of poly(A) and multiprotein machines. *Trends Genet.* 15:25–28.

Condon C *et al.* 1995. Control of rRNA transcription in *Escherichia coli. Microbiol. Rev.* 59:623–645.

Ptashne M. 1992. *A Genetic Switch Phage λ and Higher Organisms,* 2nd ed. Cell Press & Blackwell Scientific, Cambridge, MA.

Uptain SM, Kane CM, Chamberlin MJ. 1997. Basic mechanisms of transcript elongation and its regulation. *Annu. Rev. Biochem.* 66:117–172.

Chapter 10

Abundance of RNAs in Eukarya

Overview

Eukarya, like bacteria, regulate the amounts of rRNA, tRNA, and mRNA by controlling rates of synthesis and degradation. On the synthesis side, control is exerted in transcription and splicing. Degradation is controlled by RNases, and is influenced by RNA processing.

The rate-limiting binding step of transcription is complicated both by the inaccessibility of DNA in chromatin and by large genome size. Transcription goes hand in hand with the disassembly of chromatin, and a large genome requires a large number of regulatory DNA sequences and transcription factors. Chromatin disassembly is effected by the chemical modification of histones and by the actions of chromatin remodeling complexes.

RNA Degradation

Stable versus Unstable RNAs

Like bacteria, eukarya have much more stable RNA than unstable RNA. The cytoplasmic RNA of a human cell is $\approx 80\%$ rRNA, $\approx 15\%$ tRNA and other small RNAs, and $\approx 1\%$ to 5% mRNA. In growing eukaryal cells, 50% to 70% of transcription is from class I genes (big rRNAs), 20% to 40% of transcription is from mRNA genes, and the remaining 10% is from class III genes (tRNA and other small RNAs). The rarity of mRNA compared with the rate of transcription by RNA polymerase II reflects the short half-life of eukaryal mRNAs.

mRNA Degradation

Capping and polyadenylation increase RNA's resistance to degradation. These modifications are significant in mRNA, because virtually all mRNAs are capped, and most are polyadenylated. In multicellular eukarya mRNAs are more stable than they are in bacteria; eukaryal mRNA half-life is in the range 0.5 to 10 h, though some mRNAs in some cells may be even longer-lived.

The rather large variability in mRNA stability is principally due to sequences in the 5' and 3' untranslated regions, where endonucleases make their cuts. This is significant, as small changes in half-life rapidly result in large changes in abundance. The half-life of a species of mRNA may change with physiological and developmental state. Typically, mRNA is degraded in a stepwise fashion by endo- and exonucleases. The poly(A) tail is degraded by 3'→5' exonucleases first. Next, the 3' untranslated region (3' UTR, the region downstream of the polypeptide-coding region), the 5' UTR, and the coding region are cleaved by endonucleases. Following cleavage of the 5' UTR, mRNA is susceptible to exonucleases. Alternatively, a decapping enzyme can remove the 7mGpppN cap, making the mRNA vulnerable to 5'→3' exonucleases.

Variation in stability of mRNAs is influenced by several factors:

- Polyadenylated mRNA is more stable than non-polyadenylated mRNA.
- Poly(A)-binding protein protects against degradation.
- Proteins that bind to sequences within the 3' UTR, especially AU-rich sequences, also protect against degradation.

RNA Surveillance

Aberrant mRNAs with internal nonsense codons, either from mutation or from mistakes in transcription, are degraded by an **RNA surveillance system**. How RNA surveillance works is not yet known. One hypothesis is that the ribosome stalls at an early STOP codon, surveillance proteins assemble and scan the mRNA downstream for sequences such as a splice junction, and a decapping enzyme removes the 5' cap. Following this, RNases degrade the mRNA.

Regulation of Transcription

Regulation of transcription in eukarya is a vast subject, but here it suffices to describe the basic roles of chromatin, histones,

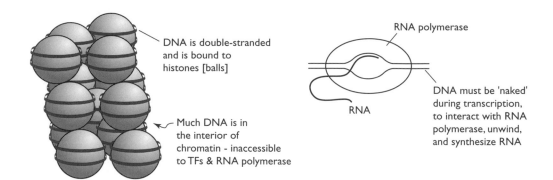

DNA is double-stranded and is bound to histones [balls]

Much DNA is in the interior of chromatin - inaccessible to TFs & RNA polymerase

RNA polymerase

RNA

DNA must be 'naked' during transcription, to interact with RNA polymerase, unwind, and synthesize RNA

and transcription factors. Keep in mind that chromatin contains a lot of protein – the ratio of protein to DNA in chromatin ranges from 1:1 to 2:1, depending on the kind of cell.

Fig 10.1 Chromatin must be disassembled before its DNA can be transcribed.

Chromatin Structure and Histones Repress Transcription

Most of the potentially active genes in a eukaryal genome are located in euchromatin, which in the interphase chromosome is a 30-nm-thick chromatin fiber arranged in loops fixed to the nuclear matrix. Genes in euchromatin fibers are not transcribed because the DNA is inaccessible to RNA polymerases (Figure 10.1).

Protein interactions pack nucleosomes in chromatin. Core nucleosomes directly contact each other, mostly via the tails of histones. Histone H1, which binds to both core nucleosomes and to the linker region of DNA between them, stabilizes the chromatin fiber by interacting with other histones, especially H2A. Nonhistone proteins may also play a role in forming chromatin and maintaining its structure.

Unmodified histones have higher binding affinity for DNA than do transcription factors. In transcriptionally active regions, (1) gene-activating proteins, rather than histones, are bound to enhancers; (2) histones are extensively acetylated, phosphorylated, and ubiquitinated; and (3) there are $\sim 10^2$-bp stretches of DNA not associated with protein. Nucleosomes can be phased near genes; i.e., they are bound to certain sequences. When positive transcription factors bind to enhancers, they displace phased nucleosomes. The mechanisms of phasing and nucleosome displacement are unknown but likely depend on sequence-specific DNA-binding proteins, since histones themselves bind to the sugar-phosphate backbone and do not bind preferentially to specific sequences of DNA.

Covalent Modification of Histones

During its lifetime, any given histone molecule may be modified repeatedly by the covalent addition/removal of side groups to specific amino acids of the basic tails. Histones may be acetylated, phosphorylated, methylated, or ubiquitinated. Ubiquitin is a 76-amino acid, 9-kDa protein that joins covalently with many other proteins; in addition to participating in chromatin remodeling, it tags proteins for proteolysis.

TATA-associated factor (TAF) can modify histone H1 by adding acetyl groups and ubiquitin (Figure 10.2). Ubiquitination and acetylation of histone H1 promotes disassembly of chromatin and precedes removal of core nucleosomes from DNA. Acetylation of core histones by **histone acetyltransferases** (HATs) and phosphorylation by kinases render them less basic and hence reduce their binding affinity for DNA; these modifications promote nucleosome disassembly. Methylation of core histones by methyltransferases and deacetylation of histones by **histone deacetylases** (HDs) silence chromatin.

Acetylation of histones at specific positions lowers the histone's binding affinity for DNA. Especially important is the acetylation of lysines in the N-terminal tails of H2A, H2B, H3, and H4 by HATs; the acetyl groups are removed by HDs. Acetylation of histones at a gene locus helps the transcription of that gene, because transcription factors displace acetylated histones at enhancers or upstream activating sites.

Phosphorylation and ubiquitination of core histones also promote the displacement of histones by transcription factors. Phosphorylation and ubiquitination of histone H1 facilitate its departure from chromatin and hence favor the formation of beads-on-a-string nucleosomes. In some instances, methylation of histones inhibits phosphorylation and promotes gene silencing.

Noncovalent Modification of Chromatin

Large protein multimers called **remodeling complexes** alter nucleosome structure and are essential to gene expression. They do not modify histones or DNA covalently; they contain contractile proteins and ATPases. When a remodeling complex is bound to a nucleosome, transcription factors have increased access to the DNA of that nucleosome. A well-known example (there are homologous proteins in yeast, humans, and fruit flies) is the ≈1.2-MDa SWI/SNF remodeling complex. SWI/SNF attracts HATs to

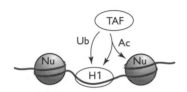

Fig 10.2 TAF modifies H1 by acetylation and ubiquitination. Nu = nucleosome, Ac = acetyl groups, Ub = ubiquitin.

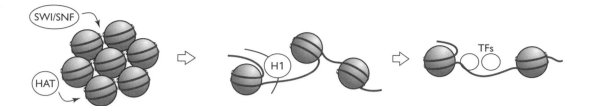

chromatin. Transcription factors bind to upstream control elements exposed by the remodeling of chromatin (Figure 10.3).

Fig 10.3 Chromatin fibers are disassembled in stages. UCE = upstream control element; TF = transcription factor; HAT = histone acetyltransferase.

Example: Expression of β-Globin Genes in Red Blood Cells

Erythrocytes (red blood cells) first develop during a human embryo's first month. Erythrocytes are nearly filled with hemoglobin, so the genes encoding hemoglobin must be active in this cell population throughout life. This is accomplished differently at the two hemoglobin loci.

The globin gene family occupies three chromosome loci in humans: the β-globin cluster (ε, $^G\gamma$, $^A\gamma$, δ, and β) on chromosome 11, the α-globin cluster (ζ [zeta], α_2, α_1) on chromosome 16, and myoglobin on chromosome 22. The α and β genes are expressed in red blood cells in a developmental pattern. Embryos (0 to 8 weeks) express the ζ and ε genes; the fetus expresses the α_2, α_1, $^G\gamma$, and $^A\gamma$ genes; the adult expresses the α_2, α_1, δ, and β genes – 97% of adult hemoglobin is a tetramer of two α and two β polypeptides. All erythroid-specific genes have one or more copies of identical enhancers recognized by erythroid-specific transcription factors; e.g., a transcription factor called GATA-1 recognizes the tissue-specific enhancer AGATAG.

In addition, chromatin of the entire β region is opened up in erythroid cells, via action of a **locus control region (LCR)**, a 15-kb region 5 kb upstream of the ε gene (Figure 10.4). The LCR contains five sites that are hypersensitive to deoxyribonuclease in erythroid cells (HS sites). In the LCR, chromatin is decondensed; the five HS sites are places where histones are not bound to DNA. The α cluster lacks an LCR. More than 30 distinct LCRs are known in vertebrates.

Fig 10.4 The locus control region of the β-globin cluster. ∇=non-tissue-specific enhancer; $\uparrow$=tissue-specific enhancer; $\psi\beta$=pseudogene.

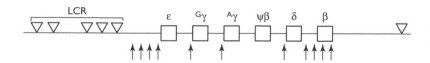

Fig 10.5 Chromosome inactivation by heterochromatin formation.

Active chromatin

Barr body

Heterochromatin

Condensed chromatin is transcriptionally silent. The most highly condensed chromatin is heterochromatin. Early in the development of female mammals, one of the two X chromosomes becomes heterochromatin throughout most of its length in somatic cells; it is genetically inactivated (Figure 10.5). Therefore, most genes in the heterochromatized copy of the X chromosome are not transcribed, leaving only one active copy of most genes on the X chromosome. This is a way of rendering XX females equivalent to XY males in the level of mRNA from X-linked genes (Chapter 25). The condensed X chromosome of females is called a **Barr body**; this heterochromatin is **facultative** (switched on or off during the life cycle), in contrast to **constitutive** (permanent) heterochromatin. In mammals, all but one X chromosome is turned into a Barr body, so that X0 females lack a Barr body, XXY males have a Barr body, and XXX females have two Barr bodies.

When a chromosomal rearrangement moves a gene close to a region of constitutive heterochromatin, the expression of that gene may vary from cell to cell. In such rearrangements, the boundary of heterochromatin is not exactly fixed, so that in some cells the gene is "mistakenly" heterochromatized and therefore is repressed, but in other cells it remains in euchromatin and is transcribed. This phenomenon, known as position-effect variegation, depends on the actions of many different proteins. Normally, these proteins, some of which bind to specific DNA sequences, precisely regulate the boundaries of heterochromatin. A chromosome rearrangement misplaces boundary-signaling DNA sequences.

Transcription Factors and Regulatory DNA Sequences

Proteins That Regulate Transcription

Each class of gene (class I, rRNA; class II, mRNA; class III, tRNA and other small RNAs) is regulated differently by a different set of general transcription factors. Specific transcription factors are

Class	RNAs produced	General transcription factors	N of subunits	(Total) MW
				(kDa)
I	Large rRNAs	SL1, UBF	6	450
II	mRNAs, some snRNAs	TFIIA, B, D, E, F, & H	≈28	≈1500
III	tRNAs	TFIIIB & C	9	730
	5S RNA	TFIIIA & C	7	575
	snRNAs	—	—	—

especially important and diverse in class II genes, which typically have many regulatory DNA sequences.

In contrast to bacterial activators and repressors, which all bind to DNA, some general transcription factors of eukarya bind only to other DNA-binding regulatory proteins. Also, many more transcription factors regulate expression of a gene in eukarya, compared with a bacterial gene (Figure 10.6).

The transcription of class II (protein-coding) genes requires a massive complex of many proteins. Specific transcription factors bind to a gene's enhancers or to regulatory sequences upstream from the promoter and fairly close to it. The general transcription factor TFIID is worth singling out. TFIID is a multimer containing TATA-binding protein, which binds to the TATA box, and ≈10 associated proteins called TAFs. The ≈250-kDa TAF multimer contains a HAT enzyme and an enzyme that adds ubiquitin to histones.

Fig 10.6 General transcription factors in vertebrates.

Regulatory DNA Sequences

Regulatory DNA sequences are binding sites for proteins that regulate transcription. Regulatory sequences of eukarya fall into three groups: (1) promoters, to which RNA polymerase holoenzyme binds; (2) enhancers and upstream activating sites (UASs), which help to activate transcription; and (3) silencers and upstream repressing sites (URSs), which help to repress transcription.

Promoters and upstream activating sequences (UASs) of class I and class II genes lie approximately in the region 100 bp upstream of the transcription start site, while promoters of most class III genes are internal to the coding sequence.

Enhancers are binding sites for positive transcription factors. They are short DNA sequences located upstream or downstream of the start site, sometimes close by and sometimes as far away as 85 kb. They can be found within genes as well as outside them.

Some enhancers, called **response elements**, are binding sites for hormone-bearing proteins. Enhancers commonly promote transcription in a specific tissue or at a specific developmental stage.

Like enhancers, UASs also up-regulate transcription, but they are located a short distance upstream of the promoter. A UAS that commonly occurs 80 to 100 bp upstream of the start site is the GC box (GGGCGG); transcription factors that bind to it stimulate constitutive transcription. Most UASs, though, are specific to one or a few genes. UAS-binding proteins stimulate the binding of RNA polymerase holoenzyme to the promoter.

Silencers are short DNA sequences that down-regulate transcription. They are predominantly binding sites for repressors, proteins that decrease the rate at which RNA polymerase binds to the promoter. Like enhancers, they can be located upstream or downstream of the start site, sometimes at a considerable distance, and they may be located within the gene. Most silencers act in a tissue-specific or stage-specific manner.

URSs are short DNA sequences that repress transcription; they are located a short distance upstream of the start site. URS-bound proteins inhibit the binding step of transcription by interfering with the binding of general or specific positive transcription factors or by modifying chromatin structure.

Coordinate Gene Transcription

Groups of related genes may be transcribed together. The operon is one system for this, although operons are less common in eukarya than in bacteria, and differ in some respects. The LCR is another system, an example being the β hemoglobin cluster.

An excellent and well-understood example of coordinate gene expression is the eight-gene *Gal* operon of *Saccharomyces cerevisiae* (Figure 10.7). An activator (GAL4, encoded by *Gal4*) and a repressor (GAL80, encoded by *Gal80*) regulate transcription of six structural genes of the *Gal* operon, which encode five enzymes and a permease. The eight genes are located on four chromosomes. The permease transports galactose, and the five enzymes control the metabolic pathway leading to glucose-6-phosphate. Each of the six structural genes has two to four copies of the 17-bp *Gal*-specific UAS, where GAL4 binds. GAL80 has an allosteric binding site for

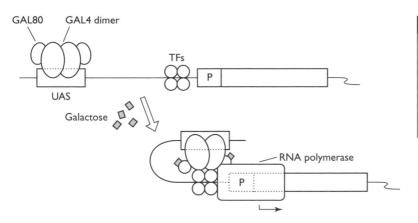

Fig 10.7 GAL4, GAL80, and galactose control the expression of the six structural genes of the *Gal* operon in yeast. UAS = upstream activating sequence, TFs = general transcription factors.

the inducer, galactose. When the inducer is absent, GAL80 binds to GAL4, blocking its ability to stimulate transcription. When the inducer is present, GAL80 does not bind to GAL4, and GAL4 then binds both to the UASs and to general transcription factors bound to the promoter, activating transcription. The yeast *Gal* operon is under catabolite repression, as is the *lac* operon of *Escherichia coli*; the glucose-driven repressor, CRP, binds to GAL4, which then loses its binding affinity for the UASs.

Methylation and Demethylation of DNA

Methylation of cytosines of the sequence CpG is widespread in the genomes of mammals, fungi, and flowering plants (Figure 10.8). Early in human development, most of the CpG sites located within transposons become methylated in somatic cells, and a small fraction of CpG sites elsewhere in the genome become methylated. Methylation in promoter regions silences transcription. DNA methylation contributes to X inactivation in female mammals.

In somatic cells of adult mammals, about 60% of the CpG sites throughout the genome contain methylated C, but in premeiotic

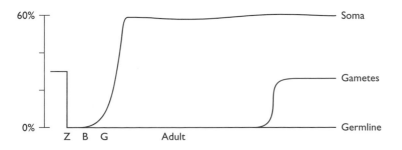

Fig 10.8 Schedule of genomic methylation at CpG sites (after Jaenisch 1997). Z = zygote (fertilized egg), B = blastula, G = gastrula.

germline cells the genome is unmethylated. Many sites are methylated during gametogenesis, and this methylation persists until the blastula stage, when nearly all methylation is erased. During gastrulation the genome again becomes massively methylated.

In mammals, DNA methylation silences parasitic genes (transposons), genes on the inactive X chromosome, and **imprinted** genes. The first of these, silencing transposons, appears to be the main function of DNA methylation. A mammalian genome typically contains $\sim 10^6$ transposons, whose gene expression would be deleterious, if not lethal. Second, methylation of cytosines is an essential step of inactivating the X chromosome in female mammals. Third, imprinting is the silencing, via DNA methylation, of the copy of a gene from only one parent; the imprinted gene is not transcribed during early development. Only a few genes are imprinted in eggs and sperm, but failure to imprint these few genes is lethal.

Methylated CpGs within transposons comprise 95% of all methylation of the genome. Methylated CpG sites within imprinted genes apparently make up substantially less than 1% of all methylated sites in the genome. The remaining methylated sites are at unknown locations. It is possible that these sites represent transcriptional silencing of genes and that gene silencing by DNA methylation contributes significantly to differential gene expression during development. It is also possible that the residual 5% of methylated sites have no biological significance.

Different tissues have different patterns of methylation, but it is not known whether this is related to tissue-specific gene expression. Abnormal methylation of tumor suppressor genes promotes tumor development and is an important cause of cancer.

Further Reading

Eloranta JJ, Goodbourn S. 1996. Positive and negative regulation of RNA polymerase II transcription. In: Goodbourn S (ed.) *Eukaryotic Gene Transcription.* IRL Press at Oxford University Press, Oxford.

Hanna-Rose W, Hansen U. 1996. Active repression mechanisms of eukaryotic transcription repressors. *Trends Genet.* 12:229–234.

Hilleren P, Parker R. 1999. Mechanisms of mRNA surveillance in eukaryotes. *Annu. Rev. Genet.* 33:229–260.

Jaenisch R. 1997. DNA methylation and imprinting: why bother? TIG 13: 323–328.

Jones PA, Takai D. 2001. The role of DNA methylation in mammalian epigenetics. *Science* 293:1068–1070.

Lee TI, Young RA. 2000. Transcription of eukaryotic protein-coding genes. *Annu. Rev. Genet.* 34:77–137.

Li Q *et al.* 1999. Locus control regions. *Trends Genet.* 15:403–408.

Ogbourne S, Antalis TM. 1998. Transcriptional control and the role of silencers in transcriptional regulation in eukaryotes. *Biochem. J.* 331: 1–14.

Ottolenghi S. 1992. Developmental regulation of human globin genes. In: Russo VEA, *et al.* (eds.) *Development: The Molecular Genetic Approach.* Springer-Verlag, New York.

Wolffe AP, Wong J, Pruss D. 1997. Activators and repressors: making use of chromatin to regulate transcription. *Genes Cells* 2:291–302.

Zawel L, Reinberg D. 1995. Common themes in assembly and function of eukaryotic transcription complexes. *Annu. Rev. Biochem.* 64:533–561.

Chapter 11

Protein Synthesis

Overview

The two kinds of final gene products, RNA and protein (polypeptides and peptides), are encoded polymers, and just as a gene's nucleotide sequence specifies an RNA molecule's nucleotide sequence during transcription, so does an RNA molecule's nucleotide sequence specify a polypeptide's amino acid sequence during **translation**, the synthesis of a polypeptide in the ribosome. The newly made polymer often undergoes enzymatic processing.

Coding in translation is less direct than in transcription, for amino acids do not pair with the nucleotides that encode them. Instead, tRNA is an intermediary, base-pairing with mRNA as it carries an amino acid to a growing polypeptide. Outside the ribosome, amino acids bond covalently to the appropriate tRNAs (**aminoacylation**); inside the ribosome, **peptidyltransferase** adds one amino acid at a time to a polypeptide chain.

This chapter breaks down protein synthesis into three phases: (1) the enzymatic aminoacylation of tRNAs; (2) translation, accomplished by ribosomes, mRNA, tRNA, and various protein factors; and (3) posttranslational processing of polypeptides (cutting, folding, covalent modification, and sometimes splicing).

Transfer of Amino Acids: A Cellular "Bucket Brigade"

A polypeptide grows by successive addition of amino acids to its carboxyl end, through the formation of peptide bonds. However, because peptides are not complementary to nucleotide chains,

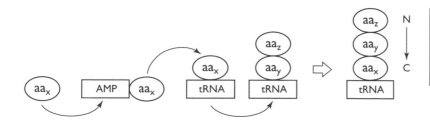

Fig 11.1 Amino acids are transferred from AMP to tRNA to polypeptide. Successive amino acids (aa) are given subscripts x, y, and z.

polypeptide synthesis requires special machinery. Each incoming amino acid is "handed off" from molecule to molecule before the peptide bond is formed. An amino acid first makes a covalent bond between its carboxyl carbon and the 5' α phosphate of ATP. This carbon is then transferred to the 3' or 2' oxygen of a tRNA. Finally, this carbon is transferred to the nitrogen of the next in-coming amino acid, to make a peptide bond.

During translation, a tRNA is attached to the carboxyl end of the polypeptide chain; this is the **peptidyl tRNA**. Every time a peptide bond is formed, the incoming tRNA (**aminoacyl tRNA**) replaces the peptidyl tRNA, which then exits the ribosome (Figure 11.1).

Aminoacylation of tRNAs

Each species of tRNA – each kind of tRNA having a particular anticodon – carries an amino acid attached covalently to the adenosine of the CCA (3' acceptor) end. Aminoacylation of the acceptor end of a tRNA molecule is a two-step reaction catalyzed by an **aminoacyl-tRNA synthetase** (Figure 11.2). These synthetases are a family of enzymes, each adding a particular amino acid and each recognizing a set of one or more tRNAs that accept the same amino acid – **isoaccepting tRNAs** – by forming weak bonds with nucleotides of the tRNA. Aminoacyl-tRNA synthetases (60 to 230 kDa), which can be monomers, homodimers, homotetramers, or heterotetramers, catalyze two reactions. First, the carboxyl carbon bonds covalently to the α phosphate of ATP, releasing diphosphate. Second, the carboxyl carbon is transferred from the 5' phosphate of AMP to the 3' oxygen of the tRNA's accep-tor end. An aminoacyl-tRNA synthetase proofreads, hydrolyzing the bond between the tRNA and an incorrectly placed amino acid.

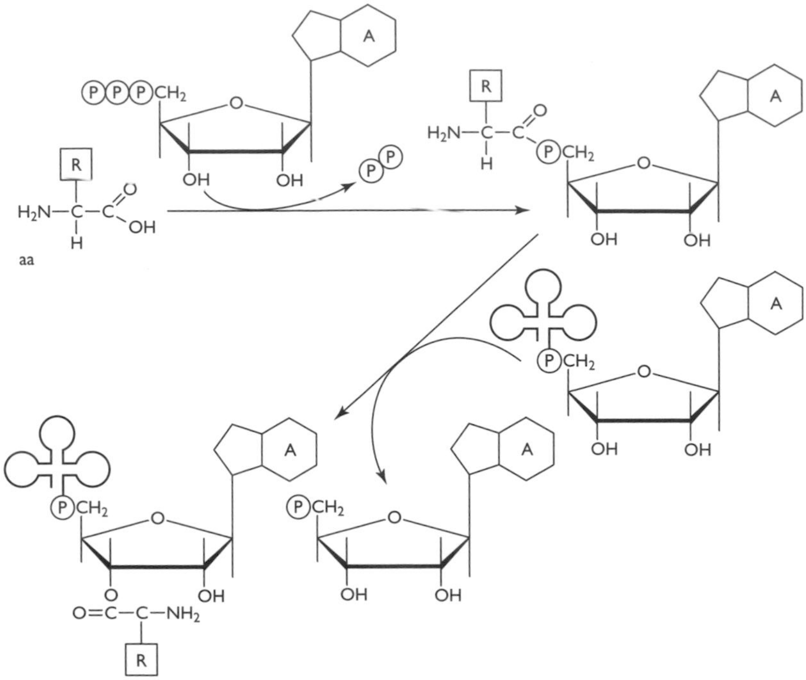

Fig 11.2 Aminoacylation of a tRNA.

aa

The Ribosome

The ribosome looks like two irregular boulders touching at many points and separated by a big space; their surfaces do not fit hand in glove. In bacteria, the 70S ribosome ($\approx$2.8 MDa, 21 × 29 nm) has two subunits: 50S ($\approx$1.8 MDa; made of 23S and 5S RNA + 34 proteins) and 30S ($\approx$1.0 MDa; made of 16S RNA + 21 proteins) (Figure 11.3). The surfaces of both subunits have ridges, valleys, knobs, and pockets. In the space between subunits, there are three distinct sites for tRNAs to occupy: the A (aminoacyl) site, the **P** (peptidyl) site, and the **E** (exit) site. The A and P sites are shared by the 50S and the 30S subunits, while the E site is a pocket in the 50S subunit alone (Figure 11.3). When a tRNA molecule sits in the A or P site, its anticodon, nestled in the 30S subunit, base pairs with a codon of mRNA.

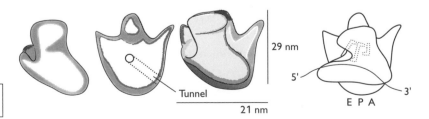

Fig 11.3 70S ribosome with 30S and 50S subunits.

Three Phases of Translation

It is useful to break translation into three phases:

• Initiation: mRNA and initiator tRNA bind to a ribosome
• Elongation: a polypeptide chain grows by the addition of amino acids one at a time
• Termination: polymerization stops and the polypeptide and mRNA are released from the ribosome.

Translation in Bacteria

Initiation: How Translation Begins

During this first stage of translation, a stable complex forms: ribosome + mRNA + charged (amino-acid-carrying) initiator tRNA. Three initiation factors, IF1, IF2, and IF3, help form the complex. Translation can be initiated at several locations on a bacterial mRNA molecule, if it encodes more than one polypeptide; the frequency of initiation on that mRNA molecule is about once per second.

First, IF1 and IF3 bind to the 30S subunit and help mRNA bind to the 30S subunit. During this step the Shine-Dalgarno sequence, a 13-nt sequence in the 5′ untranslated region 7 to 12 nt upstream of the start codon, base pairs with a complementary, anti-Shine-Dalgarno sequence near the 3′ end of 16S rRNA. Next, IF2 helps the **initiator tRNA** (carrying fmet = formylated methionine) base pair with the start codon of mRNA, usually AUG and sometimes GUG. A 50S subunit then binds to this complex. GTP is hydrolyzed to GDP and P_i on IF2 and the IFs are released, leaving a 70S ribosome bound to mRNA and initiator tRNA in the P site (Figure 11.4).

Elongation: How the Polypeptide Chain Grows

Elongation is a cyclic growth phase. During each turn of the cycle, an aminoacyl tRNA carries an amino acid into the ribosome, and a peptide bond is formed between the incoming amino acid and the amino acid at the carboxyl end of the growing polypeptide. The tRNA attached to the polypeptide is called the peptidyl tRNA. Growth of the polypeptide is N to C (amino terminus to carboxyl terminus). Elongation requires two elongation factors, EF-Tu and EF-G. At the end of the initiation phase and at the beginning of each elongation cycle, the peptidyl tRNA occupies the P site and

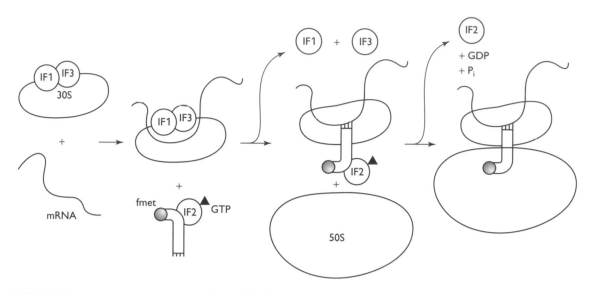

Fig 11.4 Initiation of translation.

the A and E sites are empty; the codon for the next amino acid sits in the 30S A site.

At the start of elongation, an aminoacyl tRNA enters the A site, accompanied by EF-Tu bound to GTP. If the tRNA's anticodon is not complementary to the mRNA's codon, the tRNA leaves the ribosome, but if the anticodon and codon base pair, then the GTP is hydrolyzed and EF-Tu/GDP exits. Next, peptidyltransferase, which is an integral part of the 50S subunit, catalyzes the formation of a peptide bond between the carboxyl group attached to the tRNA in the P site and the amino group of the tRNA in the A site; this leaves an uncharged tRNA in the P site and a peptidyl tRNA in the A site. Instead of being a protein, peptidyltransferase is a ribozyme, part of 23S rRNA. EF-G/GTP helps the uncharged tRNA of the P site move to the E site and the peptide-charged tRNA in the A site move to the P site, emptying the A site. The uncharged tRNA exits the ribosome from the E site, and the mRNA translocates three nucleotides relative to the ribosome. During translocation, the GTP is hydrolyzed to yield EF-G/GDP + P_i. The elongation cycle is then repeated until a stop codon (one that does not base pair with any tRNA's anticodon) enters the A site, whereupon translation stalls owing to the failure of a tRNA to enter the A site stably (Figure 11.5).

The main energetic cost of translation lies in the elongation cycle (2 GTPs/amino acid), double the cost of charging tRNAs with amino acids (1 ATP/amino acid). The average rate of elongation is about 15 amino acids per second. Translation is relatively

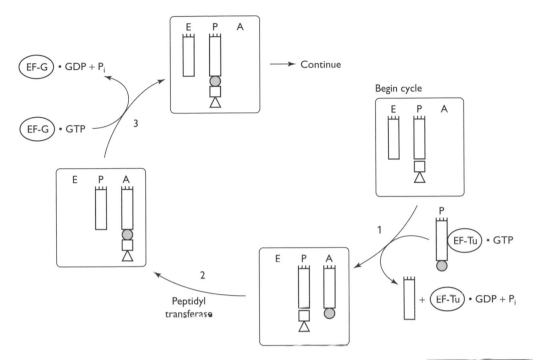

Fig 11.5 The elongation cycle of translation. Adapted from Burkhardt et al. 1998.

inaccurate, the error rate being 10^{-3} to 10^{-4} per amino acid. Both EF-Tu and the 16S rRNA in the 30S subunit contribute to the fidelity of translation, which in turn depends on the ability of the protein-synthesizing machinery to discriminate between correct and incorrect codon-anticodon pairing.

Termination: How Translation Stops

When a stop codon (UAA, UAG, or UGA) in the mRNA enters the A site, translation stalls because no tRNA base pairs with a stop codon. A release factor, (RF1 or RF2) bound to GTP enters the A site and binds to the stop codon. RF1 recognizes UAA and UAG, and RF2 recognizes UAA and UGA. A third release factor, RF3, is a GTPase that facilitates the dissociation of the two ribosomal subunits; RF3 interacts with both the ribosome and the RF in the A site. In termination, the peptidyl tRNA is cleaved from the polypeptide, and the parts of the translation complex dissociate (Figure 11.6).

The Genetic Code. With rare exceptions, the translation machinery reads adjacent 3-nt codons in mRNA by simple rules. Nearly always, nucleotides are neither skipped nor used twice. Because of these rules, the reading frame is set by the initial AUG codon;

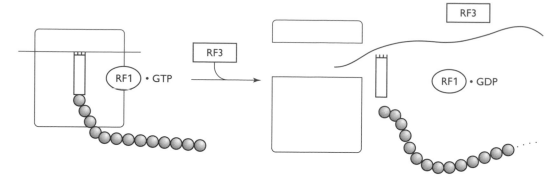

Fig 11.6 Termination of translation.

subsequent triplets are in phase with it. The genetic code is a nearly universal correspondence between codons and amino acids Figure 11.7.

The reading frame of mRNA is the phase of the triplets "read" as codons. An **open reading frame** (ORF) is accordingly a sequence in mRNA or in the nontemplate DNA strand of a gene, lacking a stop codon.

Stop codons are also known as **nonsense** codons; the other 61 are **sense** codons. Several amino acids correspond to more than

UUU Phe (F)	UCU Ser (S)	UAU Tyr (Y)	UGUCys (C)
UUC Phe (F)	UCC Ser (S)	UAC Tyr (Y)	UGC Cys (C)
UUA Leu (L)	UCA Ser (S)	UAA STOP	UGA STOP/Sec[1]
UUG Leu (L)	UCG Ser (S)	UAG STOP/Pyr[2]	UGG Trp (W)
CUU Leu (L)	CCU Pro (P)	CAU His (H)	CGU Arg (R)
CUC Leu (L)	CCC Pro (P)	CAC His (H)	CGC Arg (R)
CUA Leu (L)	CCA Pro (P)	CAA Gln (Q)	CGA Arg (R)
CUG Leu (L)	CCG Pro (P)	CAG Gln (Q)	CGG Arg (R)
AUU Ile (I)	ACU Thr (T)	AAU Asn (N)	AGU Ser (S)
AUC Ile (I)	ACC Thr (T)	AAC Asn (N)	AGC Ser (S)
AUA Ile (I)	ACA Thr (T)	AAA Lys (K)	AGA Arg (R)
AUG Met (M)	ACG Thr (T)	AAG Lys (K)	AGG Arg (R)
GUU Val (V)	GCU Ala (A)	GAU Asp (D)	GGU Gly (G)
GUC Val (V)	GCC Ala (A)	GAC Asp (D)	GGC Gly (G)
GUA Val (V)	GCA Ala (A)	GAA Glu (E)	GGA Gly (G)
GUG Val (V)	GCG Ala (A)	GAG Glu (E)	GGG Gly (G)

[1]In the rare cases where UGA → selenocysteine ratherthan STOP, a special sequence is present in the mRNA downstream of this codon.
[2]Very rarely in some archaea UAG → pyrrolysine.

Fig 11.7 Codon assignment: 1st nucleotide is 5′, 3rd nucleotide is 3′.

one codon. This property of a code is **redundancy**, which is frequently called **degeneracy**. Codons that correspond to the same amino acid are **synonymous**.

In rare instances, one codon has two meanings within a life form: the codon assignment is **ambiguous**. The dual meaning of UGA (normally STOP, rarely Sec) is an example of ambiguity. Three proteins with selenocysteine are known in *Escherichia coli*, and 12 are known in mammals. In one case in mammals, UGA is reprogrammed as tryptophan. Also, AUG in the first codon position of mRNA corresponds to initiator tRNA, which carries formylated methionine in bacteria, whereas internally AUG codes for methionine and GUG codes for valine.

The genetic code is nearly universal, which is to say that there is little variation in codon assignment among life forms. However, in *Mycoplasma capricola*, UGA = tryptophan, and in many ciliates, UAA and UAG = glutamine. There are other exceptions in various mitochondria.

Only two amino acids, methionine and tryptophan, have a single codon. For all other amino acids except leucine, serine, and arginine, the first two positions of the codon are invariant within amino acids. Leucine, serine, and arginine have six codons each. The third (3′) nt in the codon can vary within most amino acids. The 5′ nt of the anticodon has freedom of movement called **wobble**, which allows one anticodon to pair with more than one codon. Wobble in the anticodon allows for pairing between a given anticodon and several different codons (Figure 11.8). For example, CAU and CAC are codons for histidine; a tRNA with the anticodon GUG could pair with either codon, owing to wobble at the 5′ G. Without wobble pairing, translation would work only if organisms had 61 different kinds of tRNA, corresponding to the 61 sense codons (Figure 11.9). Typically, organisms make do with about 40 species of tRNA.

Programmed Translational Frameshifting. The frequency of spontaneous **frameshifts** – shifts from "reading" adjacent

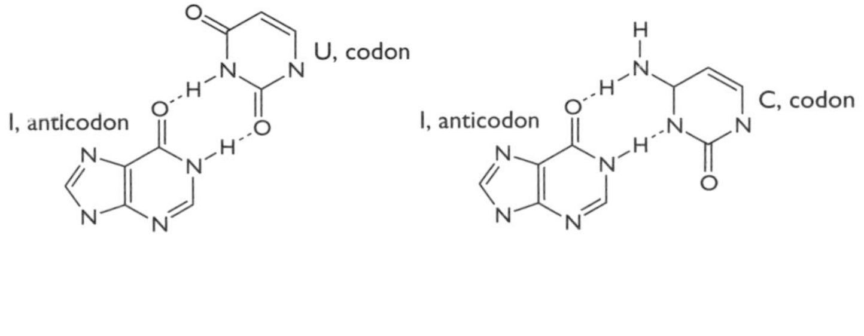

Fig 11.8 Examples of wobble pairing at the 3′ position of a codon.

Fig 11.9 Base-pairing in the 3rd nt of a codon, to allow for fewer than 61 tRNAs.

3′ nucleotide in codon	5′ nucleotide in anticodon
A	U or I
G	C or U
C	G or I
U	A, G, or I

codons – is very low, about 3×10^{-5}. In a handful of genes, though, frameshifting is normal and may occur a high percentage of the time in translation of these genes' mRNAs. Either two different polypeptides are encoded by the mRNA, or else only the frameshifted polypeptide is encoded. The *dnaX* gene of *E. coli* encodes both the γ and τ polypeptides of DNA polymerase III by translational frameshifting.

Most translational frameshifting entails single-nucleotide shifts – slipping back 1 nt (−1 frameshifting) or skipping 1 nt (+1 frameshifting). In a few cases, about 50 nt are skipped – translational hopping. Translational frameshifting is based on nearby sequences in the mRNA, which interact with the ribosome.

Translation in Eukarya

In eukarya, translation of nuclear mRNAs occurs in the cytoplasm. Ribosomes, tRNAs, and mRNA are exported from the nucleus to the cytoplasm. In eukarya, the 80S ribosome ($\approx$4.5 MDa, 22 × 32 nm) has two subunits: 60S ($\approx$3.0 MDa; made of 28S, 5.8S, and 5S RNA + 45 to 50 proteins) and 40S ($\approx$1.5 MDa; made of 18S RNA + 30 to 35 proteins). Translation in the 80S eukaryal ribosome is similar to that in the 70S bacterial ribosome. The names of eukaryal IFs, EFs, and RF are preceded by "e," for "eukaryal" – for example, eIF1. Initiation requires at least 11 proteins, compared with 3 proteins in bacteria. Some eIFs attach to mRNA and some to the 40S ribosomal subunit; eIF2 attaches to initiator tRNA, which carries methionine. The mRNA is aligned on the 40S subunit by mRNA's cap; there is no eukaryal equivalent to the bacterial Shine-Dalgarno sequence. Each mRNA has a single initiator codon (AUG or GUG). The ribosome "scans" the mRNA for the initiator codon. The eIFs are released and the 60S subunit attaches to the complex to begin the elongation phase. Elongation is similar to that in bacteria. An eRF recognizes the three stop codons.

Posttranslational Processing

Processing Common to all Organisms

The N terminus of some polypeptides is clipped off, and some polypeptides are cut into pieces by proteases. Secondary structures, mainly α-helices and β-ribbons, form because of interactions among amino acids. Polypeptides then assume tertiary structures, folding into characteristic shapes with the help of **chaperonins**, an important kind of **heat shock protein**. Heat shock proteins help polypeptides fold in cells that experience an increase in temperature. The polypeptides of multimeric proteins bind to each other with weak bonds; in some cases, disulfide bonds form between cysteines within or between polypeptides.

Some polypeptides have intervening sequences of amino acids, or **inteins**, which are removed by the action of the intein itself. The remaining N- and C-termini of the **extein** are then spliced. Most inteins are also DNases that promote the insertion of a copy of the gene that encodes the extein-intein-extein into another site in the genome. Over 20 different intein-containing proteins are known in bacteria, archaea, and eukarya.

Posttranslational Processing in Eukarya

In eukarya, there are added complexities. Proteins destined for secretion by the cell have a highly hydrophobic **signal sequence** at the N terminus. The signal sequence easily enters the lipid membrane; it is removed by a peptidase as the polypeptide is secreted. Chain folding typically begins when the N terminus emerges from the ribosome and ends after the termination of translation.

Proteins with destinations of the cytoplasm, nucleus, mitochondria, or chloroplasts are synthesized in free ribosomes in the cytosol. The mitochondrial or chloroplast proteins contain signal sequences at the N termini that recognize organelle membranes. Systems of proteins in the outer and inner membranes of the mitochondrion or chloroplast transport proteins from the cytoplasm to the interior of the organelle. After entering the organelle, a protein's signal sequence is cut off.

Proteins that migrate to cellular membranes or are secreted are synthesized on the rough endoplasmic reticulum. As a polypeptide is synthesized there, the hydrophobic signal sequence of the N terminus binds to a signal recognition particle made

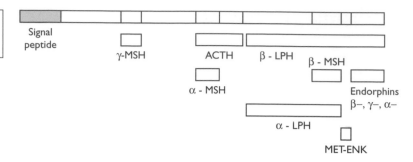

Fig 11.10 Cleavage of POMC into peptide hormones.

of 7SL RNA and several proteins. The signal recognition particle helps to transfer the protein into a system of organelles (endoplasmic reticulum, vesicles, and Golgi) where the signal sequence is clipped off and the polypeptide folds into its characteristic secondary structure. Mature proteins are then transported to the appropriate membranes or are secreted by exocytosis.

Many eukaryal proteins are extensively modified by the addition of acetyl groups, methyl groups, phosphates, and sugars (glycosyl groups) to specific amino acids. These modifications are performed enzymatically.

Peptides are made by proteolysis of a precursor polypeptide. In the pituitary gland, one 267-amino acid polypeptide, pro-opiomelanocortin (POMC), is precursor to 10 different peptide hormones, such as adrenocorticotropic hormone (ACTH) and lipotropin hormone (LPH). Specific proteases cleave POMC, and different cells in the pituitary gland make different peptides from this polypeptide (Figure 11.10).

Further Reading

Burkhardt N et al. 1998. Ribosomal tRNA binding sites: three-site models of translation. *Crit.Rev. Biochem. Mol. Biol.* 33:95–149.

Farabaugh PJ. 1996. Programmed translational frameshifting. *Annu. Rev. Genet.* 30:507–528.

Frank J. 1998. How the ribosome works. *Amer. Sci.* 86:428–439.

Green R, Noller HF. 1997. Ribosomes and translation. *Annu. Rev. Biochem.* 66:679–716.

Nakamura Y, Ito K. 1998. How protein reads the stop codon and terminates translation.*Genes Cells* 3:265–278.

Nissen P *etal.* 2000. The structural basis of ribosome activity in peptide bond synthesis. *Science* 289:920–930.

Chapter 12

DNA Replication

Overview

To reproduce, a cell must first copy its genome via DNA replication. In DNA replication, double-stranded DNA (dsDNA) is copied to yield two daughter molecules, each identical to the parent molecule. As in transcription, the new polymer grows $5' \rightarrow 3'$. Unlike transcription, whose product is a free, single-stranded RNA molecule, replication yields two double-stranded B-form DNA helices. DNA replication is **semiconservative**, as each daughter helix consists of one parental strand and one newly synthesized strand.

In cells, a large complex of proteins, the **replisome**, replicates DNA. The replisome includes DNA-dependent DNA polymerase, which, unlike RNA polymerase, cannot begin with a single nucleotide, but rather adds nucleotides to the $3'$ end of a **primer**, a short chain of RNA synthesized by **primase**. DNA synthesis is faster and much more accurate than RNA synthesis.

Chain Growth

In DNA synthesis, as in RNA synthesis, the polymer is built by making a phosphodiester bond between the α phosphate of a nucleotide triphosphate (dNTP) and the $3'$ oxygen of the growing polynucleotide chain, lengthening the chain by one and producing diphosphate. Required are the four dNTPs (dATP, dGTP, dCTP, and dTTP), a template strand, a primer (a short nucleotide chain that base pairs with the template) to which nucleotides are added, and DNA polymerase. Each of the two strands of a DNA double helix is a template for the synthesis of a new, complementary strand. The template strand's nucleotides base pair with the nucleotides being added to the 3'-OH end of the primer.

Fig 12.1 The synthesis of one new chain of DNA. Thin lines = DNA, box = RNA primer.

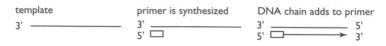

template	primer is synthesized	DNA chain adds to primer
3' ————————	3' ————————	3' ———————— 5'
	5' ☐	5' ☐———————→ 3'

DNA polymerases work by adding nucleotides to a preexisting strand; replication, like simple water pumps, must be primed. During replication the cell makes short RNA primers to which DNA polymerase can add DNA nucleotides (Figure 12.1).

Replication in Bacteria

DNA Polymerases

Escherichia coli and its relatives have three DNA polymerases (Figure 12.2): pol I, pol II, and pol III. Pol III is the main replication enzyme, pol I removes and replaces primers on the discontinuous strand, and both pol I and pol II help repair damaged DNA.

All bacteria have a pol III, but many lack pol I or pol II. DNA polymerases commonly have $3' \rightarrow 5'$ exonuclease activity to remove mispaired nucleotides. Many other proteins participate in replication, with functions such as unwinding the double helix, relaxing supercoiling during the unwinding process, stabilizing single-stranded DNA (ssDNA), synthesizing primers, and sealing gaps (Figure 12.3).

Unwinding the Double Helix and Relaxing Supercoiled Regions

To be replicated, dsDNA must be melted (unwound) to yield single strands. In *E. coli*, after chromosomal proteins are removed from the DNA a helicase unwinds the helix at an average rate of 5500 rpm! Helicase, which is associated with primase, uses one or two ATPs, GTPs, or CTPs to melt each base pair. Unwinding the double helix induces positive supercoiling, which is relaxed by a topoisomerase. *E. coli* has several topoisomerases, but only one of

Fig 12.2 The three DNA polymerases of *E. coli*.

Characteristic	pol I	pol II	pol III
Number of subunits	1	1	24 in holoenzyme
Total mass (kDa)	103	90	924
Molecules/cell	≈400	unknown	≈10
Exonuclease activity	$3' \rightarrow 5'$ and $5' \rightarrow 3'$	$3' \rightarrow 5'$	$3' \rightarrow 5'$

Protein (subunits)	Mass (kDa)	Gene(s)	Function
Helicase (B₆)	300	*dnaB*	Melts dsDNA, 5'→3'
Gyrase (A₂B₂)	374	*gyrA* *gyrB*	Relaxes DNA
Primase	64	*dnaG*	Primer synthesis
DnaC (C₆)	174	*dnaC*	Helps helicase
DnaT (T₃)	66	*dnaT*	Part of primosome, helps primase assembly
SSB (tetramer)	76	*ssb*	Single-stranded DNA binding protein, sustains unwinding, restricts priming, helps renaturing, inhibits nucleases
PriA	76	*priA*	Primosome assembly, 3'→5' helicase
PriB	12	*priB*	Primosome assembly
PriC	23	*priC*	Early primosome assembly
Ligase	75	*lig*	Seals DNA nicks

them, gyrase, functions during replication. Behind the helicase, tetramers of ssDNA-binding proteins bind to the single strands to stabilize them.

Fig 12.3 Other proteins that function in DNA replication of the *E. coli* chromosome.

Primosome

The primosome, a 280-kDa heptamer, synthesizes a 10- to 12-bp RNA primer. The primosome contains primase, which synthesizes the primers, and a 5'→3' helicase, which unwinds the double helix.

Leading Strand, Lagging Strand

The two strands of DNA have opposite polarity, one pointing 3'→5' and the other pointing 5'→3', if the double helix is viewed from one end. As in transcription, the polymerization reaction is always 5'→3'. Synthesis of one new strand, called the **leading strand**, proceeds continuously toward the double-stranded region; its template is oriented 3'→5' looking down the molecule in the direction of unwinding. Synthesis of the other new strand, called the **lagging strand**, proceeds away from the double-stranded region. Synthesis is discontinuous on the lagging strand because the double helix must unwind before a primer can be synthesized. Each discontinuously synthesized piece is ≈2 kb long; these pieces are Okazaki fragments. As a primer-containing Okazaki fragment is being synthesized on the lagging strand, the double helix

Fig 12.4 Replication fork.

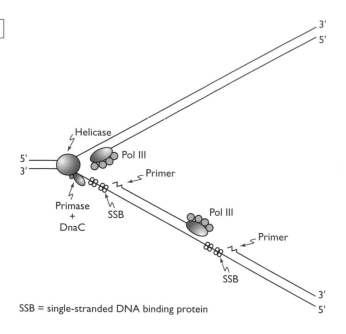

SSB = single-stranded DNA binding protein

unwinds further, and another fragment is synthesized by pol III. The DNA polymerase III holoenzyme has two parts, one that synthesizes the leading strand and one that synthesizes the lagging strand. In Figure 12.4, the two parts of the DNA polymerase III enzyme are depicted as physically separate.

Removing and Replacing Primers, Sealing Gaps

Pol III holoenzyme stops synthesis and unbinds when it reaches the RNA-containing end of another fragment. Pol I binds at the junction and removes the ribonucleotides $5' \rightarrow 3'$ one at a time (pol I has exonuclease activity).

A ligase then displaces pol I and joins the two adjacent fragments (Figure 12.5). In bacteria, DNA ligase uses NAD^+ for energy; in T4 phage and eukarya, it uses ATP.

DNA synthesis has been described here as if DNA polymerase moves along the DNA molecule. There is considerable evidence that the opposite is true: DNA polymerase is fixed in position in the cell membrane, and DNA moves past it during synthesis. A second important feature of DNA synthesis is that the leading-strand and lagging-strand DNA polymerases are really two adjacent parts

Fig 12.5 Replacing primers and ligating fragments.
1 – DNA polymerase I removes primer (zigzag) with $5' \rightarrow 3'$ exonuclease activity and adds deoxynucleotides (thick line) with polymerase activity.
2 – Ligase joins two adjacent nucleotides to seal the gap.

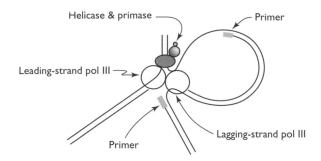

Helicase & primase

Primer

Leading-strand pol III

Lagging-strand pol III

Primer

Fig 12.6 Lagging-strand and leading-strand pol III enzymes work together.

of one complex (Figure 12.6). Because of this, the lagging strand's template makes a loop.

Errors and Proofreading

Nucleotides mispair spontaneously in every possible combination, at frequencies ranging from $\sim 10^{-3}$ (G·T) to $\sim 10^{-5}$ (G·A, C·C). This is a serious problem, because mispaired bases can lead to mutations (permanent alterations in the DNA). However, when replication is complete, mispaired bases occur at a much lower frequency, $\sim 10^{-10}$. Mispairing is reduced in three stages.

- Before the polymerization reaction, DNA polymerase assists pairing, which reduces mispairing 100-fold.
- After the polymerization reaction, DNA polymerase proofreads bases for fidelity of pairing. During proofreading, the $3' \rightarrow 5'$ exonuclease removes and replaces 90% to 99.5% of the mismatched pairs.
- After replication, mismatch repair systems replace 99% of the remaining mispaired nucleotides.

Is the overall repair rate, due to assisted pairing, proofreading, and postreplication repair, as good as it can be? No! For example, antimutator strains of bacteria exist, whose error rates in proofreading DNA are orders of magnitude lower than those of wild, natural strains. The main tradeoffs appear to be the immediate costs of accuracy and the evolutionary costs of a very low mutation rate (without mutations, adaptation to new environments is impossible) versus the benefits of a low mutation rate (most mutations are deleterious).

Replication Enzymes of Archaea

Archaeal chromosomes consist of DNA bound to histones. The histones must be removed from DNA before replication and added back to the newly synthesized daughter DNA molecules. Nothing is known about this process. The archaeon *M. jannaschii* appears to have a single, monomeric DNA polymerase, which shows homology to eukaryal polymerases and to pol II of bacteria. *Methanococcus jannaschii* also has two proteins homologous to eukaryal replication factor complex proteins and a suite of replication enzymes like those in bacteria – helicase, gyrase, and ligase. Presumably primase activity is present in other replication enzymes. The DNA polymerases of archaea that live at extremely high temperatures are widely used to synthesize DNA *in vitro* in a procedure called polymerase chain reaction (PCR) (Chapter 28).

Replication Enzymes in Nuclei of Eukarya

Eukaryal chromosomes, found in the cell's nucleus, are made of chromatin, DNA complexed with histones and other proteins. Before DNA can be replicated, the chromosomal proteins are removed. Chromatin re-forms after replication.

Eukarya have several nuclear DNA polymerases; four are known in yeast and five in mammals (Figure 12.7). It is thought that polymerases δ and ε direct replication of the leading strand and that polymerase α directs replication of the lagging strand because only α has primase activity. Early-branching eukarya appear to have several DNA polymerases. One class of DNA polymerase is unique to eukarya: telomerase, which is needed in most taxa to replicate telomeres (Chapter 13).

Name	Mass	Subunit size(s)	Characteristics
α	350	50, 60, 70, 170	Replication, primase activity, no exonuclease activity[1]
δ	173	48, 125	Replication, $3' \rightarrow 5'$ exonuclease
ε	270	55, 215	Replication, $3' \rightarrow 5'$ exonuclease
β	37	37	Repair only, no exonuclease activity
Telomerase[2]	127	127	Ribonucleoprotein

[1]DNA pol α of *Drosophila* is a $3' \rightarrow 5'$ exonuclease.
[2]Human telomerase reverse transcriptase, or TRT.

Fig 12.7 DNA polymerases of mammals (nucleus); mass in kDa.

Auxiliary replication proteins such as ligases, helicases, topoi-somerases, and exonucleases are less well known in eukarya than in bacteria. Two mammalian ligases have been identified. SSB-like proteins have been found in humans, rats, cows, and lilies. $5' \rightarrow 3'$ helicases have been identified in humans, mice, and yeast. Type II topoisomerases, which break both strands of dsDNA, are known from several eukarya. They relax DNA but do not introduce negative supercoiling. However, type I topoisomerase, which breaks one strand of dsDNA, seems to function in replication rather than type II. Eukarya have several $3' \rightarrow 5'$ exonucleases in addition to DNA pol δ and ε, and more than one $5' \rightarrow 3'$ exonuclease. DNA polymerase and other replication proteins are bundled into a replisome in eukarya, as they are in bacteria.

Replication Enzymes of Mitochondria and Chloroplasts

Mitochondria and chloroplasts possess replication enzymes encoded by the organelle genome. DNA polymerase of mitochondria, designated Pol γ, is typically an $\approx$160-kDa heterodimer. Pol γ is a $3' \rightarrow 5'$ exonuclease but not a primase. The DNA polymerase of chloroplasts is similar to Pol γ. In the mitochondria of mammals, DNA replication does not produce lagging strands, as parental strands are replicated full circle, each from a different origin.

Replication Enzymes of dsDNA Viruses and Plasmids

Some dsDNA viruses of bacteria and animals encode their own DNA polymerases (e.g., T4 and T7 phages, herpes, vaccinia, and hepatitis B), while others use host enzymes (e.g., ϕX174, polyoma, simian virus 40). Typically, viral polymerases are monomers with $3' \rightarrow 5'$ exonuclease activity but without primase activity. Primers are synthesized either by host primosome (ϕX174) or by host RNA polymerase (T4 and T7 phages). The DNA polymerase of T7 phage contains T7-encoded polymerase (80 kDa) and host-encoded thioredoxin (12 kDa). The T7 enzyme is homologous to pol I of *E. coli*, although it lacks $5' \rightarrow 3'$ exonuclease activity. T7 phage also encodes a helicase and a $5' \rightarrow 3'$ exonuclease. Plasmids are replicated by cellular replication machinery, but in many cases RNA polymerase synthesizes the primer.

Modification of DNA Bases

In bacteria and eukarya, methylases modify newly replicated DNA by adding methyl groups to bases, usually to adenine or cytosine. The roles of methylation in bacteria are as follows:

- To protect DNA from endonucleases that cleave incompletely methylated sites (in bacteria)
- To mark newly synthesized strands for repair of mismatched bases (in bacteria)
- To inactivate transposons (in eukarya)

The DNA of T-even phages contains hydroxymethyl-dC instead of dC; after DNA synthesis, some nucleotides are glycosylated by phage-encoded β-glycosyltransferases. In phage Mu, a glycinamide is added to about a seventh of the dA's. Some phages of *Bacillus subtilis* use dU or hydroxymethyl-dU instead of dT.

Speed of Replication

The speed of DNA replication in bacteria, archaea, mitochondria, chloroplasts, and viruses depends on the speed of chain elongation. In *E. coli,* the speed of chain elongation is $\approx$1000 nt s^{-1}. In archaea and eukarya, the speed of chain elongation is typically $\approx$50 nt s^{-1}.

Further Reading

Edgell DR, Doolittle WF. 1997. Archaea and the origins of DNA replication proteins. *Cell* 89:995–998.

Hozak P, Cook PR. 1994. Replication factories. *Trends Cell Biol.* 4:48–51.

Kornberg A, Baker T. 1991. *DNA Replication.* WH Freeman, San Francisco.

Marians KJ. 1992. Prokaryotic DNA replication. *Annu. Rev. Biochem.* 61: 673–719.

Waga S, Stillman B. 1998. The DNA replication fork in eukaryotic cells. *Annu. Rev. Biochem.* 67:721–751.

Chapter 13

Chromosome Replication

Overview

The mode of chromosome replication varies from life form to life form and fits with chromosome structure and with the course and tempo of reproduction.

Chromosome replication begins at **origins of replication** and stops at termination sequences; small chromosomes typically have a single origin. Replication may be bidirectional or unidirectional. Linear chromosomes may have interior or terminal origins. Most linear chromosomes have **telomeres**, which are terminal sequences that are replicated in special ways. The large, linear chromosomes of eukarya are divided into many **replicons** (simultaneously replicated regions), each having an origin at its center.

Chromosomes made of RNA or single-stranded DNA replicate by exceptional mechanisms.

Replication of Circular Chromosomes

Theta Replication
Many circular chromosomes – bacterial chromosomes, most plasmids, and some dsDNA viral chromosomes – replicate by the theta mechanism, so named because the shape of partly replicated chromosomes resembles the Greek letter theta. Theta replication is usually bidirectional, but in some chromosomes it is unidirectional – e.g., colE1 phage.

Replication begins at the chromosome's origin (Figure 13.1). In *Escherichia coli* the origin is called *oriC*, which consists of 245 bp and contains short repeated sequences: four copies of a 9-bp sequence and three copies of a 13-bp sequence. First, about 30 copies

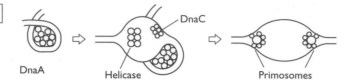

Fig 13.1 Initiation at oriC.

of initiator protein (dnaA) bind to the four 9-bp sequences of oriC. DnaA helps the DNA to melt, making a single-stranded region. Helicase + dnaC bind to the three 13-bp sequences and open up the double-stranded DNA (dsDNA) further, after which two primosomes are assembled and bidirectional replication starts (Figure 13.2). The replication origins of many bacterial and phage chromosomes resemble *oriC*.

Replication of the *E. coli* chromosome stops at the termination region 180° around the circle from the origin. This region contains six widely dispersed 23-bp termination (*ter*) sequences, to which the 36-kDa Tus protein binds. Tus protein interferes with DNA helicases. At the end of theta replication, the two daughter chromosomes are circles of dsDNA connected like links of a chain. A topoisomerase cuts both circles, releasing two separate chromosomes whose ends are ligated to make them whole.

Rolling Circle Replication

Rolling circle replication is named for its shape and apparent movement. It is common in small circular viral chromosomes and circular plasmids made of dsDNA. In *Trypanosoma*, the larger mitochondrial chromosomes (maxicircles) undergo rolling circle replication. In eukarya, genes may be amplified this way; amplification is the synthesis of extra copies of one nuclear gene.

Origins for rolling circle replication are short sequences containing cleavage sites where a single strand of DNA is nicked as replication begins. One of the DNA strands, an unnicked circle, serves only as a template, while the nicked strand serves both as primer for its own continuous replication and as template for

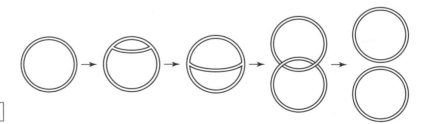

Fig 13.2 Theta replication.

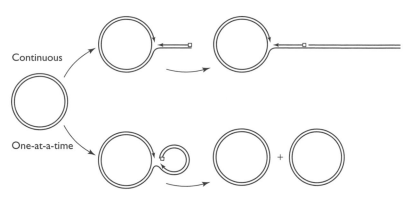

Fig 13.3 Rolling circle replication: continuous (λ) versus one-at-a-time (φX174).

the discontinuous replication of the other strand. Either primase or RNA polymerase makes the primers for the discontinuously synthesized strand. Helicase unwinds the double helix. The cell's DNA polymerase (pol III in *E. coli*) catalyzes DNA synthesis. Some yeast plasmids replicate by a double rolling circle mechanism, in which the circular plasmid assumes a dumbbell shape. The plasmid is nicked at two sites, and replication proceeds from both of these sites.

In some cases (e.g., phage λ), synthesis proceeds around the circle many times, producing a large concatemer (tandem, multiple copies of the chromosome). In other cases (e.g., phage φX174), two circular copies are completed, and then rolling circle replication starts afresh on the daughter chromosomes (Figure 13.3).

D-loop Replication

The mitochondrial DNA (mtDNA) of vertebrates replicates by the D-loop mechanism, in which each strand is synthesized unidirectionally from a different origin: origin O_H for copying the purine-rich heavy (H) strand and origin O_L for copying the pyrimidine-rich light (L) strand (Figure 13.4). At O_H a structure called the D-loop (displacement loop) forms; this permanent feature of vertebrate mtDNA is made of triple-stranded DNA (H-DNA). It does not include any gene. The third strand of the D-loop is an $\approx$0.5-kb primer for copying the heavy strand of mtDNA.

Fig 13.4 D-loop replication of the vertebrate mitochondrial chromosome.

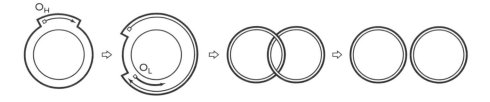

This 0.5-kb primer consists mostly of DNA, with a short stretch of RNA at the 5′ end. Mitochondrial RNA polymerase synthesizes the RNA part, and DNA polymerase γ synthesizes the rest of the primer. During most of a mitochondrion's existence, the primer sits in readiness for the completion of chromosome replication. When replication resumes, DNA polymerase γ extends the D-loop primer. The light strand begins to be copied only after DNA is unwound at O_L, more than halfway around the chromosome. Then, a primase synthesizes a primer, which is extended by DNA polymerase γ. Synthesis of each single strand continues until the entire chromosome has been replicated.

Replication of Linear Chromosomes

Large Linear Chromosomes

A large eukaryal chromosome may have thousands of origins, one origin per replicon. The number of simultaneously replicating replicons per chromosome varies within an organism: in rapidly dividing cells, a chromosome may have many more replicons and therefore more origins than it does in slowly dividing cells. In yeast, replicons are 30 to 40 kb. In multicellular organisms, the size range of replicons is 10 to 200 kb. In yeast, origins contain AT-rich autonomous replication sequences (ARSs), each containing an 11-bp sequence. The origin recognition complex, a 400-kDa hexamer, initiates replication by binding to the ARS. Eukaryal chromosomes undergoing replication have many bidirectionally growing "bubbles" (Figure 13.5). Replication in eukaryal replicons is bidirectional, with leading and lagging strands.

Small Linear Chromosomes

Small linear chromosomes – some viral chromosomes and some plasmids – have a single interior origin. In the phage T7 chromosome, for example, replication is initiated when RNA polymerase synthesizes a long primer for the continuous strand

Fig 13.5 Adjacent replicons make growing replication bubbles.

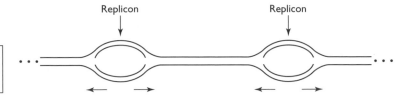

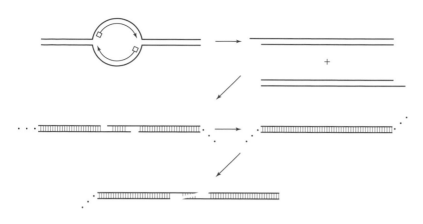

Fig 13.6 Bubble replication in phage T7.

(Figure 13.6). Primase synthesizes primers for the discontinuous strands. Short stretches at the 5′ ends are single stranded where primers are removed, but two daughter chromosomes anneal because the ends contain repeated sequences. After ligation, an endonuclease cuts the DNA, leaving 3′ overhangs at the ends.

Replication of Telomeres

The ends of a linear chromosome cannot be replicated with normal replication machinery, because each strand of the double helix is incomplete at its 5′ end after the primer has been removed (Figure 13.7).

Telomeres, present at the ends of most linear chromosomes, overcome the problem of incomplete 5′ ends. The synthesis of a linear chromosome's ends can be circumvented in three general ways:

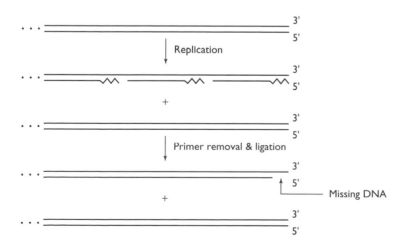

Fig 13.7 Normal replication cannot copy the ends of a linear DNA molecule.

- Extending the 3′ ends by a telomerase enzyme, and then filling in the short 5′ ends
- Attaching a primer covalently to the end of the chromosome
- Using the parent DNA molecule as a primer.

If a chromosome's termini are not completely replicated, then the chromosome shortens each reproductive cycle. One hypothesis for maintenance of telomeres without exact replication of the ends is replicative transposition of subterminal repeats.

Telomerase

Telomerase, a ribonucleoprotein enzyme, elongates the overhanging 3′ end of the telomere in eukarya (Figure 13.8). The enzyme was discovered in ciliates, whose telomeres contain tandem repeats, such as 5′GGGGTT3′ in *Tetrahymena*. Telomerase is a reverse transcriptase carrying its own RNA template for DNA. After the 3′-terminal strand has been lengthened, the "normal" replication machinery extends the 5′ terminal strand to complete replication. The telomerase mechanism leads to indeterminacy in the size of the telomeric terminus. Most mammalian somatic cells lack telomerase (the gene is not expressed), and in them the erosion of telomeres cannot be counteracted. Consequently, chromosomes in these cells become shorter each time they are replicated.

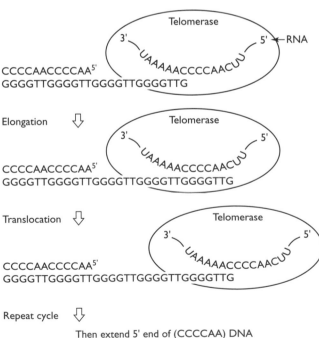

Fig 13.8 Replication of telomeres with telomerase.

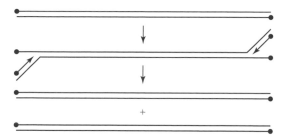

Fig 13.9 Replication of linear plasmids in *Streptomyces*, primed by protein. • = protein covalently bonded to the 5′ end.

Covalently Attached Primers

Several viruses (e.g., phage φ29 of *Bacillus subtilis* and adenovirus of mammals) use the second mechanism, a special protein primer, which remains attached covalently to each 5′ end of their linear, dsDNA chromosomes. Linear plasmids and linear main chromosomes of bacteria (e.g., *Streptomyces*) possess terminal inverted repeats and terminal proteins; it seems likely that the terminal proteins are primers in DNA synthesis (Figure 13.9). Terminal inverted repeats favor the formation of hairpin loops in the longer 3′ end, to which the terminal protein may bind. In both cases, the terminal protein acts as an initial primer for replication, so that the 5′ ends are replicated completely from the start. Primers of extrachromosomal origin are usually proteins, although they could be made of nucleic acid or of nucleic acid and protein. Retroviruses use host tRNA as an initial primer for DNA synthesis.

Parent DNA as Primer: Rolling-Hairpins

A third way to solve the problem of replicating the ends of linear, dsDNA chromosomes is to use part of the parent DNA molecules as primers. An example is the rolling-hairpin mechanism, which depends on terminal palindromic sequences that tend to form hairpin loops (Figure 13.10). This model has been proposed for the ends of some eukaryal chromosomes. Vaccinia (cowpox) chromosomes have terminal, inverted repeats that form closed terminal loops. To replicate, the molecule is nicked, the 3′ end is extended, new hairpin loops form, the chromosome is replicated, and the daughter molecule is cut to yield two new chromosomes.

Gene Amplification

Sometimes multiple copies of a gene are synthesized independently of chromosome replication; this is **gene amplification**

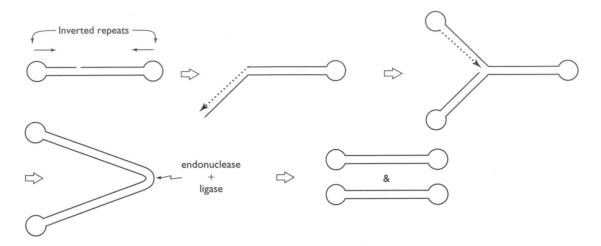

Fig 13.10 Rolling-hairpin model, using parental DNA as a primer.

(Figure 13.11). Some chorion (egg layer) genes of *Drosophila*, rRNA genes of *Xenopus* and *Tetrahymena*, and drug-resistance genes of tumor cells undergo amplification. The chorion genes of *Drosophila* are thought to undergo bidirectional bubble synthesis, making a nested loop structure. Amplified loci are adjacent to ARSs. In the macronuclei of *Tetrahymena*, genes are located in microchromosomes. The microchromosomes containing rRNA genes are amplified between 10- and 100-fold. Gene amplification is regulated differently from and independently of genome replication.

Replication in RNA Viruses

The genomes of RNA viruses consist of one or more RNA chromosomes. The genomes fall into three groups, based on the genetic material: double-stranded RNA (Figure 13.12), plus-strand RNA (equivalent to mRNA), and minus-strand RNA (complementary to

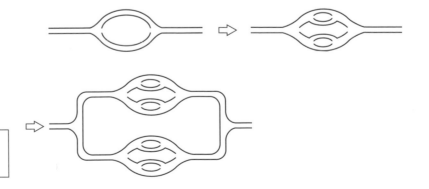

Fig 13.11 Gene amplification by bubble synthesis.

Fig 13.12 Double-stranded linear RNA replicating bidirectionally.

mRNA). Except for retroviruses (RNA viruses that make a DNA copy of the genome), all three types use virally encoded RNA-dependent RNA polymerases – replicases – to replicate chromosomes. Replicases do not proofread, and therefore replication of RNA virus genomes is error-prone. Replicases can direct the transcription of viral mRNAs and in that role are called transcriptases.

The tiny genome of retroviruses, such as HIV, is replicated in an amazingly complex fashion. Inside the retrovirus virion are two copies of the single-stranded RNA (+)-strand chromosome, tRNA from the previous host, and replication enzymes, including reverse transcriptase (115 kDa). To begin replication, the 3′ end of the tRNA molecule base pairs with a site on the RNA chromosome, and reverse transcriptase extends the primer to the 5′ terminus. In subsequent steps, the DNA translocates, ribonuclease H degrades RNA, and reverse transcriptase extends primers. The final product of replication is dsDNA, which moves into the nucleus and inserts into a host chromosome by a cut-and-paste mechanism.

Further Reading

Chen CW. 1996. Complications and implications of linear bacterial chromosomes. *Trends Genet.* 12:192–196.

Falaschi A. 2000. Eukaryotic DNA replication. *Trends Genet.* 16:88–92.

Kornberg A, Baker T. 1991. DNA replication. WH Freeman, San Francisco.

Shadel GS, Clayton DA. 1997. Mitochondrial DNA maintenance in vertebrates. *Annu. Rev. Biochem.* 66:409–433.

Toone WM *et al.* 1997. Getting started: Regulating the initiation of DNA replication in yeast. *Annu. Rev. Microbiol.* 51:125–149.

Zakian VA. 1996. Structure, function, and replication of *Saccharomyces cervisiae* telomeres. *Annu. Rev. Genet.* 30:141–172.

Chapter 14

Molecular Events of Recombination

Overview

The chromosomes of all life forms naturally undergo recombination, a physical exchange between two DNA molecules that results in new, recombinant DNA molecules. Recombination plays many roles: producing new combinations of gene copies, introducing novel sequences into chromosomes, making novel genes, regulating gene expression, and facilitating some DNA repair. Recombination has a huge impact on the biology of sexually reproducing eukarya.

There are three main kinds of recombination in nature: (1) **general recombination**, an exchange between homologous DNA molecules, in which the location of the exchange site is not restricted, (2) **site-specific recombination**, an exchange between nonhomologous DNA molecules and occurring only at specific short sequences, and (3) **transposition**, the movement of a transposon.

This chapter focuses on the molecular events of recombination, not on its biological consequences, analysis, or practical uses, all very important in genetics.

General Recombination

Homologous DNA molecules (different copies of a double helix, such as the two copies of chromosome 1 that you received from your mother and father), when they are together inside a cell, exchange parts by general recombination (Figure 14.1). In general recombination, portions of double-stranded DNA (dsDNA)

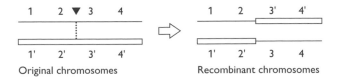

Original chromosomes → Recombinant chromosomes

Fig 14.1 A simple exchange between linear homologous chromosomes. ▼ = point of exchange. Each chromosome is one dsDNA molecule. Markers 1–4 and 1′–4′ are different copies of homologous genes on the two chromosomes.

chromosomes having extensive homology pair with the help of DNA-binding proteins. At least some homologous sequences in the two DNA molecules are aligned precisely. The two DNA molecules exchange strands in a multistep process catalyzed by many enzymes.

The Aviemore Model of General Recombination

General recombination is not identical in all life forms, and there are many theories of molecular mechanisms. One theory of how recombination often works is the Aviemore model (Figure 14.2), summarized in the following six points.

- Proteins help short homologous sequences of two dsDNA molecules to pair.
- A strand of one duplex is nicked, and the strand invades the other duplex.
- The displaced strand switches places with the nicked strand and is itself nicked.
- Both nicked strands are ligated to free ends of the recipient molecule, making an X-shaped **Holliday junction** (named after

Fig 14.2 Recombination by the Aviemore model; ▽ = site of nick.

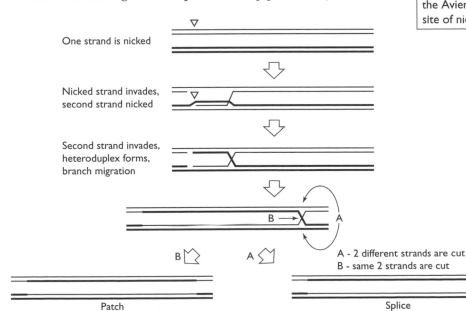

One strand is nicked

Nicked strand invades, second strand nicked

Second strand invades, heteroduplex forms, branch migration

B → A

A - 2 different strands are cut
B - same 2 strands are cut

B

A

Patch   Splice

the biologist who proposed it) connecting the two DNA duplexes. A heteroduplex is made of strands donated by two different DNA molecules.

- The heteroduplex region grows as the hybrid branch structure moves.
- Again, two strands are nicked and joined, producing either a splice or a patch. In spliced chromosomes, sequences flanking the hybrid portion recombine, while in a chromosome with a patch, flanking sequences do not recombine.

Pause and consider the implications of the Aviemore model. The original DNA molecules are homologous; that is, the paired chromosomes are different copies of one ancestral molecule. Homologous DNA from independent sources (e.g., mother and father) differ in a small percentage of their nucleotides. Recombinant DNA differs from the original (parental) DNA in nucleotide sequence. An Aviemore-type exchange has only two possible outcomes: either the recombinant DNA molecules have patches, or they have splices. Patches and splices are $\sim 10^3$ nucleotide pairs long.

The heteroduplex DNA of a patch or splice may contain mismatched pairs (pairs other than A·T and G·C). This happens when the two recombining DNA molecules differ in that part of their sequences. Suppose one heteroduplex patch contains a single mismatched nucleotide pair, A·C, and the other heteroduplex patch contains G·T (Figure 14.3).

DNA molecules are not supposed to have mismatched pairs such as A·C and G·T. How is this unhappy situation resolved? There are two possibilities: (1) repair enzymes replace the mismatched pairs with matched pairs, or (2) the mismatched pairs stay as they are until the next round of DNA replication. Possibility (1), repair, on average produces 50% recombinant molecules; possibility (2),

Fig 14.3 Aviemore-type exchange yields patches or splices with mismatched bases. The two homologous DNA molecules are drawn with either thick or thin lines for their two strands. Three genes are numbered in the diagram. To distinguish the copies of these genes, the thin set is marked with primes 1', 2', and 3'.

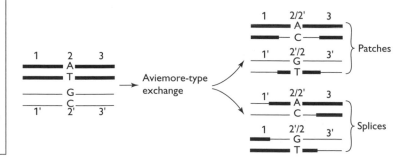

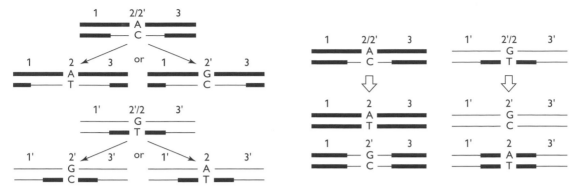

1. Repair of mismatched pairs

2. No repair, then replication

no repair, produces exactly a 2:2 ratio of recombinant and nonre-combinant molecules after DNA replication. Repair enzymes do not have intelligence, so a mismatched pair is as likely to be re-placed with the "incorrect" pair as it is to be replaced with the "correct" pair: A·C → G·C is as likely as A·C → A·T, and G·T → A·T is as likely as G·T → G·C (Figure 14.4).

When one nucleotide sequence replaces another in a heterodu-plex made during general recombination, the gene containing the changed nucleotide sequence is said to be converted. For alleles differing by a nucleotide substitution, **gene conversion** is usually unbiased: the two copies of a gene are equally likely to be converted. The rate of gene conversion is $\sim 10^{-3}$ per gene during meiosis, the part of sexual reproduction when haploid cells are made from diploid cells. Therefore, gene conversion is a significant source of recombinants in sexually reproducing eu-karya. Gene conversion is equally frequent in patches and splices. Splices, though, make for recombination between genes on either side of the heteroduplex, unlike patches.

If two copies of a gene differ in size, because nucleotides were either inserted into that gene or deleted from it, then gene con-version will be biased towards the larger copy. Thus, in an organ-ism having both a normal and a shortened copy of a gene, the normal copy replaces the shortened copy in a majority of gene conversion events. This happens because repair of a heteroduplex DNA molecule containing a single-stranded loop usually entails filling the gap opposite the loop, not removing the unpaired nu-cleotides of that loop.

To summarize, Aviemore-type recombination yields 50% patches and 50% splices. Gene conversion occurs in 50% of patches

Fig 14.4 The fate of mismatched pairs in a heteroduplex DNA molecule. Patches are shown here, but the fate of mismatched pairs is the same for splices.

and 50% of splices containing mismatched sequences, whether or not the mismatched sequence is repaired. Genes on either side of the heteroduplex recombine if a splice is made but not if a patch is made. However, a chromosome containing a converted gene is recombinant for that gene; i.e., recombination has occurred between the converted gene and other genes on the chromosome.

Protein (structure)	Mass (kDa)	Functions
RecA (monomer)	38	ssDNA binding
RecBCD (heterotrimer)	327	exonuclease, endonuclease, ATPase, helicase; binds to χ sequences
RuvA (tetramer)	88	binds to Holliday junctions
RuvB (hexamer)	148	helicase
RuvC (dimer)	38	endonuclease, cleaves Holliday junctions

Fig 14.5 Enzymer of recombination in *E. coli*.

Enzymes of Recombination in Bacteria

The enzymology of general recombination has been worked out only for *Escherichia coli*; other bacteria have similar enzymes. For eukarya, the molecular details of general recombination (arguably a more important process in eukarya than in bacteria) are not known. In *E. coli*, over 30 different proteins participate; the foremost are in Figure 14.5.

Another important component of recombination in *E. coli* is an 8-bp sequence called χ ("chi," 5'GCTGGTGG3'). χ is evenly spaced in the chromosome and occurs every ≈ 5 kb, about ≈ 13 times as frequently as it would by chance.

General recombination in *E. coli* is a three-phase process: (1) prepairing, when DNA is prepared for exchange; (2) pairing, in which strands are transferred between DNA duplexes, making hybrid structures; and (3) postpairing, in which hybrid structures are resolved into separate DNA molecules.

Prepairing. RecBCD binds to dsDNA, initiates unwinding of the helix, and, as it unwinds, digests the strands by the action of its endonuclease and exonuclease activities (Figure 14.6).

Looking in the direction of unwinding, the 3'-end strand is degraded more rapidly than the 5'-end strand. When RecBCD reaches χ, it pauses, nicks one strand 4 to 6 nt 3' to the 3' G of χ, loses its exonuclease activity (RecD), and continues to unwind the double helix. RecA then binds to the shorter, 3'-ending strand.

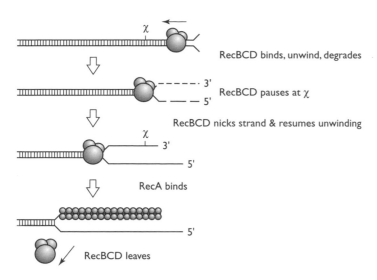

RecBCD binds, unwind, degrades

RecBCD pauses at χ

RecBCD nicks strand & resumes unwinding

RecA binds

RecBCD leaves

Fig 14.6 Prepairing in general recombination of E. coli.

Pairing. The RecA-coated strand of DNA "searches" for a homologous sequence within another DNA duplex, and the ssDNA invades that duplex to pair with a complementary sequence. The other DNA duplex is an "invader"; e.g., it may have been delivered by conjugation (Chapter 18). A Holliday junction results (Figure 14.7).

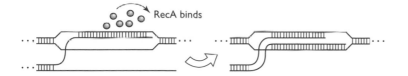

RecA binds

Fig 14.7 Pairing in general recombination of E. coli.

Postpairing. A tetramer of RuvA binds to the Holliday junction. A donut-shaped hexamer of RuvB binds to each side of RuvA. RuvA and RuvB promote branch migration of the Holliday junction. After branch migration, a dimer of RuvC then binds to the Holliday junction and nicks strands at these sites. Either splices or patches are produced (Figure 14.8).

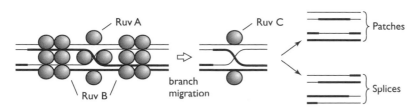

Ruv A

Ruv C

Patches

branch migration

Ruv B

Splices

Fig 14.8 Post pairing in general recombination of E. coli.

Nonreciprocal Exchange

When pairing is not precise at the site of recombination, one recombinant chromosome may suffer a deletion (be missing some of

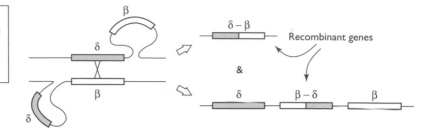

Fig 14.9 Unequal crossing-over between human globin genes. The δ-β recombinant makes for anemia.

its DNA sequences) and the other chromosome may contain a duplication (extra DNA sequences). Nonreciprocal exchange is called **unequal crossing-over** (Figure 14.9). Duplications and deletions in hemoglobin genes, which have profound medical consequences, were produced by unequal crossing-over. Unequal crossing-over has also produced duplications and deletions in human genes coding for red and green color vision.

Site-specific Recombination

Unlike general recombination, site-specific exchange is restricted to a few chromosomal **loci** (locations). Examples include the integration of plasmids or phage chromosomes into a cell's essential chromosomes, the inversion of chromosome segments to control expression of alternative genes, and the assembly of mammalian immunoglobin genes during development. A site-specific **recombinase** binds to a short DNA sequence on the donor molecule and to a homologous sequence on the target chromosome; this enzyme also cleaves and ligates the DNA. In an intermediate step, covalent bonds are formed between DNA and the recombinase. Two large families of site-specific recombinases have been investigated, the **integrase family** and the **invertase-resolvase family**. Integrases cut two strands at a time and produce Holliday intermediates with very short heteroduplexes. Resolvases cut all four strands at one site and splice the two duplexes.

Integrases
An example of integrase-mediated recombination is integration of the λ phage chromosome into the host chromosome, an essential step in the lysogenic (noninfectious) stage of λ bacteriophage. Both the λ chromosome and the host chromosome possess homologous sequences where recombination takes place, called attachment (*att*) sites – *attP* in the phage chromosome and *attB* in

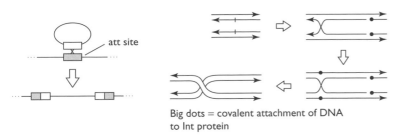

att site

Big dots = covalent attachment of DNA to Int protein

Integration of λ into the E. coli chromosome

Details of the exchange

Fig 14.10 Integration of the λ phage chromosome into the E. coli chromosome.

the bacterial chromosome. To begin integration, 21-bp sequences shared by *attP* and *attB* pair. Three bacterial DNA-binding proteins, IHF, Xis, and Fis, assist the phage integrase, λ **Int** (40 kDa). Four Int monomers interact to cut the two *att* sites. In the first step, Int nicks both chromosomes and temporarily forms a DNA-protein intermediate by a covalent bond between each 3′ phosphate of DNA and a tyrosine of Int. After a short branch migration, Int cuts the second two strands in the same way, yielding a recombinant DNA molecule, free of proteins (Figure 14.10).

Invertase–Resolvase

Invertase–resolvases catalyze inversions of a DNA segment. They cut all four strands before the exchange between DNA molecules. In the reactions, each strand of DNA is connected to an invertase-resolvase protein via a 5′-phosphoserine bond. An example is the *Salmonella* Hin invertase, which inverts sequences in two genes coding for flagellar proteins, activating one gene and inactivating the other.

Immunoglobins in Mammalian B Cells

B cells within the mammalian immune system produce immunoglobins, which are encoded by *Ig* genes that exist only in B cells. What!? Genes that exist in somatic cells but are absent from gametes? Strange, but true: all the starting materials – prefab pieces from which immunoglobin genes are to be made – are inherited in both sperm and egg, but functional *Ig* genes are assembled from these DNA sequences only in B cells.

Early in the development of a B cell, two different *Ig* genes are assembled by site-specific recombination at two separate locations in the genome. One *Ig* gene (light) is assembled from segments

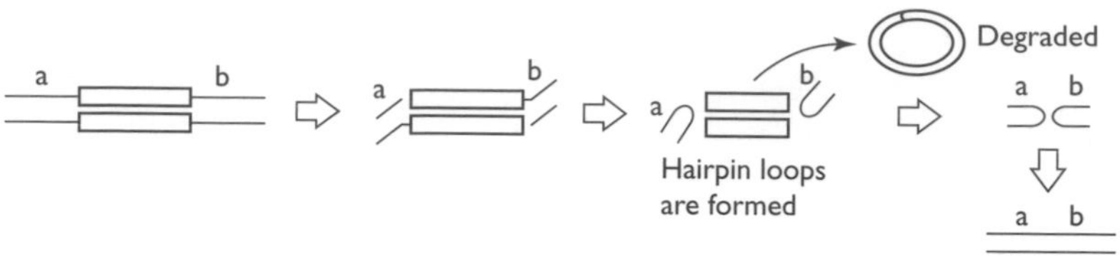

Fig 14.11 Ig segments (a and b) recombine to make a functional gene.

called C (constant), V (variable), and J (joining); the other *Ig* gene (heavy) is constructed of C, V, and J segments, and an additional gene-segment called D (diversity). A light gene encodes a short polypeptide, and a heavy gene encodes a long polypeptide; the light and heavy polypeptides form a tetrameric Ig protein. In pre-B cells, before the assembly of an *Ig* gene, each kind of segment (C, V, J, ...) exists in multiple tandem repeats.

Short, conserved sequences that signal recombination flank each copy of each segment. The segments (C, V, ...), one copy of each, come together at random and they recombine. The recombinase for *Ig* gene assembly appears not to make covalent bonds with DNA during recombination, in contrast to the bacterial and phage recombinases.

More information about *Ig* genes and the prefab DNA sequences used to construct them is presented in Chapter 24. The focus here is the nature of the recombination event. A hypothesis of how it works is cartooned in Figure 14.11.

Chromosome Rearrangement in Ciliate Macronuclei

Ciliates (e.g., *Paramecium*) possess two micronuclei, which contain the stably inherited but transcriptionally inactive genome, and a macronucleus, which contains thousands of tiny, gene-sized, transcriptionally active chromosomes. Macronuclei develop from micronuclei, and one feature of that development is the rearrangement of the large, micronuclear chromosomes, which are cut into gene-sized pieces; telomeres are added to each gene-sized piece of DNA, making thousands of tiny chromosomes. Furthermore, pre-gene DNA undergoes site-specific recombination to make new, functional genes.

Transposition

Transposons and transposable bacteriophages move by transposing donor sequences into a distant target site, either within the same chromosome or in a different chromosome. Hundreds of different transposons have been identified, and probably millions or more of them exist. DNA transposes by three different recombination mechanisms:

- Nonreplicative transposition, a cut-and-paste process
- Replicative transposition, an integrate-and-resolve process
- Retrotransposition, which works via an RNA intermediate

Some **transposases**, the enzymes catalyzing transposition, utilize random target sites, as with phage Mu, while others display preference for specific sequences, as with *P* elements in *Drosophila*.

Nonreplicative Transposition

Nonreplicative transposition entails excising a segment of DNA from a donor site, cutting open a target site, and inserting the segment into the target site (Figure 14.12). *Tn5*, a 5.7-kb transposon of *Klebsiella pneumoniae*, is a good example of this. *Tn5* encodes a transposase and proteins conferring antibiotic resistance. Transposase cuts both strands at the terminal inverted repeats of *Tn5* and ligates the nicks at the donor site. Transposase also cuts both strands at the target site, where *Tn5* is inserted into target DNA.

Nonreplicative transposons in eukarya include *P* elements of *Drosophila* and *Ds* and *Ac* elements in *Zea mays*. *P* and *Ac* elements are autonomous: they encode transposases that can catalyze the excision of the element from a donor site and its insertion into a target site. *Ds* elements and some defective *P* elements are nonautonomous; to transpose they require enzymes encoded by autonomous elements. Transposons sometimes insert into genes, mutating them. Excision can be precise or imprecise. After precise excision, the exised gene regains its functionality, but after imprecise excision, sequences in the exised gene are deleted, often rendering it nonfunctional.

Replicative Transposition

In replicative transposition a new copy of the transposon is made when it is inserted in a new site (Figure 14.13). An example is

Fig 14.12 Nonreplicative transposition. Parallel lines are the two strands of DNA; little boxes are targets of transposase.

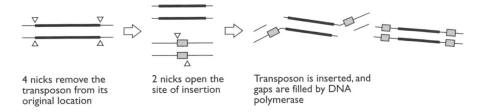

4 nicks remove the transposon from its original location

2 nicks open the site of insertion

Transposon is inserted, and gaps are filled by DNA polymerase

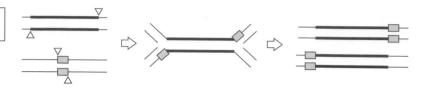

Fig 14.13 Replicative transposition.

phage Mu of *E. coli*, which usually transposes by a replicative mechanism. To do so, a distant insertion site is brought near an integrated copy of Mu. Mu-encoded transposase nicks single strands at each end of Mu and makes a staggered cut at the target site. Both strands prime DNA synthesis, yielding two copies of Mu, one at the new location.

Retrotransposition

Retrotransposons are transcribed into mRNA whose proteins include reverse transcriptase and integrase. Retrotransposons resemble retroviruses, which are ssRNA viruses whose dsDNA intermediates insert into host chromosomes. Perhaps retrotransposons evolved from retroviruses. Many retrotransposons, like retrovirus chromosomes, have long terminal repeats (LTRs) at both ends. Even retrotransposons lacking LTRs are quite similar in sequence to retroviruses.

Retrotransposons move by the same mechanism as retroviruses. Reverse transcriptase synthesizes a DNA copy of the mRNA. Occasionally, one of these DNA copies is inserted into the host chromosome, with the help of the integrase protein. Integrase makes staggered cuts at the target site and nicks the retrotransposon, so that both donor and target have short overhangs at the 5′ ends. The retrotransposon inserts into the target site, as in replicative transposition.

Retrogenes are a separate category of transposable elements found in eukarya. Retrogenes are intronless pseudogenes that can transpose. They do not code for reverse transcriptase or integrase; indeed, since they are pseudogenes, they do not code for RNA. Their transposition depends on the reverse transcriptase of retroviruses or retrotransposons. Other elements that may transpose are the *Alu* sequences of primates, 7SL RNA genes, and fold-back elements. A fold-back element has a noncoding central region flanked by terminal inverted repeats.

Further Reading

Haren L, Ton-Hoang B, Chadler M. 1999. Integrating DNA: Transposases and retroviral integrases. *Annu. Rev. Microbiol.* 53:245–281.

LLoyd RG, Low KB. 1996. *Homologous Recombination.* In: Neidhardt (ed.). *Escherichia coli* and *Salmonella* Cellular and Molecular Biology, 2nd ed., ASM press, Washington, DC. 2236–2253.

Myers RS, Stahl FW. 1994. χ and the RecBCD enzyme of *Escherichia coli. Annu. Rev. Genet.* 28:49–70.

Nash HA. 1996. Site-specific recombination. In: Neidhardt (ed.). *Escherichia coli* and *Salmonella* Cellular and Molecular Biology, 2nd ed., ASM press, Washington, DC. 2363–2377.

Chapter 15

Micromutations

Overview

No life form is a perfect machine, and from generation to generation, a genome may be subject to change – mutation. A mutation is any newly arisen, stably inherited alteration of a life form's genome. A mutation is not a transitory change, and it is not the same as damage, although damage to DNA may lead to mutation. A mutation may not impact the functioning or even the structure of gene products, but if it does, the effects can range from beneficial to fully lethal.

Micromutations are small, affecting at most a single gene and usually involving fewer than $\sim10^3$ nucleotides. In this chapter, a classification of micromutations is followed by descriptions of their causes and effects. Large mutations – chromosomal abnormalities – are sufficiently different from micromutations that they are taken up later (Chapter 21).

Types of Micromutation

The five kinds of small mutation are **substitution**, **deletion**, **insertion**, **duplication**, and **inversion**. Most mutations are point mutations (single nucleotide change), and the most frequent kind of point mutation is a nucleotide substitution, also commonly called a base substitution. Micromutations can occur anywhere in a chromosome.

Substitution
Substitution is nucleotide replacement (Figure 15.1). There are two types of nucleotide substitution, **transition** and **transversion**. A transition is substitution of a purine nucleotide for a purine

		Pu = purine, Py = pyrimidine	
(normal)	..CAT..	Transitions	Transversions
	↓	Pu ↔ Pu	Pu ↔ Py
(substitution)	..GAT..	or Py ↔ Py	
		A ↔ G	A ↔ C, A ↔ T
		T ↔ C	G ↔ C, G ↔ T

Fig 15.1 Example of substitution.

nucleotide or substitution of a pyrimidine nucleotide for a pyrimidine nucleotide. A transversion is a substitution of a purine nucleotide for a pyrimidine nucleotide, or vice versa. Substitutions occur throughout chromosomes.

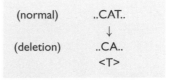

(normal)	..CAT..
	↓
(deletion)	..CA..
	<T>

Fig 15.2 Example of deletion.

Deletion

Deletion is loss of a nucleotide or a string of nucleotides (Figure 15.2). Deletions vary in size over a wide range. Deletion of nucleotides from an open reading frame, if the number of deleted nucleotides is not a multiple of three, causes a shift in the reading frame of translation, a **frameshift mutation**. Frameshift mutations cause downstream codons to be altered, resulting in amino acid substitutions or early chain termination if one of the three STOP codons occurs.

	A
(normal)	..CAT..
	↓
(insertion)	..CAAT..

Fig 15.3 Example of insertion.

Insertion

Insertion is the addition of a segment of DNA from elsewhere (Figure 15.3). Insertions can be as small as one nucleotide, or as big as a transposon or viral chromosome inserted at the target site. Insertion of nucleotides into an open reading frame can shift the reading frame, if the number of inserted nucleotides is not a multiple of three.

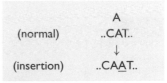

(normal)	..CAT..
	↓
(duplication)	..CATCATCAT..

Fig 15.4 Example of duplication.

Duplication is a special case of insertion (Figure 15.4); a sequence contains two or more copies of a segment, which may be a few nucleotides or gene sized. Often, the duplicated copies occur side by side, as **tandem duplications**.

(normal)	..CAT..
	↓
(inversion)	..TAC..

Fig 15.5 Example of inversion.

Inversion

Inversion is reversal of a segment of nucleic acid (Figure 15.5). Inversions of a few nucleotides are very rare.

The Consequences of Micromutations

Mutations in protein-coding genes have a wide range of effects, depending on size and where the mutation occurs. Mutations in the region that encodes an mRNA leader can affect ribosome binding or, in eukarya, transport from nucleus to cytoplasm. Changes at or near intron-exon junctions can alter splicing. Mutations that alter polyadenylation signals can interfere with polyadenylation. Mutations within exons cause either amino acid substitution, or premature chain termination (if the substitution produces a stop codon), or no change in protein product. A **synonymous substitution** produces a codon change but not an amino acid substitution (Figure 15.6). This can happen because of the redundancy of the genetic code.

Deletion or insertion often causes frameshift, which may lead to a stop codon and chain-termination (Figure 15.7). Below, reading is left to right.

...GGG... → ...GGC... ;
both codons → glycine

Fig 15.6 Synonymous substitution.

		delete T		
gene	...GTG T̲GG ATA GGG...			...GTG GGA TAG GG...
mRNA	...gug ugg aua ggg...	⇒		...gug gga uag gg...
polypeptide	...val–trp–ile–gly...			...val–gly–STOP.

Fig 15.7 Deletion of one nucleotide in a protein-coding regions shifts the reading frame.

Mutations occurring at regulatory sites can result in altered rate or control of transcription (Figure 15.8). For example, RNA polymerase may no longer bind to a mutated promoter. The mutation of a tRNA could result in a global amino acid substitution or even recognition of a stop codon by the tRNA.

Mutation in	Possible result
enhancer	misregulation of transcription
promoter	misregulation of transcription
exon	amino acid substitution, premature stop codon, downregulation of translation, inefficient transport of mRNA from nucleus, failure to polyadenylate RNA
exon–intron junction	incorrect splicing → reduced or missing final gene product

Fig 15.8 Some consequences of point mutations in a protein-coding gene.

Causes of Micromutations

Nucleotide Substitutions

Substitutions are produced from localized damage to DNA or mispaired bases. Three causes of substitution are considered in turn:

- **Mispaired bases**
- **Modified or missing bases**
- **Cross-linkage**, usually between adjacent pyrimidine bases

Misrepair of these premutations is often a key step in making mutations, but the mechanics of repair and misrepair are described in Chapter 16.

Mispaired Bases

Copy errors — mistakes made spontaneously during replication – account for a big fraction of naturally occurring base substitutions. The average rate of copy errors in organisms is very low, $\sim 10^{-10}$ per nucleotide.

Suppose a C mispairs with an A during replication and that the mistake is not fixed by DNA polymerase, proofreading, or repair. One of the two daughter DNA molecules will contain an A·C premutation. After the next round of replication, one granddaughter DNA molecule will contain a mutation, and the other three granddaughter DNA molecules will contain the original base pair (Figure 15.9).

Mispaired bases can occur within heteroduplex regions formed during an Aviemore-type exchange during homologous recombination. Repair of such mispaired bases results in gene conversion, the substitution of one version of a gene for a different version,

Fig 15.9 Example of mispaired base.

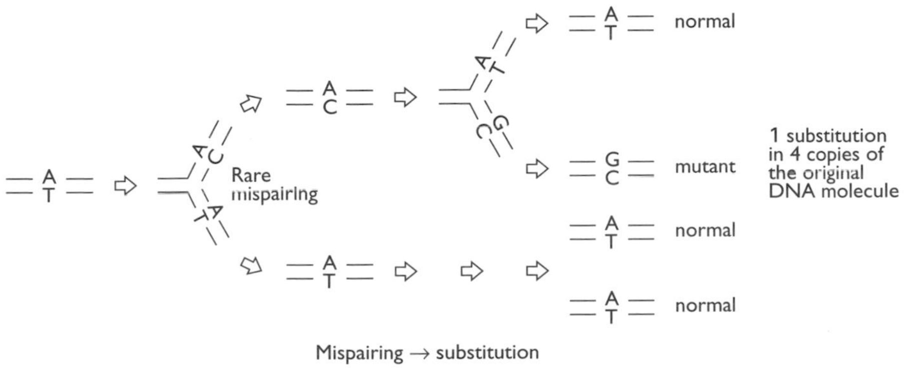

Mispairing → substitution

following recombination. Gene conversion differs from *de novo* mutation in two ways: a preexisting version of a gene is copied, and repair of mispaired nucleotides is not biased toward one strand or the other, so that conversion does not yield a net increase in the frequency of any version of the gene in a population of organisms.

Base Modification

Hydroxylation and deamination of DNA bases are two common kinds of premutation in nature (Figure 15.10). Modified bases pair differently, leading to base substitution in a newly synthesized strand during replication. Bases can be hydroxylated by exposure to hydroxyl radicals. Ultraviolet (UV) radiation and oxidizing chemicals are major sources of hydroxyl radicals; they are also produced during oxidative metabolism. Spontaneous deamination of cytosine, 5-methylcytosine, or adenine produces uracil, thymine, or hypoxanthine, respectively.

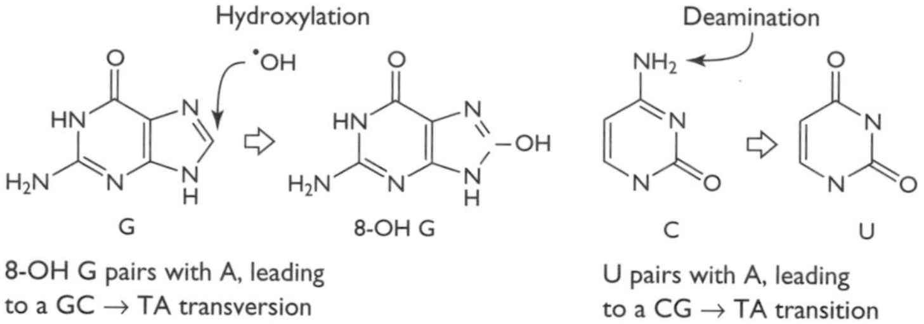

Fig 15.10 Examples of hydroxyletion and deamination.

8-OH G pairs with A, leading to a GC → TA transversion

U pairs with A, leading to a CG → TA transition

Loss of Bases

Apurinic and apyrimidinic (AP) sites occur spontaneously due to acid-catalyzed hydrolysis; the sugar-phosphate backbone remains intact (Figure 15.11). Nuclear DNA in a human cell loses 5,000 to 10,000 purines and 200 to 500 pyrimidines per cell cycle.

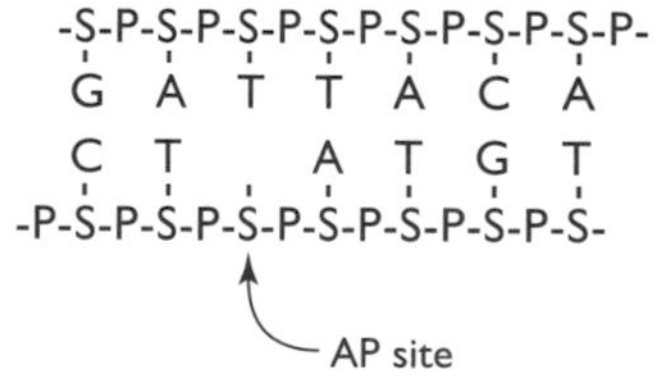

Fig 15.11 Example of loss of bases.

Intrastrand Cross-Links

Why do people and other organisms exposed to strong sunlight need sunscreen? The UV portion of sunlight produces premutations called pyrimidine dimers – cross-linked adjacent cytosine or thymine bases within a single strand of DNA. About 85% are cyclobutane dimers and the remainder are 6,4 dimers (Figure 15.12).

Other intrastrand cross-links between adjacent pyrimidines form when DNA is exposed to chemical mutagens. Replication is blocked at all intrastrand cross-links, because normal base pairing is impossible with cross-linked pyrimidines. Intrastrand

...GATTACCATTACTGGCG...   ∧ = dimer within strand
...CTAATGGTAATGACCGC...

Fig 15.12 Example of cyclobutane dimer and 6, 4 dimer.

Cyclobutane TT dimer   6, 4 CT dimer

cross-links, including pyrimidine dimers, are sometimes repaired by error-prone excision repair, leading to base substitutions, most commonly C→T or CC→TT transitions.

DNA absorbs UV light peaking at a wavelength $\lambda \approx 255$ nm. The ozone layer of earth's atmosphere filters out most light with $(\lambda) < \approx 280$ nm, but the longer UV that reaches the earth's surface nonetheless produces pyrimidine dimers. Biologists classify UV light into three groups by wavelength, UV-A (315 to 400 nm), UV-B (280 to 315 nm), and UV-C (100 to 280 nm). The amount of solar UV that reaches the earth's surface is high in the UV-A range, decreases sharply in the UV-B range, and drops to nearly zero by 290 nm. Sunlight reaching us contains nearly no UV-C. Pyrimidine dimers are produced by UV-C and UV-B; OH radicals are produced by UV-B and UV-A; most pyrimidine dimers in cells are produced by UV-B (Figure 15.13).

Other Damage to Chromosomes Induced by UV Light and Oxidants

Cross-links between the two strands of DNA also occur and can be induced by radiation, free radicals, some carcinogenic chemicals, and some agents used in chemotherapy for cancer. UV light also makes cross-links between DNA and DNA-bound proteins such as

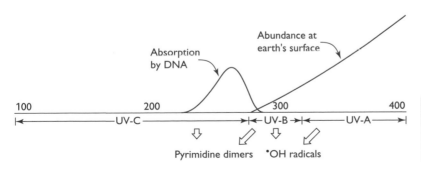

histones. However, because interstrand and DNA-protein cross-links are repaired by mechanisms that are relatively error-free, they are not a significant source of new mutations. UV light also damages sugars; about 20% of the UV-induced DNA damage is breakage or modification of deoxyribose.

Deletions and Insertions

Three important natural causes of deletions and insertions are:

* **Transposition of transposons**
* **Primer-template misalignment** during replication
* **Unequal crossing-over** during general recombination.

Deletions and insertions are produced less frequently than substitutions, in nature.

Transposition of Transposons

When a transposon moves to a new site, the new site immediately mutates by virtue of containing an insertion. Replicative transposition causes no mutation at the original site, but nonreplicative transposition mutates the original site from which the transposon excises because that site then lacks the transposon (Figure 15.14). Precise excision of the transposon restores the original site to its original sequence – the sequence before the transposon's original insertion. Imprecise excision produces an insertion or a deletion at the original site, compared with the preinsertion sequence, and therefore may render the transposon-losing gene nonfunctional.

Primer-Template Misalignment during Replication

Accurate replication depends on perfect alignment of template and primer strands. Each nucleotide in the newly synthesized

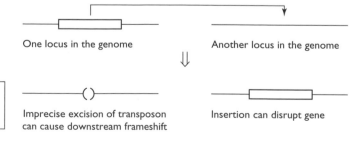

One locus in the genome   Another locus in the genome

Fig 15.14 Nonreplicative transposition can produce two mutations.

Imprecise excision of transposon can cause downstream frameshift

Insertion can disrupt gene

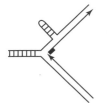 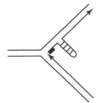

Hairpin/loop in template strand causes shortened new strand

Slippage & hairpin/loop formation in new strand causes lengthened new strand

Fig 15.15 Misalignment of tandem repeats in DNA replication → deletions/ duplications.

strand must correspond to one and only one nucleotide in the template strand; otherwise, the number of nucleotides added to the newly synthesized strand will be greater or fewer than in the template strand.

Tens of genes in the human genome are known to contain perfectly tandemly repeated microsatellites, especially trinucleotide repeats such as CGG and CAG. When too many repeats are present in such a gene, it becomes nonfunctional. This kind of mutation is responsible for several genetic diseases, e.g., Huntington disease (late-onset, progressive neurodegeneration), fragile-X syndrome (mental retardation, male sterility, and other traits), and myotonic dystrophy (late-onset, progressive wasting of the muscles, and other symptoms). Normally, repeat-containing genes have a small number of repeats (<20), but the number of repeats varies among individuals. The source of variations in repeat number may be (1) hairpin formation in the template or newly synthesized strand, or (2) slippage of DNA polymerase along the template strand (Figure 15.15).

Unequal Crossing-Over

Nonreciprocal homologous recombination, or unequal crossing-over, can produce duplications and deletions of various sizes (Figure 15.16). This error can be a source of gene duplication as well as expansions and contractions of tandemly repeated satellites, microsatellites, and minisatellites.

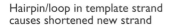 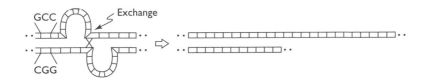

Fig 15.16 Unequal crossing-over → DNA sequence duplication.

Mutation in Different Life Forms

The intrinsic rate of mutation for any life form depends on the accuracy of replication machinery, the accuracy of repair, and exposure to mutagens. As a rule of thumb, the natural mutation rate for an average-sized gene in organisms is $\sim10^{-5}$ per replication cycle. The rate of mutation appears to be higher in mitochondria and chloroplasts than in bacteria or eukarya. Viruses typically have much higher mutation rates.

In multicellular sexual species (e.g., animals) mutations in germ line cells (gametes) can be passed to the diploid offspring, whereas mutations in somatic cells can be passed only to daughter cells within the body. Germ line mutations, not somatic mutations, are the raw material for evolution of species. Some somatic cells are exposed to more mutagens (e.g., UV light) than germ line cells; consequently, the mutation rate in those somatic cells is higher. Somatic mutations are a major cause of cancer, a type of genetic disease in which tumor cells proliferate in an uncontrolled way. According to one theory, accumulated somatic mutations are one cause of aging.

Further Reading

Brash DE. 1997. Sunlight and the onset of skin cancer. *Trends Genet.* 13: 410–414.

Dizdaroglu M. 1993. Chemistry of free radical damage to DNA and nucleoproteins, p. 19–35. In: Halliwell B, Aruoma O (eds.). *DNA and Free Radicals.* Ellis Horwood, New York.

Marmett LJ, Plastaras JP. 2001. Endogenous DNA damage and mutation. *Trends Genet.* 17:214–219.

Pearson CE, Sinden RR. 1998. Trinucleotide repeat DNA structures: dynamic mutations from dynamic DNA. *Curr. Opin. Struct. Biol.* 8:321–330.

Samadashwily GM *et al.* 1997. Trinucleotide repeats affect DNA replication *in vivo. Nature Genet.* 17:298–303.

Chapter 16

Repair of Altered DNA

Overview

Every cell's genome is vulnerable to change. DNA-modifying chemicals and electromagnetic radiation continually damage it, and the inherent inaccuracy of DNA synthesis erodes its perfect replication. There are two main lines of defense against DNA damage: prevention and repair. Preventive mechanisms (e.g., natural sunscreens, enzymatic degradation of mutagens) go beyond the scope of this book. The enzymes of DNA repair and their modes of action are described here.

Systems of Repair

Every organism has multiple systems to repair premutational alterations to DNA – both damage and mispaired nucleotides. For example, *Drosophila* has nearly 90 DNA repair enzymes. Systems of repair, some of which are universal among organisms and some of which are not, fall into broad categories:

- **Direct repair** – the sugar-phosphate backbone remains intact (this is the only nonsurgical option for repairing DNA damage)
- **Repair of apurinic and apyrimidinic sites (AP repair)** – excision of a short sequence of one strand containing an AP site (nucleotide missing its base)
- **Mismatch repair** – a region containing a mismatched base is removed and replaced, usually repairing a newly synthesized strand
- **Excision repair** – uses **exinucleases**, which are endonucleases that nick a single strand of DNA on both sides of an altered site

• **Recombination repair** – uses recombination to fill gaps in DNA
• **End-joining repair** – joins double-strand breaks in DNA

Direct Repair

The two kinds of repair of altered nucleotides, which does not involve breaking the sugar-phosphate backbone of DNA, are **photoreactivation** of pyrimidine dimers and **dealkylation of alkylated bases**.

Photoreactivation
Bacteria and many eukarya possess DNA **photolyases**, which directly repair cyclobutane or 6–4 pyrimidine dimers (different enzymes for the two kinds of dimer). Photolyases have noncovalently bound **chromophores** (light-absorbing pigments), typically $FADH_2$ plus a second chromophore. Photoreactivation to repair a cyclobutane dimer occurs in steps (Figure 16.1): (1) Independently of light, photolyase ($\approx$60 kDa) binds with high affinity to the site of a pyrimidine dimer. (2) A chromophore other than $FADH_2$ absorbs a photon (300 to 500 nm) and transfers the energy to $FADH_2$. (3) $FADH_2$ transfers an electron to the pyrimidine dimer, making a dimer radical. (4) The dimer radical splits into two pyrimidines, and the electron jumps back to $FADH_2$.

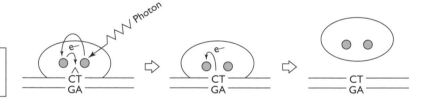

Fig 16.1 Photoreactivation of cyclobutane pyrimidine dimers.

Dealkylation of Alkylated Bases
DNA inside cells becomes alkylated by various mechanisms; especially damaging is alkylation of the O^6 of guanine or the O^4 of thymine, which leads to mispairing during replication. Several DNA alkyl transferases, present in bacteria and eukarya, remove alkyl groups from DNA bases or even from the sugar-phosphate backbone. An example is the Ada protein of *Escherichia coli*, a DNA methyltransferase that restores alkylated bases, such as O^6-methylguanine or O^4-methylthymine, to normal (Figure 16.2). The 39-kDa Ada (<u>ad</u>aptive response) protein transfers a methyl group from a nucleotide to one of its own cysteine residues in

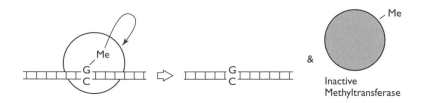

Fig 16.2 Demethylation repair by DNA methyltransferase (Ada protein).

Inactive
Methyltransferase

an irreversible reaction, destroying its own enzyme activity; it is therefore called a **suicide enzyme**.

AP Repair

An AP site is a nucleotide lacking a base. **AP endonucleases** initiate repair of AP sites by nicking a strand of DNA (Figure 16.3). Most AP endonucleases cleave DNA on the 5′ side of the AP site, yielding a 3′-OH end on the adjacent nucleotide, and a 5′ terminal phosphate on the AP-containing piece. After the AP endonuclease acts on the 5′ side, an exonuclease removes a short stretch of nucleotides, beginning with the AP site, a DNA polymerase (pol I in *E. coli*) fills the gap, and a ligase joins the broken ends. Another class of endonuclease that cleaves DNA at an AP site is **AP lyase**, which cuts DNA 3′ to the abasic site, producing a 2′-3′ double bond on the abasic sugar and a 5′ terminal phosphate. An exonuclease removes a short stretch starting with the abasic nucleotide, and DNA polymerase (pol I in *E. coli*) fills the gap.

DNA **glycosylases** produce AP sites by removing damaged or modified bases. There are many specific DNA glycosylases, which are small (15 to 40 kDa) and do not require exogenous energy. AP repair follows removal of a base by glycosylase, and, in some cases, one enzyme may act both as a glycosylase and as an AP endonuclease or a lyase.

Like bacteria, mammals have several enzyme systems of base excision/AP site repair.

Mismatch Repair

Three systems of enzymes in *E. coli* repair mismatched base pairs, which are produced by errors of replication, recombination, or

Endonuclease   Exonuclease   DNA polymerase
                                            & ligase

Fig 16.3 AP repair by AP endonuclease.

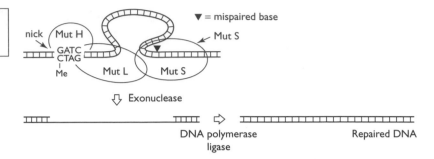

Fig 16.4 Methyl-directed, strand-specific mismatch repair.

spontaneous deamination of 5-methylcytosine (5-MeC) (making a GT pair). Only one of these, the methyl-directed system, is described here (Figure 16.4). It acts on newly replicated, hemimethylated DNA and repairs seven of the eight kinds of mismatched pair (all except C·C) by excising nucleotides from the unmethylated strand and filling in the gap. Humans and other eukarya have a strand-specific mismatch repair system that corrects all eight mismatches.

In *E. coli*, Dam methylase methylates adenine in every GATC sequence a few minutes after replication. By random chance alone, GATC occurs on average about every 250 nucleotides. During the short time between replication and methylation, the old (template) strand is methylated and the newly synthesized strand is unmethylated. Methyl-directed mismatch repair enzymes act on the newly synthesized, unmethylated strand, not on the methylated strand.

At least eight proteins participate; specific to the process are MutH, MutL, and MutS. MutS, a 97-kDa ATPase, binds to the mismatch, and MutH, a 25-kDa endonuclease, binds to GATC. MutL, 70-kDa molecular matchmaker, binds to MutS, activating its ATPase. ATP hydrolysis causes MutS to change shape and bind to MutH, activating the endonuclease of MutH. MutH nicks the unmethylated strand 5′ to GATC. An exonuclease excises the nucleotides (up to several kilobases) to a point about 100 nucleotides (nt) beyond the mismatch, and pol III synthesizes a new strand. At the end, a ligase seals the break.

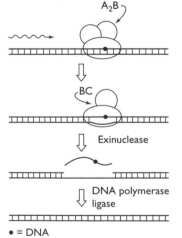

Excision Repair

Fig 16.5 Excision repair in *E. coli*.

Nucleotide excision and replacement, commonly called excision repair, restores to normal a wide variety of alterations to DNA

(Figure 16.5). The key player is an excision endonuclease, more commonly called **exinuclease**, which removes oligonucleotide segments containing an altered site. Exinucleases recognize pyrimidine dimers, AP sites, deaminated bases, oxidized bases, mispaired bases, and bulges due to deletions or insertions. Excision repair appears to be ubiquitous: bacteria, humans, and yeast have systems of excision repair.

The exinuclease of E. coli, UvrB (78 kDa), acts in concert with UvrA (103 kDa ATPase) and UvrC (69 kDa). An A_2B trimer binds non-sequence specifically to DNA, scans the duplex for distortions, and forms a stable complex of A_2B-DNA at an altered site. UvrA then dissociates, leaving UvrB covering about 20 bp including the altered site, which is unwound by 5 bp. When UvrC binds to the UvrB-DNA complex, UvrB is enabled to cut the fifth phosphodiester bond 3' to the damaged base(s) and the eighth phosphodiester bond 5' to the damage. A 12- or 13-nt fragment is thus excised (13 if the damage is a pyrimidine dimer). After UvrC dissociates, DNA polymerase (pol I) fills the gap, and a ligase seals the break.

Human exinuclease has ~10 subunits; it incises an $\approx$28-nt fragment. DNA polymerase δ/ε then fills the gap. Defects in any of seven genes encoding enzymes for excision repair cause a genetic disease, xeroderma pigmentosum (XP). Most of the seven genes, denoted *XP-A* through *XP-G*, have homologs in yeast. More than 20 genes encode proteins of the excision-repair system in humans.

Postreplication Recombination Repair

During replication in E. coli, lesions that block base pairing, such as pyrimidine dimers, interrupt DNA synthesis. DNA polymerase stops synthesis, relocates to the other side of the lesion, and resumes synthesis, leaving a single-strand gap in one of the two daughter molecules. Recombination between the two daughter molecules fills the gap opposite the lesion and leaves a gap in the other daughter molecule, which is repaired by pol I and ligase (Figure 16.6). RecA, a recombination enzyme, binds to single-stranded DNA (ssDNA); then the RecA-ssDNA complex associates with double-stranded DNA (dsDNA), and homologous strands pair. Endonucleases and ligases complete the exchange.

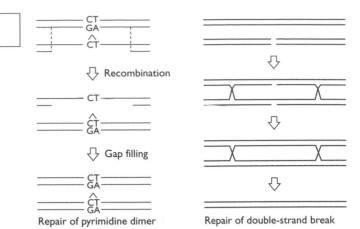

Fig 16.6 Recombination repair.

Recombination repair circumvents blocked base pairing, but it does not remove the lesion. Another process, such as photoreactivation or excision repair, then repairs the lesion.

General recombination can repair a double-strand break in DNA. In this case, unlike the example of pyrimidine dimers, recombination actually effects repair.

End-Joining Repair of Double-Strand Breaks

In eukarya, an important mechanism for repairing double-strand breaks is **end-joining repair**. The protein components of the end-joining repair system include (1) Sir2, Sir3, and Sir4, which also bind to telomeres; (2) Ku80-Ku70, a 150-kDa heterodimer with protein kinase activity; and (3) an end-joining multimer that includes endonuclease, exonuclease, and ligase (Figure 16.7). The Sir proteins protect chromatin from degradation, and Ku protein may bind the end-joining enzymes.

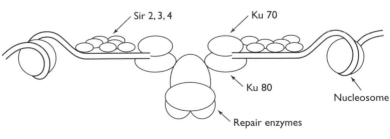

Fig 16.7 End-joining complex.

End-joining complex

Systems Related to DNA Repair

The SOS Response in Bacteria

Mutagens such as strong UV light may damage DNA faster than normal systems can repair it. Pyrimidine dimers and broken DNA strands interrupt DNA replication, because the replisome stalls when it encounters these lesions. In bacteria, repeated stalling of the replisome is a first step in triggering the SOS response, which is to transcribe a set of about 20 genes – the SOS regulon – whose products include DNA repair enzymes and factors that promote error-prone DNA synthesis. A regulon is a set of functionally related, coordinately expressed operons. Gene products of the SOS regulon include (1) enzymes for excision repair and recombination repair, and (2) two proteins that bind to pol III holoenzyme, causing it to fill spots opposite pyrimidine dimers without base pairing. While error-prone DNA synthesis has the disadvantage of increasing the rate of mutation, it has the advantage of allowing replication to proceed under conditions that normally prevent it. The SOS response is widespread in bacteria but absent in eukarya.

Transcription-Repair Coupling

During transcription, lesions such as pyrimidine dimers in the template strand of a gene can block RNA polymerase. They are repaired faster than similar lesions in nontranscribed genes. RNA polymerase itself, stalled at a DNA lesion, interferes with exinuclease, thus blocking excision repair. Transcription-repair coupling factor facilitates repair at this site (Figure 16.8). Transcription-repair coupling factor, a 130-kDa protein in *E. coli*, binds to the DNA-RNA polymerase complex, displaces RNA polymerase, binds to UvrA in the A_2B exinuclease and facilitates the dissociation of UvrA, leaving a UvrB-DNA complex ready for activation by UvrC.

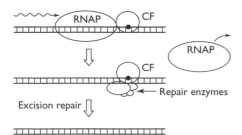

Fig 16.8
Transcription-repair coupling.
RNAP = RNA polymerase,
CF = Transcription-repair
coupling factor, • = lesion.

Humans have a similar transcription-repair coupling factor, a 160-kDa protein encoded by a gene named CSA (Cockayne's syndrome, gene A). Mutations in this gene can cause Cockayne's syndrome, whose symptoms include sensitivity to UV light as well as serious developmental and neurological abnormalities. Human transcription-repair coupling factor interacts with RNA polymerase II and exinuclease.

Further Reading

Haber J. 2000. Partners and pathways repairing a double-strand break. *Trends Genet.* 16:259–264.

Kanai S *et al.* 1997. Molecular evolution of the photolyase-blue-light photoreceptor family. *J. Mol. Evol.* 45:535–548.

Lehmann AR. 1995 The molecular biology of nucleotide excision repair and double-strand break repair in eukaryotes. *Genet. Engin.* 17:1–19.

Marnett LJ, Plastaras JP. 2001. Endogenous DNA damage and mutation. *Trends Genet.* 17:214–220.

Sancar A. 1995. DNA repair in humans. *Annu. Rev. Genet.* 29:69–105.

Sutton MD *et al.* 2000. The SOS response. *Annu. Rev. Genet.* 34:479–495.

Chapter 17

Reproduction of Bacteria

Overview

At each turn of a **cell cycle**, a bacterium or archaeon converts molecules taken from the environment into its own components and divides into two daughter cells. Each daughter cell receives a copy of the genome as well as copies of plasmids and nongenetic components. Mitochondria and chloroplasts proliferate by a similar division process, and, although they are organelles within eukarya, they reproduce rather than being made *de novo* by the cell. A population of genetically identical life forms made by this process is a **clone**.

Under poor or stressful environmental conditions, some bacteria form **spores** – specialized resting cells – by the process of sporulation. A few species of bacteria develop into **multicellular organisms** containing different cell types. Both sporulation and the formation of multicellular bacteria are simple examples of cell differentiation, which happens as a result of regulated changes in gene expression.

Bacterial Reproduction

Cell Cycle

The cell cycle of bacteria has two phases, **C** (chromosome copying) and **D** (division). On average, cells double in size between divisions. The cell cycle is tightly regulated and division is precise, ensuring that the chromosomes, cytoplasm, and cell envelope are synthesized at the right rate and are equally apportioned between the two daughter cells. During the C phase the chromosome(s) and plasmids replicate, and components of the cytoplasm such as ribosomes, tRNAs, and enzymes are approximately doubled in

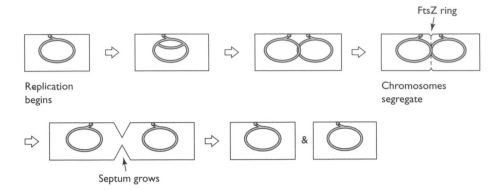

Replication begins

FtsZ ring

Chromosomes segregate

Septum grows

| Fig 17.1 | Bacterial cell cycle. |

number. During the D phase, the cell envelope grows, daughter chromosomes segregate (move apart), a contractile ring pinches the cell in two, and a septum forms between the incipient daughter cells.

In slowly growing cells, chromosome replication and cell division are distinct and separate, with division following replication, whereas in rapidly growing cells, replication and division overlap. The cell cycle is short in favorable environments and long in unfavorable environments. For *Escherichia coli* at 37°C, the minimum duration of the cell cycle is ≈20 minutes (Figure 17.1). The minimum division time (D) is also ≈20 minutes. However, replication of the main chromosome takes at least 40 minutes. This apparent paradox is resolved by the simultaneous replication of at least two copies of the chromosome; replication must overlap completely with the division phase, so that each daughter cell receives an entire chromosome. When the cell cycle lasts 60 minutes or more, replication precedes division with no overlap, and only one copy of the chromosome need be present at the beginning of the cycle.

C Phase. Chromosome Replication. Circular bacterial chromosomes undergo theta replication, beginning at the origin. The initiation of replication is rate limiting.

In *E. coli* and *Bacillus subtilis*, two distantly related bacteria, both the essential chromosome and some of the plasmids are attached to the cell envelope. The prereplication attachment site is central, near the place where the septum will grow to divide the cell. The essential chromosome and plasmids replicate independently of each other, and their copy number is regulated differently.

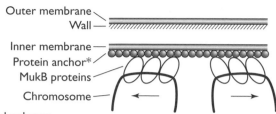

* Hypothetical polymer

Fig 17.2 Segregation of chromosomes. MukB proteins move along the protein anchors, pulling the daughter chromosomes in opposite directions, as indicated by the arrows.

D Phase. Cell Division. Two main processes of cell division are chromosome segregation (separation and movement of daughter chromosomes) and cytokinesis (splitting of the cell into daughter cells). Chromosome segregation and cytokinesis are coordinated, but this coordination can go awry in mutants for the structural proteins assisting either process.

In *E. coli*, a protein needed for the segregation of daughter chromosomes is MukB (177 kDa, the largest known protein in this bacterium). MukB and other proteins attach to the chromosome near *oriC* and move with *oriC* to the poles of the dividing cell (Figure 17.2). MukB may generate mechanical force to help pull apart the nucleoids. Chromosomes begin to segregate as soon as the oriC region replicates. Segregation also requires topoisomerases to decatenate interconnected daughter chromosomes.

Cytokinesis requires many proteins, including the products of the *fts* (filamentous temperature-sensitive) genes (Figure 17.3). FtsZ (40 kDa) is the most abundant of these proteins; during cell division there are $\sim10^4$ copies per cell. FtsZ, a homolog of tubulin (a microtubular protein in eukarya), polymerizes and forms a midcell ring, growing from a nucleation center near the initial attachment site of the chromosome. The FtsZ ring contracts as

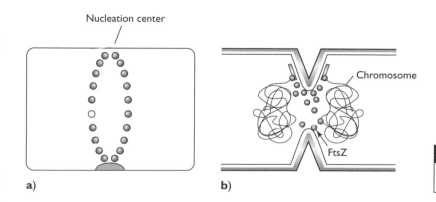

a)

b)

Fig 17.3 Cytokinesis: Z-ring formation (**a**) and septation and separation (**b**).

the cell envelope grows inward to form a septum. FtsZ is associated with many other proteins, but it seems likely that it plays an active role in septal growth. When the septum grows across the cell, the cytoplasmic membrane pinches off.

Products of the *minCDE* gene cluster regulate cell division. MinC and MinD, which are widely dispersed on the cell's inner membrane, inhibit FtsZ ring formation. MinE localizes in a ring at the equator and inhibits MinC-MinD; FtsZ therefore forms a ring near MinE. The mechanism for controlling the production and localization of MinE is unknown. The SOS response to DNA damage includes rapid synthesis in SulA protein, which blocks cell division by inhibiting formation of the FtsZ ring.

Sporulation

Some bacteria, for example *Bacillus* and *Clostridium*, form spores in conditions unfavorable to growth – for example, depletion of nutrients or water (Figure 17.4). Spores are reproductively dormant, resist extremes of temperature and harsh chemicals, and can survive for long periods. There are many reports of spores remaining viable more than 10^4 years and one report of viable spores $\approx$250 million years old (*Bacillus marismortui*, found in salt crystals). This is remarkable because the DNA, enzymes, and ribosomes must remain functional during spore dormancy, despite the susceptibility of these components to damage under normal conditions.

In *B. subtilis*, a regulon comprising more than 70 operons controls the 8-hour sporulation process. A developmentally programmed series of sporulation-specific sigma factors – σ^F, σ^E, σ^G,

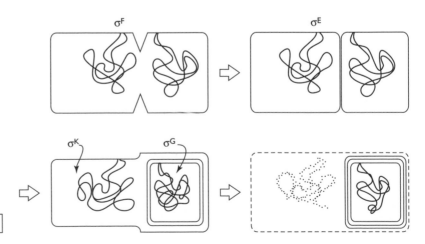

Fig 17.4 Sporulation.

and σ^K – control transcription of these genes. Under favorable conditions, a spore germinates, producing a normal cell.

Following signals to sporulate, the main chromosome replicates and the daughter chromosomes segregate into two cell compartments, the smaller of which is fated to develop into a spore. The smaller compartment is the forespore and the larger is the mother cell. At the same time, RNA polymerase holoenzyme carrying σ^H transcribes the gene coding for σ^F, the first sporulation-specific sigma factor.

σ^F directs transcription of several early sporulation genes, including the gene encoding σ^E. σ^E is active both in the forespore compartment and in the mother cell compartment. σ^G is transcribed only in the forespore compartment, while σ^K is transcribed only in the mother cell compartment. The gene for σ^K is not present in the normal genome as a functional gene but is produced in a remarkable way – by splicing together two genes, *spoIVB*, which codes for the N-portion, and *spoIIIC*, which codes for the C-portion.

As the spore develops, the mother cell engulfs it; it develops a cortex composed of a peptidoglycan-like protein, and it develops several layers of protein spore coats outside the cortex. The mother cell lyses as soon as the spore is released.

Multicellular Bacteria

Filamentous bacteria, which evolved ≈3000 million years ago and are represented by many species, have the simplest kind of multicellular arrangement (Figure 17.5). Each filament consists of cells joined end to end. Some filaments have cross-walls. Under unfavorable conditions, some cells of the filament become round, spore-like gonidia, which are released from the tips of the filament.

Myxobacteria, notably *Myxococcus xanthus*, grow in aggregates (Figure 17.6). When environmental conditions become unfavorable, specialized structures called fruiting bodies grow in the

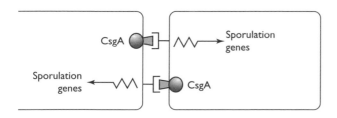

Fig 17.5 Cell–cell interactions activate transcription of developmental genes.

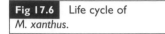

Fig 17.6 Life cycle of *M. xanthus.*

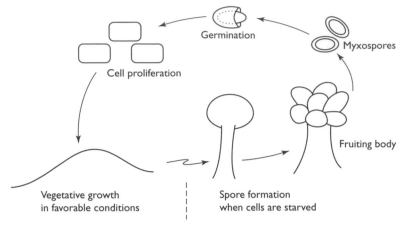

Life cycle of *Myxococcus xanthus*

aggregates. Spores develop in and are released from fruiting bodies. Cell specialization depends on signaling between cells. The intercell signals switch on transcription of certain genes in the recipient, and the gene products cause the cell to become specialized, also called **differentiated**. Signaling between cells depends on proteins displayed on the signaler's cell membrane, and receptor proteins (e.g., CsgA protein) displayed on the receiver's cell membrane. The molecular nature of signal transduction inside the receiver is unknown.

Archaea

Reproduction in archaea is less well-known than in bacteria or eukarya. Circular archaeal chromosomes are concentrated in a nucleoid region, and plasmids are commonplace. As in bacteria, the cell cycle comprises chromosome replication, chromosome segregation, and cytokinesis.

Archaeal chromosomes contain extensive regions organized into histone-containing nucleosome-like particles, reminiscent of eukarya. The single DNA polymerase of *Methanococcus jannaschii* is similar to eukaryal B polymerases; other enzymes of replication are similar to those of eukarya.

Chromosome segregation in archaea probably is similar to chromosome segregation in bacteria; some archaea have homologs MukB. *M. jannaschii* has a centromere-microtubule binding protein, whose function is unknown.

M. jannaschii has two homologs of FtsZ, a homolog of FtsJ, and three homologs of MinD, suggesting that cytokinesis is similar in archaea and bacteria. It also has a homolog of a fruit fly protein that helps regulate cell division.

Mitochondria and Chloroplasts

Mitochondria and chloroplasts undergo asexual reproductive cycles whose main events are chromosome replication, chromosome segregation, and division by splitting, as in bacteria. The cell does not and cannot construct them anew from molecular components, which is not the case with other organelles (Golgi apparatus, nucleus).

Enzymes of replication in mitochondria and chloroplasts are homologous to bacterial enzymes, reflecting their evolutionary origins, and chromosomes are attached to the inner membrane of the organelle. Chromosome replication can be more rapid than division, with the result that a mitochondrion or chloroplast may have more than one copy of the chromosome. Chromosomes segregate precisely, giving daughter mitochondria and daughter chloroplasts half the copies of the genome. How plasmids segregate is not known. Some chloroplasts have FtsZ. Genes in the nucleus apparently influence the timing of replication and organelle division; apparently, components of the molecular machines of chromosome segregation and organelle division are also encoded by nuclear genes. Recall that many genes have moved from the mitochondrial and chloroplast genomes into nuclear genomes, rendering the original bacterial symbionts genetic parasites.

Clonal Inheritance

All the life forms discussed in this chapter reproduce by splitting but with precise segregation of the two copies of the genome. Consequently, after many reproductive cycles, this process produces many identical copies of the original. This collection of progeny is called a **clone**, or population of life forms that are genetically as close to identical as the fidelity of replication and repair allows. Clones can accumulate mutations and thus produce genetically divergent subclones, but clonal reproduction is asexual, entailing no exchange of genes via union of gametes.

Plasmids of low copy number are thought to segregate precisely, as does the main chromosome, so that each daughter cell is virtually assured of receiving a copy of the plasmid. Plasmids of high copy number are thought to be apportioned in a random fashion to the daughter cells, but the high copy number gives each daughter cell a high probability of receiving at least one copy of the plasmid.

Nonchromosomal components of the cell are apportioned randomly to the daughter cells, except the envelope, which is split at the point of septation, at the cell's equator or middle. There are many copies of the nonchromosomal components, such as ribosomes, and they are distributed throughout the cell, with the result that daughter cells receive copies of each nonchromosomal component. This becomes untrue only for components that are not renewed by synthesis; in that case, the component becomes more dilute in the cell with each division cycle.

While the basic mode of reproduction of bacteria, archaea, mitochondria, and chloroplasts is clonal, there is a small amount of gene exchange between clones and recombination within each of these life forms via several mechanisms. Incoming foreign DNA may be integrated into the genome by recombination. The movement of genes from one cell to another is the subject of the following chapter.

Further Reading

Erickson H. 1997. FtsZ, a tubulin homologue in prokaryotic cell division. *Trends Cell Biol.* 7:362–367.

Hiraga S. 1992. Chromosome and plasmid partition in *Escherichia coli*. *Annu. Rev. Biochem.* 61:283–306.

Kaiser D. 1999. Cell fate and organogenesis in bacteria. *Trends Genet.* 15: 273–277.

Rothfield L et al. 1999. Bacterial cell division. *Annu. Rev. Genet.* 33:423–448.

Stragier P, Losick R. 1996. Molecular genetics of sporulation in *Bacillus subtilis*. *Annu. Rev. Genet.* 30:297–341.

Vinella D, D'Ari R. 1995. Overview of controls in the *Escherichia coli* cell cycle. *BioEssays* 17:527–536.

Vreeland RH et al. 2000. Isolation of a 250 million-year-old halotolerant bacterium from a primary salt crystal. *Nature* 407:897–900.

Chapter 18

Horizontal Gene Transfer in Bacteria

Overview

During reproductive cycles genetic information is transferred vertically from parent to offspring, and from progenitor to descendant. Genes can also be transferred horizontally between contemporaneous individuals. In this book, horizontal gene transfer is defined in a broad, inclusive way: the incorporation of DNA from any external source into the genome of the recipient life form – a cell, a virus, a mitochondrion, but excluding genomes of parasites that invade a cell. This chapter focuses on horizontal gene transfer in bacteria.

Types of Horizontal Transfer

In bacteria, there are three kinds of transfer of genes between cells:

- **Transformation** – the cell takes up exogenous DNA directly from the surrounding medium
- **Conjugation** – two cells join and a donor cell transfers DNA, usually a plasmid, to a recipient cell
- **Transduction** – bacteriophages transport DNA from on cell to another

Mitochondria transfer genes horizontally infrequently, when two genetically different mitochondria fuse and their chromosomes undergo recombination. The same is true of chloroplasts. Viruses can exchange genes by recombination, following infection of the host cell by two or more genetically different virions.

Transformation

Transformation is the alteration of a cell's genome by acquiring exogenous DNA. In nature, the acquired DNA is integrated into a main chromosome. The three major events of natural transformation are as follows:

• Binding of double-stranded DNA to receptor proteins
• Uptake of a single strand of the bound DNA
• Integration of that DNA into the genome

Transformation does not include viral infection and does not occur by cell contact. Transforming DNA outside the cell is usually double-stranded, although in some cases single-stranded DNA is effective.

Transformation occurs naturally in species of Gram-positive and Gram-negative bacteria, cyanobacteria, and mycoplasmas and rickettsias. Gram-positive bacteria (e.g., *Bacillus subtilis*) and Gram-negative bacteria (e.g., *Haemophilus influenzae*), distinguished by the Gram stain, have differently constructed cell envelopes. Gram-positive bacteria have a thick cell wall outside a single cell membrane; Gram-negative bacteria have a thin cell wall sandwiched between two cell membranes, outer and inner. Many medically significant species undergo transformation, including *Haemophilus influenzae*, *Neisseria gonorrhea*, and *Streptococcus pneumoniae*. Bacteria can rapidly acquire virulence or resistance to antibiotics via transformation.

In laboratory experiments, it is commonplace to transform bacteria, usually *Escherichia coli*, with DNA whose sequence is controlled by the experimenter. In experimental transformation, the exogenous DNA may be inserted into an independently replicating plasmid, in which form it enters the cell and is stably maintained. *E. coli* does not undergo natural transformation.

Natural Competence for Transformation

Natural competence for transformation is a physiological state in which cells bind and take up exogenous DNA. A few taxa, notably *Neisseria*, are constitutively competent, but in most taxa, some signal induces competence. A population of cells can rapidly gain or lose competence. Some species acquire competence in the slow stationary growth phase when cell density in the environment nears its upper limit, but other species may develop competence during exponential growth.

Competence may develop in response to starvation, as in *H. influenzae*, or may be signaled by peptides excreted by other cells in the growth medium, as in *S. pneumoniae* and *B. subtilis*. The response to the signal for developing competence is to switch on transcription of genes encoding the machinery that binds, takes up, and processes DNA, and that incorporates DNA into the genome. In many species, more than 40 genes are required to develop competence.

Binding of DNA to Receptors

About 50 DNA binding sites are present on the surface of competent cells of the Gram-positive species *B. subtilis* and *Streptococcus*. The binding sites are protein multimers (75 kDa in *B. subtilis*) and may be associated with a pore built of pilins and extending through the cell envelope. Large pieces ($\sim 10^4$ bp) of double-stranded DNA (dsDNA) bind with no sequence specificity. Binding is quickly followed by double-strand cleavage, yielding large fragments (7 to 30 kb); these fragments are then nicked at intervals of about 7 kb.

Competent cells of the Gram-negative bacterium *H. influenzae* have about 10 DNA-binding sites. These sites appear to be extensions of the outer membrane (vesicles). Large pieces ($\sim 10^4$ bp) of dsDNA bind and enter the vesicles. Only short, species-specific **uptake-signal sequences (USSs)** bind, so that only intraspecific DNA can participate in natural transformation. The USS from *H. influenzae* is a 29-bp sequence containing a highly conserved 11-bp sequence (consensus, AAGTGCGGTCA). There are 1465 copies of the USS in the 1.8-Mb main chromosome, on average about 1 USS per 1200 bp, a frequency $\sim 10^{14}$ higher than chance probability.

Uptake and Integration of DNA

In Gram-positive bacteria, an exonuclease digests one strand, while the other strand moves into the cell through a pore in the cell envelope, at a rate of 100 to 300 nucleotides/second (nt/s). In the cytoplasm, proteins transport this single-stranded DNA (ssDNA) to the chromosome, where it may integrate by recombination, if it is homologous to chromosomal DNA. Nonhomologous DNA is degraded. Heteroduplex molecules that result from the

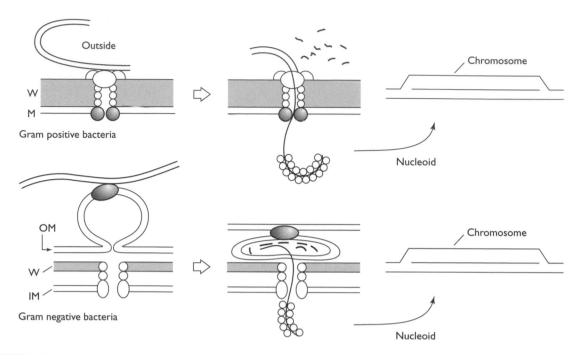

Outside

W
M

Gram positive bacteria

Chromosome

Nucleoid

OM

W

IM

Gram negative bacteria

Chromosome

Nucleoid

Fig 18.1 Binding, uptake, and integration. CM = cell membrane, W = cell wall, OM = outer membrane.

integration of the donor strand of DNA are resolved either during subsequent chromosome replication or by mismatch repair followed by replication.

In Gram-negative bacteria, the DNA-containing vesicle moves into the periplasm. One strand is cleaved repeatedly by endonucleases, and the other strand is transported into the cytoplasm $3' \rightarrow 5'$ at 500 to 1000 nt/s. Proteins transport the ssDNA to the chromosome, and homologous DNA integrates into the chromosome by recombination (Figure 18.1).

A sizeable fraction of cells in a natural population can be transformed. The main biological significance of transformation is that it produces new, potentially favorable recombinants in bacteria.

Conjugation

Conjugation is a process of one-way DNA transfer between live cells in direct contact with each other. Usually only a plasmid is transferred from donor cell to recipient cell; occasionally part of the main chromosome can also be transferred. Like natural transformation, conjugation is regulated genetically. Conjugation occurs in most Gram-negative bacteria and in some Gram-positive bacteria (*Streptomyces, Streptococcus*). It seems likely that

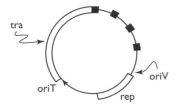

tra

oriT

oriV

rep

Fig 18.2 The F plasmid. *tra* = conjugation genes, dark rectangles = pairing regions, *rep* = genes for vegetative replication, dark trangle = *oriT*, one strand of which leads entry of plasmid into recipient cell.

conjugation occurs widely among other groups of bacteria and in archaea. The main steps of conjugation are

- Establishment of contact
- Mobilization of donor DNA
- Synthesis and transfer of donor DNA.

After conjugation, the cells separate.

The F plasmid (Figure 18.2) is frequently transferred by conjugation in *E. coli* from a donor cell possessing the F plasmid (F$^+$cell) to a cell lacking the F plasmid (F$^-$ cell). A second type of donor cell is Hfr, in which the F plasmid has been integrated into the main chromosome by site-specific recombination. The circular, 95-kb F plasmid has 60 known genes; more than 20 of these genes code for proteins related to conjugation – plasmid-incompatibility genes, genes encoding the sex pilus (a microtubule on the surface of F$^+$ cells, that adheres to F$^-$ cells), and genes coding for proteins facilitating intimate cell contact, DNA mobilization, and DNA transfer. The conjugation genes are located in a region called *tra* (transfer). F has two origins of replication, *oriV* for replication during the cell cycle and *oriT* for transfer. F also contains three insertion sequences and a transposon.

An F plasmid-containing cell has one to three F-encoded sex pili on its surface; 1 gene codes for the pilus subunit, and 11 other genes code for proteins involved in assembly of the pilus. A sex pilus is a long, hollow cylinder (outside diameter 8 nm, inside diameter 2 nm, and length $\approx$1 μm). The sex pilus binds to a receptor on the F$^-$ cell, probably a lipoprotein. Surface exclusion proteins (encoded by *traS* and *traT*) inhibit mating between two F$^+$ cells by blocking the pilus receptor from binding to the pilus. The pilus connecting the cells contracts, the cell envelopes make contact, and a pore forms between the cells. A copy of the F plasmid is transferred through the pore (Figure 18.3). Here's how.

A single strand of F plasmid DNA is nicked at *oriT*, and a helicase starts unwinding the dsDNA from the 5$'$ end of the

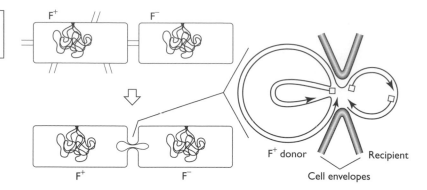

Fig 18.3 Conjugation in *E. coli* and transfer of the F plasmid.

nicked site. Single-stranded DNA binding proteins (SSBs) protect the ssDNA. After part of the ssDNA is transferred, both the transferred strand and the unnicked, nontransferred strand serve as templates for replication. In the recipient cell, synthesis is discontinuous, by cellular replication enzymes (pol III holoenzyme, topoisomerase, SSBs, pol I, ligase, etc.). Priming may be either by primosome or by RNA polymerase. In the donor cell, synthesis is probably continuous, using cellular replication enzymes; RNA polymerase synthesizes the primer. Mating pairs separate after DNA transfer.

There are two kinds of donor cell, F$^+$ and Hfr (Figure 18.4). F$^+$ cells have a free plasmid, while in Hfr cells the F plasmid is integrated into the main chromosome.

If the donor cell is F$^+$, then the recipient becomes F$^+$ upon receiving the F plasmid, immediately expresses pilus genes and surface exclusion genes. Genes of the main chromosome are transferred at a very low rate ($\sim 10^{-5}$).

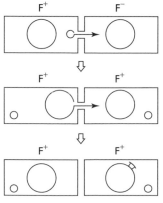

Rare recombinants

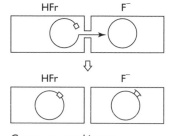

Common recombinants

Fig 18.4 Gene transfer from F$^+$ (left). Gene transfer from Hfr (right).

If the donor cell is Hfr, then a copy of the main chromosome is transferred, starting at the integration site of F, and potentially ending with the transfer of F. Transfer of the full chromosome is rare because conjugation is often interrupted by random shearing forces. In Hfr × F⁻ conjugation, genes of the main chromosome near *oriT* are transferred at a high rate (several percent), but the recipient likely remains F⁻ because of interrupted conjugation.

Plasmids abound in bacteria, and most are species specific. Some circular conjugative plasmids carry genes for drug resistance and replicate by a rolling circle mechanism. Two classes of circular plasmids, RK2 and RSF1010, are transferred by conjugation and grow in virtually all Gram-negative species; this opens the possibility for transfer of genes between distantly related species. *Agrobacterium tumefaciens* regularly transfers its 9-kb plasmid (RSF1010) into plant cells.

Transduction

Transduction, originally discovered in Gram-negative and Gram-positive bacteria but now known to occur in eukarya, too, is the transfer of DNA from the genome of a donor cell to the genome of a recipient cell, carried between the cells in a viral capsid. RNA viruses can transduce if the RNA is reverse-transcribed in the recipient cell and the DNA then integrates into the genome by recombination. When a virus transfers DNA from one cell to another, but when that DNA fails to integrate into a recipient chromosome, transduction is abortive. Abortive transduction has little or no genetic or evolutionary significance.

Bacterial transduction is either **generalized** or **specialized**. In generalized transduction, the viral capsid contains only cellular DNA – a small segment of the donor cell's chromosome. In specialized transduction, the viral chromosome has a bit of the donor cell's DNA integrated into it.

Generalized Transduction
Bacteriophage P1 can transduce *Salmonella typhimurium*. PBS1, whose genome is ≈300 kb, transfers bacterial genes in *B. subtilis* by generalized transduction.

The linear, 90-kb chromosome of phage P1 circularizes and undergoes rolling circle replication to produce a concatemer. The

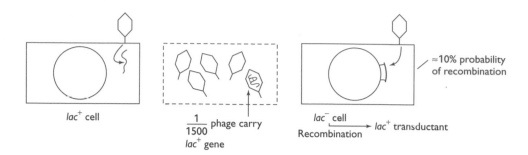

lac^+ cell

$\frac{1}{1500}$ phage carry lac^+ gene

lac^- cell

$\xrightarrow[\text{Recombination}]{}$ lac^+ transductant

≈10% probability of recombination

Fig 18.5 Generalized transduction by P1 phage.

P1 chromosome is packed into the phage capsid by the headful mechanism. A P1 endonuclease cuts the concatemer at a *pac* (packaging) site; DNA then enters the capsid until it is full (≈100 kb of DNA); and another endonuclease cuts the concatemer.

However, the chromosome of *S. typhimurium* contains many sites homologous to the P1 *pac* site, which the viral endonuclease cleaves. Phage particles attach to the cell envelope and inject their DNA into uninfected cells, and the ones containing bacterial DNA work just as well as those containing phage DNA. Suppose P1 phages infect and lyse a cell that was capable of using lactose (*lac$^+$*), and one phage particle picks up *lac$^+$*. Then suppose it injects the *lac$^+$* DNA into a *lac$^-$* cell, and that the *lac$^+$* gene replaces the *lac$^-$* gene, by recombination. Then the recipient cell becomes *lac$^+$* and can metabolize lactose.

Transduction with P1 is efficient (Figure 18.5): ≈1/30 viral capsids contain bacterial DNA instead of viral DNA, ≈2% of the bacterial chromosome fits into the capsid, and recombination of the injected DNA happens ≈10% of the time. Therefore, the chance that transformation will happen with any given bacterial gene is ≈$(1/30)(1/50)(1/10) = 7 \times 10^{-5}$. Rates of generalized transduction for a specific gene range from 10^{-8} to 10^{-5}. Like transformation, transduction is a significant source of new recombinants.

Specialized Transduction

Specialized transduction differs from generalized transduction in that the transduced DNA is covalently joined to the viral chromosome, and only a small subset of the bacterial genome can be transferred in this way. Theoretically, any temperate phage can mediate specialized transduction, although few examples are known: phage SPβ of *B. subtilis*, and phage λ of *E. coli*. A temperate phage has both lysogenic and lytic modes of reproduction.

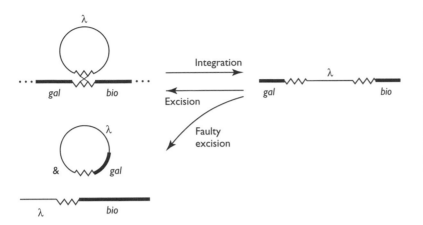

Fig 18.6 Faulty excision of a prophage can make a *gal*-carrying λ chromosome. Thin line = λ DNA; thick line = bacterial DNA; zigzag line = pairing site. Integration of λ DNA into the main chromosome requires site-specific recombination, and excision is the reverse of integration. Faulty excision occurs when the integration sites do not align properly, and cuts are made at other places.

In the lysogenic mode, the phage chromosome inserts into the host chromosome by site-specific recombination, and viral reproduction is repressed. The integrated phage chromosome is a **prophage**. When the virus switches to the lytic mode of reproduction, the prophage excises from the bacterial chromosome by the reverse of integration (Figure 18.6). Excision is nearly always precise, yielding a perfect phage chromosome. Rarely excision is imprecise, producing a phage chromosome with bacterial DNA on one end and missing viral DNA from the other end. The recombinant chromosome replicates and is packaged into capsids. In this way, genes near the prophage site undergo specialized transduction at rates as high as 10^{-5}.

Other Life Forms

In sexually reproducing eukarya, mitochondria infrequently exchange genes. During sexual reproduction, two sex cells, or gametes, fuse to produce a zygote. In most eukarya except the amitochondrate groups, both sex cells contain at least one mitochondrion. In animals, eggs are large and contain many mitochondria, while sperm are small and contain one or a few mitochondria; therefore, animals have a strong bias toward inheritance of mitochondria from the mother. In many animal species mitochondria are inherited only from the mother. From time to time, some mitochondria fuse, and their chromosomes recombine. If the mitochondria of the two sex cells differ genetically, which is often the case, and if some mitochondria are inherited from the father, then some of the zygote's mitochondria will carry some genes derived from the mitochondria of one sex cell and some genes derived from the other. In this way, genes can move between mitochondria despite their clonal mode of reproduction and even in organisms where there is a strong bias toward maternal inheritance. The frequency of horizontal gene movement in mitochondria has not been assessed quantitatively with precision.

Viruses exchange genes by recombination whenever two or more viruses simultaneously infect a host cell. This fact has been exploited successfully to study recombination in dsDNA bacteriophages such as T4. The main biological impact of recombination between homologous viral chromosomes is to make new recombinant viral types.

RNA viruses undergo recombination in cells infected by two viruses, both within and between taxa. Recombination between chromosomes of eastern equine encephalitis and sindbi virus gave rise to western equine encephalitis. New strains of influenza A virus frequently arise by recombination, and poliovirus chromosomes also recombine. RNA virus genomes recombine in two ways in doubly infected cells. First, the chromosomes of segmented viruses (kinds with multichromosome genomes) assort; e.g., one virus's copy of chromosome 1 and another virus's copy of chromosome 2 are packaged into one capsid. Second, during chromosome replication intrachromosomal recombination may occur by a template-switching mechanism called **copy choice** (Figure 18.7).

Fig 18.7 Recombination of RNA virus chromosomes. Assortment in a segmented virus (top). Copy choice recombination (bottom).

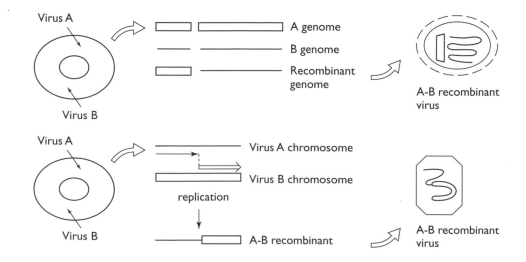

Further Reading

Dubnau D. 1999. DNA uptake in bacteria. *Annu. Rev. Microbiol.* 53:217–244.

Masters M. 1996. Generalized transduction. In: Neidhardt FC (ed.). *Escherichia coli and Salmonella Cellular and Molecular Biology*. 2nd ed., ASM Press, Washington DC.

Worobey M, Holmes EC. 1999. Recombination in RNA viruses. *J. Gen. Virol.* 80:2535–2553.

Chapter 19

Cell Cycles of Eukarya

Overview

In eukarya, as in bacteria and archaea, single cells reproduce asexually via a cell cycle. The cell cycle of eukarya, though, is more complex and elaborately regulated. The large, linear chromosomes of eukarya, whose DNA is packaged in chromatin, are replicated once per cycle, and copies are apportioned to daughter cells with great accuracy in a nuclear division phase, **mitosis**. The phases of the cell cycle and the main points of cell-cycle regulation are described in this chapter.

Keeping Track of Chromosome Number

Ploidy refers to the number of chromosome sets per cell nucleus. A **euploid** cell has an integer multiple of chromosome sets. Normal sex cells are **haploid**, having one set of chromosomes; sex cells unite to make **zygotes**, **diploid** cells with two sets of chromosomes.

In sexually reproducing eukarya, the haploid chromosome number of a species is **N**, and the diploid number is **2N**. A gamete has N chromosomes, and most somatic cells of plants and animals have 2N chromosomes. Some cells, though, are **polyploid** – i.e., they have three or more sets of chromosomes. For example, in humans N = 23; a sperm has 23 chromosomes, a skin cell has 46 chromosomes, and a liver cell has 92 or more chromosomes. Mitosis works the same no matter what the cell's ploidy.

However – and this is important – the cell's ploidy does not change with the phase of the cell cycle, even though the number of copies of the genome does change. Consider a newly formed human skin cell; it has 46 chromosomes, and each chromosome is made of one DNA molecule. There are two copies of the nuclear genome, so the amount of nuclear DNA is **2C**. Later in the cell's development, the genome replicates, after which the cell still has $2N = 46$ chromosomes, each one made of two DNA molecules, and the amount of nuclear DNA is **4C**. The equal copies of a replicated chromosome are **sister chromatids**. Sister chromatids segregate during mitosis, and, as soon as they segregate, they cease to be chromatids and are simply chromosomes.

Phases of the Eukaryal Cell Cycle

The cell cycle of actively growing and dividing eukarya has four phases (Figure 19.1):

- G_1 (gap 1), the phase following nuclear division and preceding genome doubling
- **S** (synthesis), the phase when the nuclear genome is replicated
- G_2 (gap 2), the phase in which the cell prepares for nuclear division
- **M** (**mitosis**), the phase of genome segregation; both **karyokinesis** (nuclear division) and **cytokinesis** (cell division) usually accompany mitosis.

The genome doubles in S phase and halves in M phase. Together, G_1, S, and G_2 make up **interphase**. Quiescent, nondividing cells are said to be in G_0 phase.

Interphase

Biochemically and genetically, the phases within interphase – G_0, G_1, S, and G_2 – are profoundly different, but in most cells they cannot be distinguished simply by microscopic examination (Figure 19.2). Most plant and animal cells are diploid during the

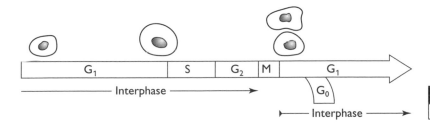

Interphase

G_1 | S | G_2 | M | G_1

G_0

Interphase

Fig 19.1 Phases of the eukaryal cell cycle.

Fig 19.2 Interphase cell.

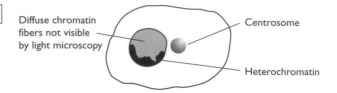

Diffuse chromatin fibers not visible by light microscopy

Centrosome

Heterochromatin

entire cell cycle, but there are plenty of examples of polyploid and haploid cells that go through cell cycles. The nature of the cell cycle is unaffected by ploidy level, but ploidy level is affected by the cell cycle (no karyokinesis→polyploid nucleus; karyokinesis without cytokinesis→multinucleate cell).

G_1 Phase

In physiological terms, G_1 is a time for "business as usual," when the cell performs the bulk of its normal functions. Proteins, ribosomes, and tRNAs are synthesized as the cell grows and carries out physiological activities.

The centrosome, an amorphous structure containing >30 different proteins essential to nuclear division, sits near the center of the cell and, in animals, plants, and fungi, to one side of the nucleus. The primary function of centrosomes is to generate the spindle during mitosis. In many animal species, the centrosome contains a pair of centrioles, which have no known function in cell division.

Each nuclear chromosome in G_1 contains one double-stranded DNA (dsDNA) molecule. Euchromatin consists of chromatin fibers arranged on the nuclear matrix, and the highly condensed heterochromatin is usually at the periphery of the nucleus. The nucleolus is prominent in the nucleus. Genes are actively transcribed.

In the context of the cell cycle, G_1 is the phase when the cell commits itself to enter S phase and hence to go through another division cycle. Commitment to proceed through another cycle occurs at the **R point** (R for restriction) in mammalian cells; this point has other names in other organisms. The R point divides G_1 into two parts, early G_1 and late G_1.

S Phase

Chromosomes replicate in S phase, when the amounts of DNA and histones in the nucleus double. At the end of S phase, each chromosome consists of two sister chromatids. In species with localized centromeres, sister chromatids are joined at their centromeres, although the double nature of each chromosome cannot be observed microscopically because chromatin fibers are thinner than the wavelength of visible light. At the centromeres are kinetochores, multiprotein structures to which spindle fibers attach in mitosis. Commonly, the components of the centrosome double during S phase, but the centrosome does not divide until M phase.

G_2 Phase

During G_2 proteins that participate in M phase are synthesized – for example, α and β tubulin, which comprise the chromosome-moving spindle.

G_0 Phase

G_0 is a state in which passage to S phase is blocked. A cell may exit the cell cycle permanently or temporarily, by moving from early G_1 to G_0. G_0 cells may be very active physiologically, as are the neurons in adult humans; our neurons seldom reenter the cell cycle.

M Phase

The main events of M phase are mitosis, karyokinesis, and cytokinesis. During mitosis, the two copies of the genome segregate. In karyokinesis, the nucleus splits into two, each daughter nucleus having a copy of the genome. At the same time (cytokinesis), the cell divides and each daughter cell receives one of the two new nuclei. Thus, a diploid mother cell is 2N and 4C, while each of its daughter cells is 2N and 2C.

Mitosis

It is convenient to divide mitosis into six stages:

- Prophase
- Prometaphase
- Metaphase
- Anaphase
- Telophase
- Karyokinesis

Mitosis can be either open or closed, depending on the species. Animals and plants have open mitosis, during which the nuclear envelope disintegrates at the beginning and re-forms at the end. Closed mitosis takes place inside the intact nuclear envelope (e.g., in yeast).

Prophase

Chromosomes, observable under the light microscope, appear in early prophase. Chromosomes become visible in early prophase when they thicken as chromatin condenses. Each chromosome consists of a pair of sister chromatids, joined along their entire length by glue-like proteins. Three such proteins are known in yeast. The two centrosomes move apart toward opposite sides of the nucleus, and microtubules begin to form between them. The mechanism of centrosome movement is unknown (Figure 19.3).

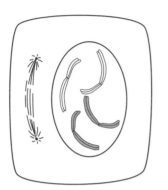

Fig 19.3 Prophase of mitosis.

Prometaphase

The spindle forms in prometaphase. The spindle is a set of microtubules, made α- and β-tubulin (55 kDa each), growing from the centrosomes; some attach to chromosomes (Figure 19.4). In open mitosis, the nuclear envelope disintegrates during prometaphase, and the spindle fibers enter the nuclear region. In closed mitosis the spindle fibers either form inside the nucleus or grow through the nuclear envelope. The number of spindle fibers that

Fig 19.4 Microtubule α- and β-tubulin subunits. Microtubule α- and β-tubulin subunits (dark and light) **(a)**. Prometaphase of mitosis **(b)**.

a)

25 nm

End view

b)

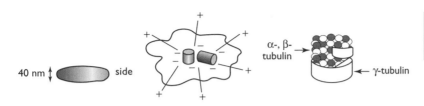

Fig 19.5 Kinetochore (left). Animal centrosome (center). Microtubule, (−) end (right).

attach to one chromosome varies among species, from two (one on each side of the chromosome) to many. In species with localized centromeres, the **kinetochore spindle fibers** attach to kinetochores. In species with holocentric chromosomes, the spindle fibers attach to several points along each chromosome. The plus ends of spindle fibers attach to kinetochores and the minus ends are embedded in the centrosome (Figure 19.5). The (−) end is attached to γ-tubulin rings in the centrosome. Spindle fibers grow (−) end → (+) end. In most species, spindle fibers of a second kind, **polar spindle fibers**, remain unattached to kinetochores and run from spindle pole to spindle pole (from centrosome to centrosome).

Spindle fibers apply force in two ways, by changing length as a result of adding or subtracting subunits at the ends, and by interacting with contractile proteins (e.g., kinesins). The minus end, attached to the centrosome, grows slowly because it is capped by proteins including γ-tubulin. The plus end, attached to the kinetochore (or functionally equivalent structure in holocentric chromosomes), can grow quickly. A net gain of subunits in a microtubule increases its length, causing it to push at both ends. A net loss of subunits in a microtubule decreases its length, causing it to pull at both ends. Kinesins exert further pulls and pushes on spindle fibers at the minus end.

Spindle fibers attach to kinetochores in random order. After attaching to one kinetochore, the singly attached chromosome jiggles toward and away from the associated pole. The jiggling motion is caused by oscillations in poleward growth and cessation in growth at the minus end of the spindle fiber, and it continues until a spindle fiber attaches to the second kinetochore, on the opposite side of the chromosome. After both kinetochores of a chromosome attach to spindle fibers, the growth of the two kinetochore spindle fibers somehow balance, eventually moving the chromosome equidistant from the two poles.

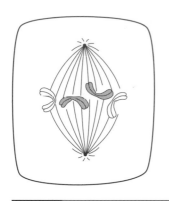

Fig 19.6 Metaphase of mitosis.

Metaphase

In metaphase, the chromosomes have been aligned midway between the spindle poles by the equal tension of the kinetochore microtubules (Figure 19.6). The chromosomes are scattered throughout the plane that bisects the cell, midway between the spindle poles. This central region is the **metaphase plate** or the equator. Metaphase has great practical significance, because cytologists observe and photograph chromosomes at this stage, in order to study the **karyotypes** of individuals (i.e., their chromosomal make-up). In metaphase, sister chromatids are still glued together, although in karyotypes they have separated as an artifact of preparing the cells for microscopic examination.

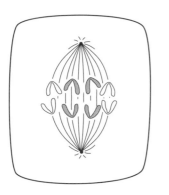

Fig 19.7 Anaphase of mitosis.

Anaphase

Anaphase begins when the centromeres pull apart from the contraction of spindle fibers (Figure 19.7). Anaphase is subdivided into two parts, A and B. During **anaphase A**, the chromatid arms become unglued by the action of **separin**, a protease, leaving sister chromatids connected only at their centromeres; more about the action and timing of separin appears in the section on regulation of the cell cycle. The kinetochore spindle fibers shorten, separating sister chromatids and then pulling the now separate chromosomes to the spindle poles. During **anaphase B**, the polar spindle fibers elongate, pushing the two spindle poles move farther apart. **Kinesins** contribute to the movement of spindle fibers.

Telophase

In telophase, the daughter chromosomes arrive at the **poles** (Figure 19.8). The kinetochore microtubules disappear and the polar microtubules achieve maximum length. Karyokinesis, the final step of mitosis, is the production of two daughter nuclei – the division of the nucleus in closed mitosis or the re-formation of the nuclei in open mitosis.

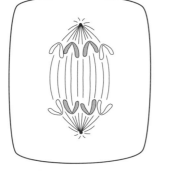

Fig 19.8 Telophase of mitosis.

Karyokinesis and Cytokinesis

Karyokinesis and cytokinesis normally occur during M phase (Figure 19.9). Karyokinesis is nuclear division, and cytokinesis is cell division. There are two major kinds of cytokinesis. Cells without walls cleave by constriction, while cells with walls cleave by cell plate formation. In animals, a contractile ring of actin forms at the middle of the cell; it causes the cell to pinch in two. Pieces of the spindle left over from mitosis may persist until the end of cytokinesis. In plants, the cell plate, consisting of cell wall and membrane, grows between the cells.

Endomitosis

A cell cycle with neither nuclear nor cell division is called **endomitosis**. One round of endomitosis doubles the chromosome number; two rounds quadruple the chromosome number. Endomitosis is widespread in plants and animals, although it is generally restricted to particular cell types. Human liver cells are at least tetraploid, the macronuclei of ciliates range from 16N to 13,000N, and the nucleus of a giant abdominal neuron in *Aplysia* has about 7.5×10^4 copies of its chromosomes. The larval salivary glands of *Drosophila* and related flies contain giant polytene chromosomes, produced by 10 or so rounds of endomitosis. In this case, identical and homologous copies are synapsed point for point, making each polytene chromosome more than 1000 times as thick as a single copy.

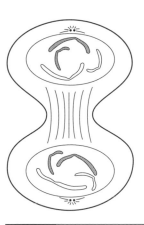

Fig 19.9 Karyokinesis and cytokinesis.

Regulation of the Cell Cycle

Growth and development, maintenance of appropriate numbers of cells in a multicellular organism, and safeguarding against genetic damage require tight regulation of passage through the cell cycle. Regulation of the cell cycle depends on the action of scores of proteins and entails a combination of physiological and genetic processes. This section is a brief summary.

The **R point** of the G_1 phase is critical to the cell cycle. The cell remains in G_1 phase until it is released from the R point. When a cell in G_1 passes the R point, it then continues through an entire cell cycle; R is the point of no return. However, the cell's progress through the cell cycle may be arrested at **checkpoints** if there are problems such as DNA damage or incomplete development of the spindle (Figure 19.10). The cell's progress halts at a checkpoint until the problem has been corrected. There are two checkpoints for DNA damage, one in G_1 and another in G_2. At the spindle checkpoint in metaphase of mitosis, each kinetochore must be attached to the spindle before entry into anaphase.

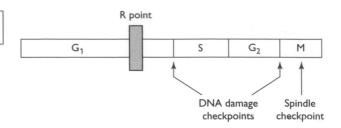

Fig 19.10 Cell-cycle R point and three checkpoints.

Many proteins, including **cyclin-dependent kinases (CDKs), cyclins**, Rb, p53, p21, E2F, and APC regulate the passage through the cell cycle. Each CDK binds to a cyclin to make an active dimer. There are three groups of CDK-cyclins, for G_1, S, and M phases. The concentrations of CDK-cyclins change systematically through the cell cycle. CDKs phosphorylate other regulatory proteins.

External signals such as growth factors trigger passage through the R point. Rb binds to E2F, a transcription factor for genes encoding the DNA-replicating machinery. Rb also stimulates histone deacetylases (HDs), which help repress transcription by promoting nucleosome-DNA binding. When growth factors bind to receptors on the cell's surface, a cytoplasmic domain of the receptor activates a cascade of intracellular signaling that mobilize CDK4-cyclinD and CDK2-cyclinE. These CDK-cyclins phosphorylate Rb, which then becomes inactivated (Figure 19.11). Inactive, phosphorylated Rb no longer binds to E2F, and E2F activates genes that code for DNA-synthesizing machinery. At the same time, phosphorylated Rb does not bind to HDs, allowing histone acetyl transferases (HATs) to derepress chromatin.

When damage to DNA is sensed at the DNA damage checkpoints, the cell cycle is arrested. DNA damage in G_1 phase induces signals that activate p53, a transcription factor. p53 activates (1) genes encoding DNA repair activities, and (2) the gene encoding p21, which inhibits CDK4-cyclinD, so that Rb continues to maintain the R point. DNA damage in G_2 phase generates

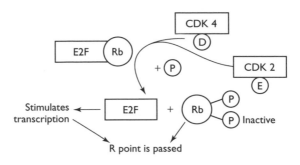

Fig 19.11 The R point is passed when protein kinases phosphorylate Rb.

intracellular signals that inhibit CDK2-cyclinB, which is necessary for progress through G_2 phase.

At the metaphase spindle checkpoint, if any kinetochore is unattached to spindle fibers, there is no tension on the chromosome. Absence of tension generates signals that inhibit APC (anaphase-promoting complex); the cell is arrested in metaphase until kinetochore-spindle attachment generates tension.

Further Reading

Kalt A, Schliwa M. 1993. Molecular components of the centrosome. *Trends Cell Biol.* 3:118–127.

Planas-Silva MD, Weinberg RA. 1997. The restriction point and control of cell proliferation. *Curr. Opin. Cell Biol.* 9:768–772.

Raff JW. 1996. Centrosomes and microtubules: wedded with a ring. *Trends Cell Biol.* 6:248–251.

Rieder CL, Salmon ED. 1998. The vertebrate cell kinetochore and its roles during mitosis. *Trends Cell Biol.* 8:310–318.

Chapter 20

Meiosis

Overview

The essence of sex is alternation between a **haploid phase**, when cell nuclei possess one set of chromosomes, and a **diploid phase**, when they possess two. Each set of chromosomes is **homologous** and usually nonidentical, having come from unrelated haploid sex cells. The transition from diploid to haploid requires **meiosis**, a process in which precisely one copy of each chromosome is apportioned to each haploid cell.

The events of meiosis determine quantitative, predictable patterns of genetic transmission from parent to offspring in sexual species. Two hallmarks of meiosis are the 1:1 segregation of gene copies and recombination of genes and chromosomes. In recombination, chromosomes and chromosome segments shuffle to make a virtually limitless number of new genetic combinations.

The first part of the chapter describes meiosis as a formal dance of chromosomes. The second part of the chapter explains the genetic consequences of meiosis – the segregation of homologous chromosomes and recombination. The third part of the chapter describes exceptional patterns of meiosis.

Recap of Ploidy and DNA Content

To recap what was explained in Chapter 19, for any eukaryon with sexual reproduction the haploid number of nuclear chromosomes is N and the diploid number is 2N; in a diploid cell the two sets of chromosomes are nonidentical and homologous. In asexually reproducing cells, chromosome number is constant through the life cycle. In sexual organisms, gametes (sperm and eggs, or their equivalents) are haploid, while zygotes (cells formed by the union of gametes) are diploid. The amount of DNA in a cell is given as the number of copies (C) of the haploid genome. A diploid human cell in G_1 has 2C DNA and in G_2 has 4C DNA.

Stages in the Diploid→Haploid Transition

Meiosis is two successive nuclear divisions, accompanied by cell divisions (Figure 20.1). A diploid cell enters meiosis from G_2. The two successive nuclear divisions are meiosis I and meiosis II; each consists of prophase, prometaphase, metaphase, anaphase, and telophase. Meiosis I is called the **reductional division**, because it reduces the number of microscopically observed chromosomes by half. Meiosis II is the **equational division**, because chromatids segregate during this division; mechanically, meiosis II closely resembles mitosis.

Meiosis is either **chiasmate** or **achiasmate**. In chiasmate meiosis, the more common type, X-shaped structures called **chiasmata** (sing. **chiasma**) form during prophase I; in achiasmate meiosis, chiasmata are absent. Chiasmata are positive indicators of chromosome exchange (recombination) in all cases that have been checked for this. Chiasmate meiosis is described here.

In the following descriptions, the main features of each stage of meiosis, as it typically appears by light microscopy, are diagrammed for a hypothetical example in which there are two chromosomes per haploid genome (N = 2).

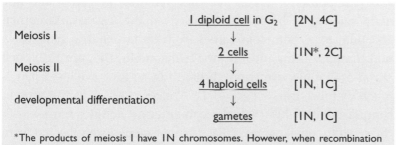

	1 diploid cell in G_2	[2N, 4C]
Meiosis I	↓	
	2 cells	[1N*, 2C]
Meiosis II	↓	
	4 haploid cells	[1N, 1C]
developmental differentiation	↓	
	gametes	[1N, 1C]

*The products of meiosis I have 1N chromosomes. However, when recombination occurs in meiosis I, the "sister chromatids" are not identical.

Fig 20.1 Meiosis I and meiosis II.

Prophase I

Understanding prophase I is the key to understanding meiosis. It is the longest stage of meiosis and very different from prophase of mitosis. Prophase I is subdivided into five distinct

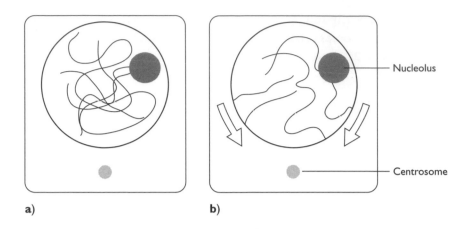

a)   b)

Nucleolus

Centrosome

parts: **leptonema, zygonema, pachynema, diplonema,** and **di-akinesis.** The adjectives derived from the names of the first four of these end in "-tene": leptotene, zygotene, pachytene, and diplotene.

Leptonema

In leptonema, chromosomes become visible as thin threads whose telomeres are attached to the nuclear membrane (Figure 20.2). In mid-leptonema of most plants and animals, one telomere of each chromosome attaches to the nuclear envelope and migrates toward the centrosome.

Zygonema

In zygonema, **homologous chromosomes** (**homologs**) thicken and pair point for point; i.e., they **synapse.** In plants and animals, one set of homologs comes from the organism's mother, and one set of homologs comes from its father. The **synaptonemal complex** forms and joins the two homologs (Figure 20.3). Each chromosome has two sister chromatids, so each synapsed pair is made of four strands (= four DNA molecules). Gigantic protein complexes ($\sim 10^5$ kDa) called recombination nodules form within the synaptonemal complex. Each synaptonemal complex has from one to a few recombination nodules. In the synaptonemal complex are three electron-dense stripes, two lateral elements next to the chromatin and a central element in the middle; transverse fibers run across the synaptonemal complex, perpendicular to the lateral elements. Synapsed chromosomes are called **bivalents** or **tetrads,** depending on whether one wishes to emphasize the fact that chromosomes are paired or that four strands are brought

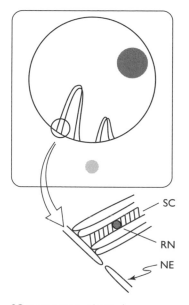

Fig 20.3 Zygonema (**a**).
Synaptonemal complex (**b**).

SC = synaptonemal complex
NE = nuclear envelope
RN = recombination nodule

TF = transverse filaments
LE = lateral elements
CE = central elements

a)

b)

together. For purposes of genetic analysis, they are tetrads. All telomeres attach to a region of the nuclear envelope near the centrosome, and centromeres point in the opposite direction. This arrangement, called the **bouquet**, is short-lived.

Pachynema

In pachynema, chromosomes thicken further and their telomeres are no longer attached to the nuclear envelope – the bouquet disappears (Figure 20.4). Recombination occurs, presumably at the sites of recombination nodules. Nearly all exchanges are between nonsister chromatids; exchanges between sister chromatids are rare.

Diplonema and Diakinesis

In diplonema, chromosomes continue to condense, and chiasmata arise in each bivalent (Figure 20.5). Most bivalents have one to three chiasmata (very rarely more). Chromosomes appear to repel each other except at the center of each chiasma. Sister chromatids remain glued together but chiasmata hold the homologs together at only one or a few points. Chiasmata are thought to initiate at sites where chromosomes have undergone exchange – i.e., at recombination nodules.

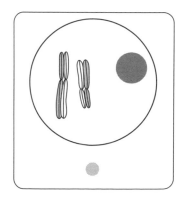

Fig 20.4 Pachynema.

Fig 20.5 Diplonema (**a**). Diakinesis (**b**).

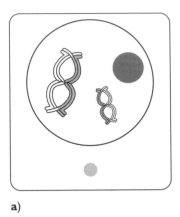
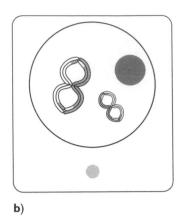

a)   b)

In diakinesis, chromosome condensation reaches a maximum, and chiasmata terminalize; that is to say, the points of connection between chromosomes move to the telomeres (Figure 20.5). In some species, chromosomes tend to lie at the periphery of the nucleus, near the nuclear envelope. The cellular motors that move chromosomes during prophase I are unknown. The centrosomes begin to move apart and the spindle begins to form. At the end of diakinesis, the nuclear envelope disintegrates and, in many species, the nucleolus disintegrates as well.

Prometaphase I to Telophase I

In prometaphase I the centrosomes move to the poles and the spindle is organized (Figure 20.6). The paired homologs attach to the spindle and move to the equator of the spindle. In metaphase I, the paired homologs are lined up on the metaphase plate (Figure 20.6). In contrast to mitosis, spindle fibers attach only to one side of each paired chromosome. Apparently, additional proteins bind to both kinetochores of each homolog, and the spindle fibers bind to these proteins. At the end of metaphase,

Fig 20.6 Prometaphase I, metaphase I, anaphase I, and telophase I.

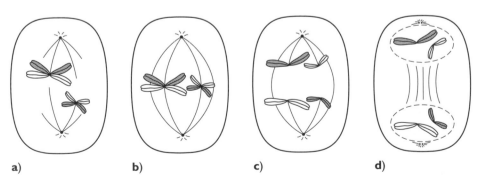

a)   b)   c)   d)

sister chromatids become unglued, just as they do in metaphase of mitosis.

During anaphase I, **homologs disjoin**, the first step of segregation (Figure 20.6). In telophase I, chromosomes are pulled to the poles, chromosomes begin to decondense, and a nuclear envelope re-forms at each pole around the chromosomes (Figure 20.6). Importantly, there are N chromosomes at each pole, a haploid number, and 2c DNA. In some species, cytokinesis follows telophase I.

Meiosis II

Mechanically, meiosis II resembles mitosis of haploid cells, except that the arms of sister chromatids remain separated in early meiosis II; they are connected only at the centromere in metaphase II. Meiosis II follows a special, post-meiosis I interkinesis, which is G_2-like. Important point: DNA does not replicate in meiosis.

In prophase II, chromosomes condense and centrosomes begin to move apart as spindle fibers form (Figure 20.7). The nuclear envelope, if it re-forms in telophase I, disintegrates in prophase II. In prometaphase II, the centrosomes move to the poles, the spindle forms, chromosomes attach to the spindle, and chromosomes move towards the equator. In metaphase II, chromosomes line up on the metaphase plate (Figure 20.7). Kinetochore fibers connect to both sides of each chromosome. In anaphase II, sister chromatids disjoin (Figure 20.7). In telophase II, chromosomes move to the poles, chromosomes begin to decondense, and a nuclear envelope re-forms around each set of chromosomes (Figure 20.7). Cytokinesis typically follows.

Fig 20.7 Prophase II, metaphase II, anaphase II, and telophase II.

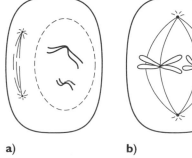

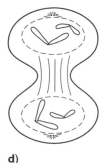

a)   b)   c)   d)

Fig 20.8 Four haploid genotypes produced by meiosis: meiosis I = MI and meiosis II = MII. Light chromosomes = maternal, dark chromosomes = paternal.

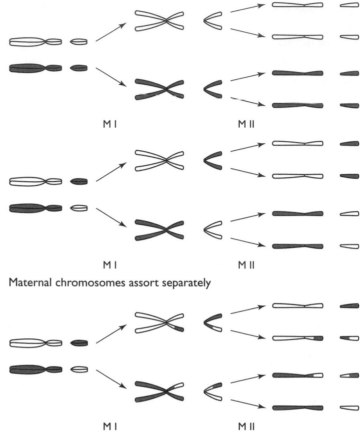

M I   M II

M I   M II

Maternal chromosomes assort separately

M I   M II

Recombination in both chromosomes

Consequences of Meiosis

Segregation, Assortment, and Recombination of Chromosomes

The diploid phase of a sexually reproducing organism begins when two different haploid sex cells fuse. In animals, the sex cells are an egg and a sperm. The genome and chromosomes of the egg are **maternal**, and the genome and chromosomes of the sperm are **paternal**. Every diploid nucleus of a diploid organism has two homologous but usually nonidentical copies of each chromosome, maternal and paternal. In meiosis, homologous chromosomes recombine during pachynema and segregate during anaphase I. Furthermore, the orientation of chromosome pairs on the spindle is random and independent. The result: a haploid cell after meiosis contains a random collection of maternal and paternal copies of the chromosomes. Homologs segregate, and so do the copies of genes located on those homologs. Nonhomologous chromosomes assort independently, as do the genes located on those nonhomologous chromosomes. (Figure 20.8)

Recombination and Chiasmata

Recombination and physical exchange between chromosomes (crossing-over) depend on chiasmata in meiosis. In every case, chiasmata correlate with both crossing-over and with recombination, furthermore, neither crossing-over nor recombination occurs in the achiasmate sex. It remains to be seen, however, whether species in which both sexes have achiasmate meiosis also lack recombination. So far as we know, whenever a crossover (physical exchange) occurs between two strands, a chiasma also occurs, and genes straddling the exchange point recombine.

Chiasmata do not form in heterochromatin, and nucleolar organizer regions (NORs) inhibit nearby formation of chiasmata. Chiasmata tend to form near the centromeres in some species, near the telomeres in other species, and at roughly equal frequencies throughout the chromosome's length in yet other species. The spatial pattern of chiasma formation matches the rates of recombination throughout euchromatic parts of the chromosomes. The frequency of recombination varies with chromosome length, correlating very well with the frequency of chiasmata. Chiasmata tend to interfere with each other, so that two chiasmata are unlikely to occur at nearby chromosomal loci (positions). A typical homologous pair of chromosomes of a chiasmate species forms one chiasma, but large chromosomes tend to form two or three. However, no matter how big the chromosome, the number of chiasmata per paired homolog rarely exceeds three. The mean number of chiasmata per cell in diplonema is usually less than 2N, the diploid chromosome number. As a result, species with small chromosomes undergo more recombination per Mb of DNA than species with large chromosomes.

Cells That Undergo Meiosis

In unicellular organisms such as yeast, meiosis is developmentally restricted; by manipulating culture conditions, an experimenter can induce or suppress meiosis. In multicellular organisms, meiosis is restricted to special reproductive cells as well as being developmentally restricted.

In animals, meiosis takes place only in germ line cells; all other cells are called somatic cells. Germ line cells differ from somatic cells:

1. Meiosis is restricted to germ line cells, which in animals are located only in gonads.
2. The germ line lineage undergoes a small number of cell division cycles before meiosis.
3. Germ line cells are developmentally flexible and can, after futilization, differentiate into all the cell types of the multicellular organism.

Variations of Meiosis

Achiasmate Meiosis

Achiasmate meiosis, which has evolved many times from chiasmate meiosis, occurs in various species of flagellates, ciliates, foraminifera, helminth and annelid worms, arachnids, crustacea, insects, and angiosperms. All the species in two orders of insects, Lepidoptera and Trichoptera, have achiasmate meiosis. In some species of animal, one sex has chiasmate meiosis, while the other sex has achiasmate meiosis (e.g., male *Drosophila* are achiasmate). Whereas in achiasmate meiosis homologs remain completely synapsed throughout prophase I, recombination appears to be absent. Otherwise, chiasmate and achiasmate meiosis are the same.

Inverted Meiosis

Most species have chiasmate or achiasmate meiosis, but a few species exhibit a third kind, inverted meiosis. In inverted meiosis, chromatids disjoin in meiosis I, and homologs segregate in meiosis II. In inverted meiosis, homologs remain paired until meiosis II by an unknown mechanism.

Further Reading

Schmekel K, Daneholt B. 1995. The central region of the synaptonemal complex revealed in three dimensions. *Trends Cell Biol.* 5:239–242.
Zickler D, Kleckner N. 1998. The leptotene-zygotene transition of meiosis. *Annu. Rev. Genet.* 32:619–697.

Chapter 21

Chromosomal Abnormalities

Overview

Most mutations are clearly either very small, involving $<10^3$ nucleotides, or big, involving many genes or whole chromosomes – chromosomal abnormalities are rearrangements (**deletion, duplication, inversion,** or **translocation**), or altered numbers of chromosomes (**aneuploidy** and **polyploidy**). Here, "chromosomal abnormality" means a mutation affecting two or more genes. This chapter deals with the causes and consequences of chromosome mutations in eukarya.

Chromosomal Rearrangement

Chromosomal rearrangements arise more rarely than do micromutations. Furthermore, having two identical copies of a chromosome rearrangement in a diploid genome is often lethal. Deletions and duplications are extremely rare in natural populations, but, surprisingly, inversions and translocations are relatively common in many species of plants and animals. The evolutionary basis of this seeming paradox is, in a nutshell, that diploid individuals having one rearranged copy and one normal copy of a chromosome often have a reproductive advantage (e.g., higher survival or fertility). Such a reproductive advantage leads to an increase in the population frequency of the chromosomal rearrangement (Chapter 39).

Rearrangements usually result from an exchange following two or more double-strand breaks in physically close chromosomes. Ionizing radiation, transposons, or oxidation by free radicals can induce double-strand breaks.

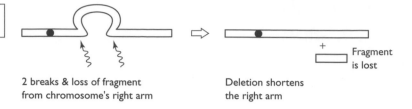

Fig 21.1 Deletion of a chromosome segment.

2 breaks & loss of fragment from chromosome's right arm

Deletion shortens the right arm

Fragment is lost

Deletion (Deficiency)

A chromosomal deletion is the loss of a segment of the chromosome (Figure 21.1). Large deletions, sometimes called chromosomal deficiencies, result from two double-strand breaks followed by loss of the segment between the breaks and a rejoining of the outside pieces. Can a chromosome break at one point and lose its end? Not for long, because the telomere would be lost. Telomeres are essential to the replication of the chromosome end, and they protect the chromosome from degradation. Most genes are essential for life, and organisms having zero copies of one or more genes usually cannot survive.

Inversion

Paracentric and Pericentric Inversions. Inversion is reversal of a segment's orientation, caused when double-strand breaks occur at two locations in a chromosome, the middle part turns around backwards, and the ends rejoin. If the inversion breakpoints are in one chromosome arm, to one side of the centromere, it is a **paracentric inversion**, but if the inversion breakpoints are in both chromosome arms, on both sides of the centromere, it is a **pericentric inversion** (Figure 21.2). Of course, these terms do not apply to organisms having holocentric chromosomes.

Consequences of Chromosome Inversions. Recombination is restricted between two homologous chromosomes if one of them

Fig 21.2 Paracentric and pericentric inversions. Letters denote chromosome segments and arrows show breakpoints. Paracentric inversion (**a**). Pericentric inversion (**b**).

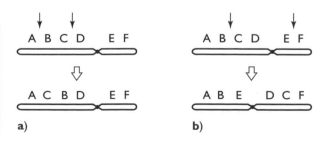

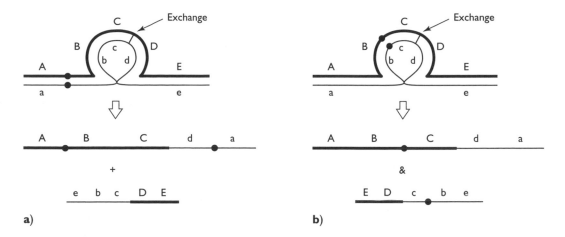

a)

b)

contains an inversion that the other does not. How inversions block recombination is shown in Figure 21.3.

Exchange within the paracentric inversion yields acentric and dicentric recombinants. The acentric chromosome is lost because it does not attach to the spindle and the dicentric chromosome is pulled apart during telophase. Gametes receiving a broken chromosome (because of the dicentric chromosome) or nothing (because of the acentric chromosome) make for massive deletions in the embryo, which then dies. No surviving offspring have a recombinant chromosome for exchanges between the inversion breakpoints. Recombination is thus suppressed.

Exchange within the pericentric inversion yields recombinant chromosomes with massive deletions and duplications. The embryo receiving one of these chromosomes cannot survive the deletions and duplications. The final result is the same as with a paracentric inversion: recombination is suppressed.

Because recombination is suppressed between inverted and noninverted chromosomes, for all exchanges between the inversion breakpoints, the genes within the inversion – which can easily contain 10^3 genes – are inherited as a block, a "supergene" not subject to recombination. Inversions do not suppress recombination outside the inversion breakpoints.

Translocation

Structure. Translocation is exchange between two nonhomologous chromosomes, following breaks in both strands of DNA: a translocated chromosome is a new combination of parts from nonhomologs.

Fig 21.3 Crossing-over between normal and inverted chromosomes in meiosis. Letters identify parts of the chromosomes; inversions contain B→D. Each paired homologous chromosome has two sister chromatids, but for simplicity only one chromatid/chromosome is shown in the diagram. Centromere = big dot. Paracentric inversion (**a**). Pericentric inversion (**b**). For clarity, only one chromatid per homolog is shown.

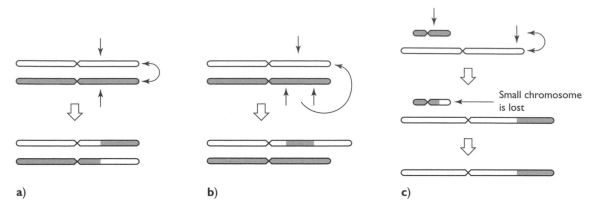

a)   b)   c)

Fig 21.4 Reciprocal translocation (**a**). Simple transposition (**b**). Fusion (**c**).

In **reciprocal translocation**, each chromosome donates and receives DNA. The simplest way for this to happen begins with a double-strand break in each chromosome; then the distal chromosome fragments switch and rejoin (Figure 21.4).

A **nonreciprocal translocation** is an exchange in which one chromosome donates material and the other receives. This can happen either by simple transposition or by fusion. Simple transposition requires three double-strand breaks, two in the donor chromosome and one in the recipient. Fusion results from a reciprocal exchange between two acrocentric chromosomes followed by loss of a translocated chromosome that contains mostly heterochromatin; heterochromatin mostly lacks essential genes. The chromosome number of chimpanzees (N = 24) and humans (N = 23) differs because of a fusion in the human lineage. Human chromosome 2 is a fusion of chimpanzee chromosomes 12 and 13.

Consequences of Translocations

Possessing one copy of a translocation can lead to partial sterility; having two copies can be lethal. An example in humans illustrates this well. Several different nonreciprocal translocations between chromosome 21 and a medium-sized chromosome (notably chromosome 14) cause familial Down syndrome, a condition whose symptoms include poor intellectual development,

Fig 21.5 Fusion of chromosome 14 and 21q → translocation 14^{21}.

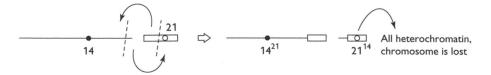

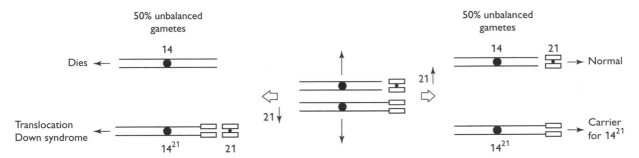

Fig 21.6 How unbalanced gametes are made in 14^{21} carriers. Line = chromosome 14; box = chromosome 21. In anaphase I of meiosis, the unpaired chromosome 21 assorts randomly with chromosome 14 or with chromosome 14^{21}.

epicanthic folds, and short life (Figure 21.5). Persons with Down syndrome have an extra copy of chromosome 21 or an extra copy of 21q. Chromosome 21 is the smallest human chromosome, and 21p is heterochromatin (and therefore not essential for survival).

A person who has one normal chromosome 14, one normal chromosome 21, and translocation 14^{21} is called a 14^{21} **carrier** (also known as heterozygote; Chapter 31) (Figure 21.6). Although 14^{21} carriers have no abnormal traits from this chromosomal makeup, a third of their children have translocation Down syndrome. Female 14^{21} carriers have an increased frequency of miscarriage in pregnancy.

Duplication

Duplications can arise when a chromosome segment translocates to a different, nonhomologous chromosome by simple transposition. Following the translocation, the donor chromosome can be lost to future generations, as multigene deletions often are. The recipient chromosome contains an insertion that contributes an extra copy, or duplication, of the inserted chromosome segment to the entire genome. Extra copies of several genes usually reduce viability.

Aneuploidy

Euploid cells have an integer number of chromosome sets per nucleus. A eukaryon with too many or too few copies of one chromosome has an **aneuploid** chromosome number. Aneuploidy of the entire organism is rare in mature animals. It is possible, but extraordinary, for an organism to be aneuploid for more than one

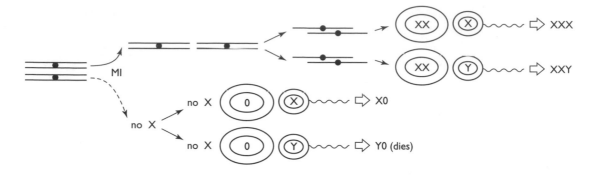

Fig 21.7 Nondisjunction of chromosome X in meiosis I (MI) → aneuploidy for X. Half the gametes have two copies of chromosome X, leading to trisomy-X and XXY children; the other half have no copy of chromosome X, leading to X0 or Y0 children.

chromosome. Sometimes aneuploidy arises in the somatic cells of a multicellular organism. This occurs frequently in cancer cells (Chapter 26).

The usual cause of aneuploidy is chromosomal **nondisjunction** – failure of homologous chromosomes to segregate in meiosis (Figure 21.7) mitosis. The daughter cells then have two or zero copies of that chromosome instead of one copy.

Figure 21.8 summarizes commonly encountered aneuploidies in humans and their frequency among live births.

The consequences of aneuploidy are profound, for an abnormal number of copies of any chromosome causes an imbalance in the amount of gene product for many genes. In the absence of a compensatory mechanism, the amount of gene product (RNA, protein) made from a gene is proportional to the number copies of that gene. In monosomy $(2N - 1)$ the amount of RNA or protein encoded by genes on the single-copy chromosome is reduced to half the normal level, and in trisomy

Fig 21.8 Common aneuploidies in humans.

Chromosomal constitution	Description	Approximate frequency per 100,000 live births
45, X	Female, monosomy X (Turner syndrome, sterile)	10
47, +21	Trisomy 21 (Down syndrome)	100[1]
47, XXY	Male (Klinefelter syndrome, sterile)	50
47, +18	Trisomy 18 (Edwards syndrome, lethal[2])	10
47, +13	Trisomy 13 (Patau syndrome, lethal[2])	20
47, XYY	XYY male (normal)	50
47, XXX	Trisomy X	50

[1] frequency increases with age of mother
[2] death may occur as late as days after birth

The names of common aneuploidies are: $2N - 1 =$ **monosomy**, $2N + 1 =$ **trisomy**, $2N - 2 =$ **nullosomy**.

(2n + 1) the amount of RNA or protein encoded by genes on the triple-copy chromosome is increased to 150% of the normal level. Some of the proteins encoded by any chromosome have regulatory functions – for example, regulating the transcription of many other genes. Abnormal expression of regulatory genes can cause the misexpression (too much or too little transcription) of many other genes scattered throughout the genome. This magnifies the primary effects of aneuploidy. Moreover, if there is only one copy of a chromosome instead of two, and if that chromosome contains less-than-fully-functional copies of its genes, then the genes' products will be deficient or missing. For these reasons aneuploidy can be lethal, and aneuploid animals are often sterile. In many plant species and a few animal species, offspring can inherit aneuploidy from their parents.

In humans, aneuploidy for all chromosomes except X, Y, and 21 is lethal past infancy, and there are particular reasons for these exceptions. Chromosome Y has only a handful of genes; chromosome 21 is the smallest chromosome and has the second smallest number of genes. Even so, only a fourth of trisomy-21 individuals survive until birth. A special mechanism compensates for extra copies of chromosome X – namely, the inactivation of all but one copy in most somatic cells – the inactive X's are turned into heterochromatin.

In fruit flies, XO (single-X) individuals are sterile, fully viable males. In mice, XO females are fertile.

Polyploidy

Polyploidy, a condition found only in eukarya, is the presence of three or more copies of the genome per cell nucleus. The Latin or Greek prefix indicates the number of sets of chromosomes (tri-, tetra-, hexa-, octo-).

Polyploidy in Cells within Multicellular Organisms
Most multicellular species have some polyploid cells. Polyploid cells arise when chromosomes replicate in the synthesis phase and identical copies of each chromosome (chromatids) separate, but the nuclear/cell division phase is not completed. If a diploid cell does not complete the nuclear/cell division phase the

 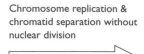 

Fig 21.9 A cell cycle without nuclear division doubles the number of chromosomes.

Chromosome replication & chromatid separation without nuclear division

resulting cell is tetraploid, and after two such modified cell cycles the resulting cell is octaploid. Other related mechanisms can achieve the same result; for example, a cell may undergo nuclear division but not cell division, and then the daughter nuclei may fuse. A diploid cell may undergo nuclear division but not cell division, making a **multinucleate** cell with two diploid nuclei. In some flies (*Drosophila* and its relatives), some cells undergo repeated cell cycles without nuclear/cell division, and the daughter chromosomes synapse (pair point for point); the resulting **polytene chromosomes** may have as many as 2000 copies stuck together (Figure 21.9). Polyteny is a special case of polyploidy. In all these situations, only two nonidentical versions of the genome are present, no matter how many copies of each version there are.

Polyploid Species

During **speciation**, a new species evolves from ancestral species (Chapter 42). In most cases one ancestral species splits into two new ones. A new species may also evolve from a hybrid between two ancestors. Usually, the new species has the same number of chromosomes as the ancestor, but it is sometimes the case that one of the new species is polyploid compared with the other and with the ancestor. The polyploid species gained one or more entire sets of chromosomes, and the evolutionary change is polyploidization. Polyploidization is common in flowering plants and rare in animals (there are a few polyploid insects, fish, and amphibians).

Polyploid species lead us to a pesky complication in terminology used for the number of chromosomes per nucleus. All sexually reproducing species alternate between N and 2N phases. That is to say, a plant or animal species may be polyploid compared with its ancestors, but its gametes are still haploid compared with the somatic cells of the same species. A system of counting chromosome sets that takes into account both reproductive cycles and evolution uses X for the number of chromosomes in one full ancestral set – the **monoploid** number – and N to denote the number of chromosomes in the gamete.

Bread wheat, for example, has 42 chromosomes in somatic cells, six copies of the ancestral grass genome (X = 7, the monoploid number). Bread wheat has 2N = 6X = 42 chromosomes; the ancestral grass plant was 2N = 2X = 14. Several wild cotton species, genus *Gossypium*, have N = X = 13 chromosomes, but the cultivated, tetraploid cotton species *Gossypium hirsutum* has N = 2X = 26 chromosomes. *G. hirsutum* evolved from wild species by polyploidization. In contrast, in our own diploid species, N = X = 23, so humans have 2N = 2X = 46 chromosomes in most somatic cells.

Two types of polyploid species, **allopolyploids** and **autopolyploids**, are recognized, based on the source of the polyploid's chromosomes. The genome of an allopolyploid species derives from two different ancestral species and therefore contains similar but not-quite-homologous sets of chromosomes from the ancestors. In contrast, the genome of an autopolyploid species contains duplicate diploid sets of chromosomes. Tetraploidy is the simplest case to describe (Figure 21.10). An allotetraploid (sometimes called **amphidiploid**) may arise from the union of unreduced (2N) gametes of closely related species with the same monoploid number. An autotetraploid may arise either from unreduced gametes of one species or from mitosis lacking karyokinesis in the early embryo.

Fig 21.10 Origins of tetraploid species. Ancestral monoploid sets of chromosomes are designated A or B. Allotetraploid (**a**). Autotetraploid (**b**).

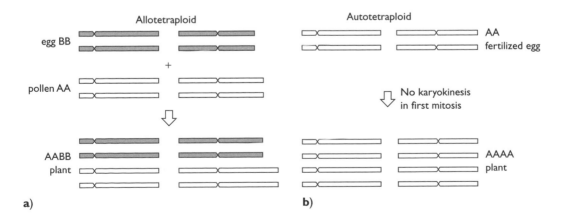

a)

b)

Further Reading

Levan A, Müntzing A. 1963. Terminology of chromosome numbers. *Portugaliae Acta Biol.* 7:1–16.

White MJD. 1973. *Animal Cytology and Evolution,* 3rd ed. Cambridge University Press, Cambridge.

Chapter 22

Life Cycles of Eukarya

Overview

Sexually reproducing organisms alternate between haploid and diploid phases during one **sexual generation**, which extends from haploid phase to haploid phase (for animals and plants, from egg to egg).

During a sexual cycle, diploid cells produce haploid cells by meiosis, and haploid cells and their nuclei produce diploid cells by **syngamy** (fusion). Sexual life cycles are broadly classified according to whether mitotic nuclear division cycles occur in the diploid phase, the haploid phase, or both. These three types are as follows:

- **Haplontic** – predominantly haploid; mitosis does not occur in the diploid phase
- **Diplontic** – predominantly diploid; mitosis does not occur in the haploid phase
- **Haplodiplontic** – mitosis occurs in both the haploid and diploid phases

Eukarya also reproduce asexually. The kinds of asexual reproduction are classified at the end of the chapter. **Parthenogenesis**, an asexual mode of reproduction that evolved from sexual reproduction, is given special attention.

Evolutionary Considerations

There are wags who never tire of asking children, "Which came first, the chicken or the egg?" While this nonsensical question has no answer, it leads to a valid evolutionary question, "What was the relationship between haploidy and diploidy in ancient eukarya?" It is reasonable to assume that mitosis evolved before meiosis, at the beginnings of eukarya, and that the earliest eukarya were monoploid. Perhaps monoploid cells fused from time to time, yielding diploid cells. In this scenario, meiosis then evolved as a mechanism that halved the chromosome number, and the resulting cells had recombinant genomes, which conferred a selective advantage. Presumably, meiosis evolved via modifications of mitosis. On this theory, monoploidy came first, but diploidy preceded the existence of sex (i.e., cycling between haploid and diploid phases).

Haplontic Sexual Cycles

The earliest sexual cycles to evolve were likely haplontic, in which only haploid cells go though mitotic cell cycles (Figure 22.1). Two haploid cells at the G$_1$ phase (each cell having 1N chromosomes and one copy of the genome [1C]) fuse, following which their nuclei fuse, making a diploid 2N, 2C cell. This cell goes through S phase to make a 2N, 4C G$_2$ cell, which immediately enters meiosis. The four haploid daughter cells reproduce by mitotic cycles until the next fusion event.

All modern haplontic organisms are simple, but not all simple organisms are haplontic. For example, ciliates such as *Paramecium* are diplontic.

Fig 22.1 Unicellular haplontic cycle (**a**). Multicellular haplontic cycle (**b**).

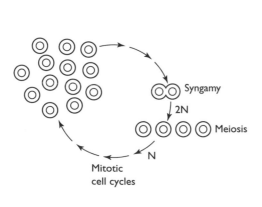

a)

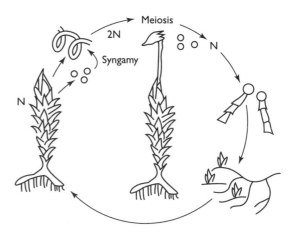

b)

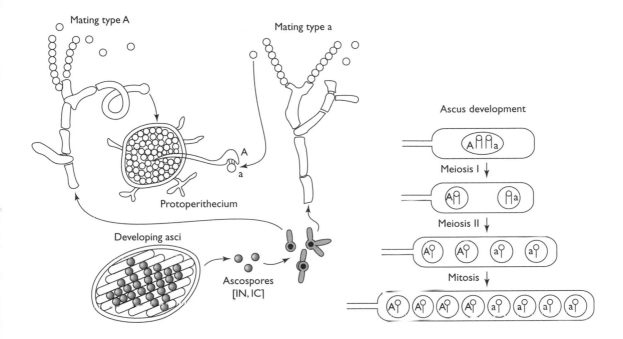

Mating type a

Ascus development

Meiosis I

Meiosis II

Mitosis

Protoperithecium

Developing asci

Ascospores
[IN, IC]

Neurospora Crassa

Fig 22.2	The life cycle of
	N. crassa.

Some fungi are haplontic – their vegetative and sexual structures contain haploid nuclei. An example is the mold *N. crassa*, whose genetics has been studied extensively (Figure 22.2).

N. crassa is an ascomycete, for it produces **asci** – sacs containing haploid spores (**ascospores**) – during meiosis. Each ascospore develops into a vegetative, asexual organism consisting mainly of branching, multicellular filaments called hyphae; each hypha is a multinucleate, walled cell. The nuclei of a hypha go through mitotic cycles, and, when a hypha grows large enough, it divides; the number of haploid nuclei per cell is indeterminate. An entire pad of hyphae is the mycelium. In a second mode of asexual reproduction, haploid spores called conidia arise by mitotic divisions at the ends of some hyphae. Conidia can disperse and grow asexually into new mycelia, each conidium producing a clone.

Conidia can also act as male sex cells. Mycelia can produce female sexual structures, called protoperithecia when they are immature and perithecia at maturity. The protoperithecium contains mating filaments called trichogynes. Fungi may be either **homothallic** (one mating type) or **heterothallic** (two mating types). *Neurospora* is heterothallic; the mating types, A and a, are determined by alleles. A conidium can fuse with a trichogyne, but they must be of opposite mating types to fuse. Conidial nuclei divide mitotically as they move along the cytoplasm of the multinucleate trichogyne. The male and female nuclei fuse to make a diploid nucleus. Each of these diploid nuclei forms a cell, the zygote, which is contained in a sac, the ascus.

The zygote immediately undergoes meiosis followed by one mitotic division, yielding eight ascospores. The ascus is a narrow sac, so that the products of a single meiosis and the subsequent mitosis line up in the exact order in which they arise. The four meiotic products, ordered tetrads, are therefore useful for analysis of segregation, assortment, linkage, and recombination. Many fundamental features of meiosis have been revealed in experiments with *N. crassa* – e.g., 1:1 segregation of alleles (homologous gene copies), recombination between a gene and the centromere of its chromosome, assortment of genes, and the occurrence of crossing-over at the four-strand stage.

Haplodiplontic Life Cycles

In haplodiplontic organisms, mitotic cell divisions occur in both the haploid and the diploid phases of the life cycle. Either or both stages can be multicellular.

Saccharomyces cerevisiae

Budding yeast – *S. cerevisiae*, a simple unicellular fungus – may undergo mitotic cell cycles during both the haploid phase and the diploid phase of the life cycle, depending on environmental conditions and genotype. Budding yeast is therefore haplodiplontic. Both diploid and haploid cells go through normal mitotic cycles, but cytokinesis is asymmetrical. One daughter cell is large and the other is small; the smaller is said to bud off the larger. Other taxa of yeast – the fission yeasts such as *Schizosaccharomyces pombe* – divide symmetrically and produce daughter cells of equal size.

Under good conditions, diploid *S. cerevisiae* proliferate by budding, but, under starvation, they undergo meiosis. In every diploid cell, the two copies of a mating type gene (*MAT*) are different – *MATa*/*MATα*. The *a* and *α* alleles segregate in meiosis to yield four haploid cells, two *a* cells and two *α* cells. The haploid cells either proliferate by mitotic cycles or else they **conjugate** (undergo syngamy). The two conjugants must be of opposite mating types,

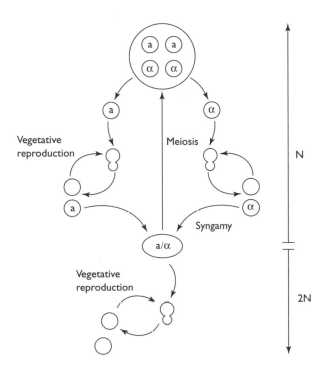

Fig 22.3 The life cycle of S. *cerevisiae*.

a and *α*. During conjugation, fusion of an *a* and an *α* cell is followed by fusion of the two nuclei, making a diploid nucleus (Figure 22.3).

Zea Mays

Many simple plants such as mosses, liverworts, and ferns are haplodiplontic. Botanists commonly refer to the haploid and diploid phases of haplodiplonts as "generations" and say there is "alternation of haploid and diploid generations." This contrasts with the genetic terminology of this book, where "generation" means the same thing in all sexually reproducing organisms; i.e., a generation is defined as one full turn of the sexual cycle, say from haploid to haploid. The half-cycle concept of haplodiplontic generation is akin to breaking up the 24-h period of the earth's rotation into two 12-h cycles, and calling the parts "day" and "night," while the universally applicable, full-cycle concept of sexual generation is analogous to calling one turn of the 24-h cycle a "day," say from midnight to midnight.

Flowering plants are also haplodiplontic, although the haploid phase is brief and entails few nuclear divisions. In both simple plants and flowering plants, the haploid phase is the **gametophyte** and the diploid phase is the **sporophyte**. Flowering plants are so

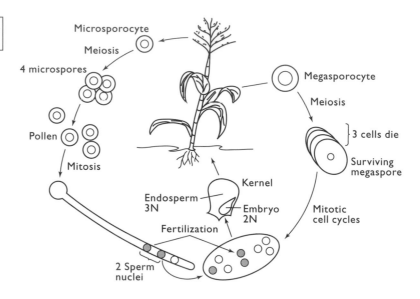

Fig 22.4 The life cycle of *Z. mays.*

important that a genetically well-known species, *Z. mays* (maize or corn), is the example here (Figure 22.4).

Z. mays is **monoecious**, which means that one individual sporophyte contains both male reproductive organs (stamens) and female reproductive organs (pistils). **Dioecious** plants such as the sugar maple, *Acer saccharum*, have stamens on male trees and pistils on female trees (trees are sporophytes).

In *Z. mays*, the diploid pistil makes an ear of corn. The pistil contains hundreds of megaspore-producing megaspore mother cells, each having a stigma; a stigma of *Z. mays* is the silk. The stamens of *Z. mays* are tassels, which grow at the top of the plant. The stamens contain pollen-producing microspore mother cells. Megaspore mother cells and microspore mother cells go through meiosis to produce haploid gametophytes – megaspores and pollen, respectively.

The haploid megaspore occupies the embryo sac. The megaspore divides mitotically to produce eight haploid cells, including three cells that contribute to the seed. Two of these cells fuse. The haploid microspore nucleus divides mitotically into two nuclei. One of the daughter nuclei becomes the large, transcriptionally active vegetative nucleus, while the other nucleus, surrounded by a membrane, becomes a generative cell. The generative cell divides mitotically to make two sperm cells. The pollen grain breaks free and air currents carry it to the corn silk, where it adheres and grows a tube into the silk. The vegetative nucleus and the sperm cells stay near the rapidly growing tip of the pollen tube. When the pollen tube reaches and grows into the embryo sac, double fertilization takes place.

One sperm cell fuses with one haploid cell of the megaspore to make the zygote, or diploid embryo nucleus, which will grow into a corn plant (sporophyte). The other sperm cell fuses with the diploid cell in the megaspore to make a triploid endosperm nucleus. This grows into parts of the seed surrounding the embryo. The sperm cell lacks mitochondria and chloroplasts, so inheritance of these organelles is strictly from the egg nuclei.

Diplontic Life Cycles

Nearly the entire life cycle of diplonts is diploid; the haploid cells of diplontic organisms do not undergo mitosis. Animals are diplontic, even though in some species most of an individual's cells are haploid; e.g., males of many wasps and bees. The life cycles of animals are amazingly diverse.

The life cycle of our own species, *Homo sapiens*, typifies diplonts. The focus of this description is on (1) meiosis and gamete development, and (2) fertilization. Two haploid gametes, egg and sperm, fuse during **fertilization**; then the two haploid nuclei fuse to make a diploid nucleus; the whole process is syngamy. The initial diploid cell is a **zygote**. The zygote goes through mitotic cell cycles, producing an embryo. Early in development, the embryo splits into two cell-lineages, **germ line** and **soma**. Germ line cells are the progenitors of gametes; soma give rise to body parts. The founder-cells of the germ line, known as gonia (= primordial germ cells), migrate to the undifferentiated gonad, where they take up residence. Thus, the gonad is a mixture of germ line and somatic cells. Gonia divide mitotically and develop into **spermatogonia** in testes and **oögonia** (the "o" with double dots is pronounced; thus ōōgōnee ah) in ovaries.

Both eggs and sperm develop by gametogenesis: oögenesis in the ovary and spermatogenesis in the testis (Figure 22.5). Gametogenesis includes mitotic cell division cycles, meiosis, and developmental changes in the products of meiosis.

Before birth, each human ovary makes $\sim 10^5$ primary oöcytes, which are not produced again by the ovary. Oögenesis begins before birth, from a lifetime supply of primary oöcytes. The primary

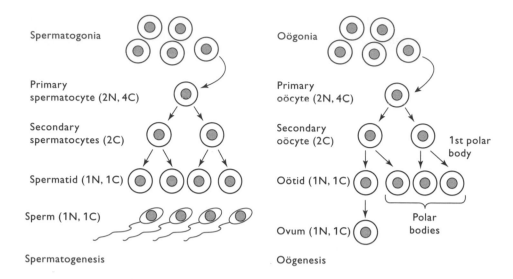

Spermatogonia

Primary spermatocyte (2N, 4C)

Secondary spermatocytes (2C)

Spermatid (1N, 1C)

Sperm (1N, 1C)

Spermatogenesis

Oögonia

Primary oöcyte (2N, 4C)

Secondary oöcyte (2C)   1st polar body

Oötid (1N, 1C)

Ovum (1N, 1C)   Polar bodies

Oögenesis

Fig 22.5 General plan and terminology of gametogenesis.

oöcytes begin meiosis synchronously, and at birth all the individual's primary oöcytes are arrested in diplonema, the fourth stage of prophase I. Each primary oöcyte grows large, develops special structures, amplifies its rRNA genes manyfold, and makes a large number of ribosomes. Somatic cells called follicle cells surround each oöcyte. At ovulation, which in humans is signaled hormonally and occurs about monthly from menarche to menopause except during pregnancy, a ripe follicle releases its primary oöcyte. The primary oöcyte moves to an oviduct (egg tube – called Fallopian tube in humans) and resumes meiosis I. A sperm fertilizes the secondary oöcyte in the oviduct, after which it undergoes meiosis II as it migrates to the uterus.

The male germ line is established in embryogenesis. Spermatogenesis takes place from puberty until old age. Primary spermatocytes are generated continuously from spermatogonia by mitotic division. During this reproductive period $\sim 10^8$ sperm are produced daily in the seminiferous tubules of the testis. Immature sperm are transported to tubes called the epididymis, where they mature into functional sperm. Sperm move from the epididymis via the vas deferens to the seminal vesicles, through the prostate gland, and from there into the urethra. Seminal fluids from accessory organs – seminal vesicles, ejaculatory ducts, and prostate gland – join the sperm to make semen. Semen is ejaculated into the female reproductive tract, and sperm move to the oviducts. Fertilization takes place in an oviduct (Figure 22.6).

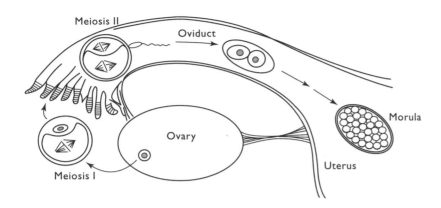

During fertilization, the sperm and egg cytoplasm mix and the sperm and egg nuclei (sometimes called **pronuclei**) fuse, so that the zygote's nucleus is 2N. The sperm carries $<10^2$ mitochondria, but the egg has $\sim 10^5$ mitochondria. This gives the inheritance of mitochondria a strong maternal bias. In addition, mitochondria of the sperm are usually destroyed in the zygote.

Fig 22.6 Ovulation and fertilization. The primary oocyte moves into the oviduct; a sperm fertilizes the secondary oocyte as it migrates; sperm and egg nuclei fuse to make a diploid zygote, which migrates toward the uterus and undergoes cell division as the embryo develops.

Asexual Reproduction

Most single-celled eukarya reproduce primarily asexually, via mitotic cell cycles. Occasionally they go through a brief sexual phase. Asexual reproduction, common in multicellular fungi, plants, and animals, can be by the following:

- Spores
- Multicellular reproductive tissues
- Broken parts of the organism
- Fission of the whole organism
- Parthenogenesis (virgin birth)

Many fungi make haploid asexual spores. Specialized tissues for asexual reproduction of plants include runners, corms, tubers, or suckers. Many plants can generate a whole organism from a broken part or cutting – leaf or branch – as can simple invertebrates such as flat worms and starfish. Cniderians (e.g., sea anemones) can reproduce by fission of the whole animal. Except for certain kinds of parthenogenesis, asexual reproduction produces clones – populations of genetically identical individuals.

Parthenogenesis

Types. Parthenogenesis, production of an embryo from an unfertilized egg, occurs widely in plants and animals. One way to classify parthenogenesis is according to the mechanism of sex determination:

- **Arrhenotoky** – unfertilized eggs develop parthenogenetically into males and fertilized eggs develop into females (many bees, wasps, and mites)
- **Thelytoky** – unfertilized eggs develop into females (aphids, a few lizards, snakes, frogs, salamanders, birds, and plants)
- **Deuterotoky** – unfertilized eggs develop into either females or males (aphids, some mites)

Parthenogenesis may also be classified according to genetic mechanism:

- **Haploid parthenogenesis** – parthenogenetically produced individuals are haploid
- **Diploid parthenogenesis** – parthenogenetically produced individuals are diploid
- **automixis** – meiosis is normal, but the diploid number is produced by fusion of two haploid nuclei
- **apomixis** – meiosis is skipped or modified, so that the egg has an unreduced chromosome number

Haploid Parthenogenesis. The honey bee, as well as other bees and wasps (order Hymenoptera), reproduces by haploid parthenogenesis: fertilized eggs develop into females and unfertilized eggs develop into males. Primary oocytes, being diploid, go through normal meiosis, but primary spermatocytes, being haploid, develop into sperm by mitotic cell division. Sisters share three-quarters of their genes with each other, versus only one half of their genes with their mother. Brothers and sisters share only one-quarter of their genes. Males do not carry lethal mutations, unlike most animals, which are diploid and do carry lethal mutations.

Automixis. Automixis – fertilization of an egg by a second polar body (the two daughter cells of the secondary oocyte), following normal meiosis – is not a common form of reproduction (Figure 22.7). The cells that fuse to make a zygote are 1N and 1C, and, because of crossing-over, they are genetically different, so

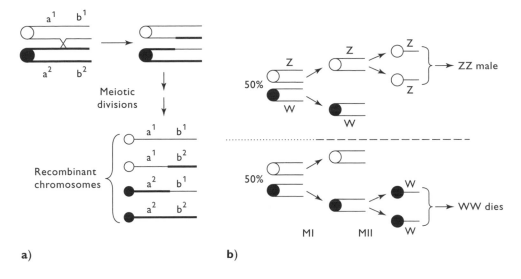

a)

b)

that the offspring do not comprise a clone. Facultative automixis (some sexual reproduction and some reproduction by automixis) is found in plants, some annelid worms, and a few avian species. The offspring of automictic birds are all males, which works in the following way. The sex of birds is determined chromosomally: females are ZW and males are ZZ. The Z and W chromosomes pair and segregate in meiosis I, so that the secondary oocyte has either Z or W; half the zygotes are ZZ and half are WW. ZZ embryos develop into males, and WW embryos die in automictic species.

Fig 22.7 Automixis does not produce clones (**a**). All-male broods of automictic birds (**b**). Egg and second polar body fuse; 50% of meioses →ZZ and 50%→WW.

Apomixis. Females reproducing by apomixis naturally clone themselves, because meiosis is skipped and therefore all the eggs are genetically identical (barring mutation) and diploid. Apomixis is found in diatoms, aphids, crustaceans, mites, ticks, lizards, frogs, salamanders, birds, and plants (notably in the grasses and the sunflower family).

There are two kinds of life cycle in aphids (a family of 4400 species): entirely apomictic and alternation of sexual and apomictic generations. The latter is more prevalent and presents an interesting biological problem – how can males be produced by apomictic parthenogenesis? Sex determination in aphids is chromosomal: females have two X chromosomes (♀♀ are XX) and males have one (♂♂ are "XO"). During the summer female aphids lack meiosis and produce apomictic female (XX) clones. At summer's end, in response to short days and cool nights, female aphids undergo "minimeiosis" that operates only on the X chromosome. The eggs made by this modified meiosis all have one X chromosome and develop into males. Females and males mate and reproduce sexually. The offspring of these matings survive the winter as eggs, finish development in the spring, and found apomictic clones the next summer.

Gynogenesis and Hybridogenesis.   Some animals can reproduce by apomixis, but eggs must be "activated" by sperm in order to develop into embryos. This is gynogenesis, found in some fish, amphibians, and plants. The viviparous "Amazon mollies" (genus *Poecilia*), commonly raised in aquaria, reproduce apomictically but mate with males of other species (they have internal fertilization). The molly eggs and foreign sperm unite, but egg and sperm nuclei do not fuse. Gynogenetic populations of some economically important fish (e.g., bass) have been made experimentally as a way of maintaining clones with desirable traits.

Hybridogenesis is the production of a diploid interspecific hybrid that discards one of the parental genomes during meiosis. It occurs in *Poecilia* and water frogs. Interspecific progeny of the frogs *Rana lessonae* (genome A) and *Rana ridibunda* (genome B) have genome AB. During meiosis, the father's genome is discarded, and the hybrid makes eggs containing only its mother's genome; genomes A and B do not recombine.

Further Reading

Blackman RL. 1987. Reproduction, cytogenetics, and development. In: Minks AK, Harmwijn P (eds.). *Aphids. Their Biology, Natural Enemies, and Control.* Elsevier, Amsterdam.

Calzada J-P, Crane CF, Stelly DM. 1996. Apomixis: the asexual revolution. *Science* 274:1322–1323.

Valero M *et al.* 1992. Evolution of alternation of haploid and diploid phases in life cycles. *Trends Ecol. Evol.* 7:25–29.

Wassarman PM. 1999. Mammalian fertilization. *Cell* 96:175–183.

Chapter 23

Reproduction of Viruses

Overview

Viruses are acellular, intracellular parasites whose reproduction depends in part on the host's genetic machinery. Viral infection may kill a cell; at the very least viruses are a burden to their host. All organisms are susceptible to viral infection, and by conservative estimate there are 10^8 different kinds of virus. The ubiquity, variety, and destructiveness of viruses have motivated study of their reproduction. In addition, viruses are extremely useful *per se* in the study of genetics (replication, mutation, recombination, and gene expression), as tools in molecular biology, medicine, and agriculture, and as "windows" into the biology of the host organisms.

   This chapter summarizes viral life cycles. A virus's host and genetic material strongly influence its mode of reproduction; accordingly, the chapter is organized both by host and by genetic material – RNA or DNA, single- or double-stranded.

A Little Terminology

The (+) strand of an RNA virus genome can be translated into proteins, and the (−) strand is complementary to the (+) strand; essentially, the (+) strand is mRNA. The (+) strand of a DNA virus corresponds to the mRNA sequence, while the (−) strand is the template for RNA synthesis. The chromosome, which may be circular or linear, may contain a small amount of protein.

Bacteriophages, the Viruses of Bacteria

Bacteriophages (also simply **phages**) are viruses that infect bacteria. Phages consist of a chromosome contained within a protein capsid, ranging in length between 10 and 300 nm.

Reproduction of the phages infecting *Escherichia coli* are described here; they appear to be representative of phages generally. Because many different phages specifically infect this bacterium, it is reasonable to suppose that the number of kinds of phages far exceeds the number of species of bacteria. This relationship appears to be paralleled by viruses infecting the other two domains of organisms.

Bacteriophages capable of killing their hosts fall into two categories, **virulent** and **temperate**. Virulent phages infect a cell, reproduce, lyse the host, and release progeny. Most temperate phages can exist in two alternative modes of life cycle, **lytic** and **lysogenic**. In the lytic mode, reproduction is like that of virulent phages. In the lysogenic mode, the phage genome is quiescent and either integrates into the host chromosome by recombination or else reproduces as a plasmid. Some phages cannot kill their hosts; their nonlethal means of egress is filaments that extend through the cell envelope.

The phage genome is a single chromosome, consisting of either single-stranded RNA, single-stranded DNA (ssDNA), or double-stranded DNA (dsDNA).

RNA Phages

RNA phages, including MS2, f2, and Qβ, have a (+) single-stranded chromosome, 3 to 4 kb long. Typically, the phage capsid contains one chromosome and a viral RNA-dependent RNA polymerase. The chromosome enters the cell through an F pilus, a plasmid-encoded microtubule (inner diameter = 2 nm), that extends from the cell envelope. Upon entry, the chromosome is translated into coat protein, replicase, lysin, and adsorption protein. Viral proteins help regulate translation. In phage MS2, the replicase forms a multimer with three cellular proteins – ribosomal protein S1, EF-Tu, and EF-Ts – to become fully active. The (+) strand is a template for the synthesis of a (−) strand and vice versa. Many copies of the virion assemble within minutes of initial entry. Lysin facilitates cell lysis by an unknown mechanism, releasing hundreds of mature viruses. The chromosome lacks telomeres, as replication of linear RNA does not present the same difficulty as replication of a linear DNA chromosome.

Single-Strand DNA Phages

These include the filamentous phages (M13 and Ff) and small polyhedral phages (e.g., ϕX174). Filamentous phages do not lyse the host, but the polyhedral ssDNA phages do kill the host. The genome size of ssDNA phages is 5 to 7 kb.

The chromosome of filamentous phages enters through an F pilus and progeny virions exit directly through the cell envelope without lysis. Host enzymes replicate the chromosome. First RNA polymerase synthesizes a short primer; then the replisome completes synthesis of the (−) strand. The resulting dsDNA is called the **replicating form (RF)**. Many copies of RF are made by rolling circle replication from the original RF. Viral RNAs are transcribed from these RFs. Finally, many copies of the (+) strand are synthesized by rolling circle replication from the RFs. Phage M13 is useful as a cloning vector; i.e., it is used by experimenters to produce copies of DNA.

The 5.4-kb chromosome of ϕX174 encodes 11 polypeptides, an efficiency due to overlapping mRNAs and overlapping reading frames within mRNAs. Some DNA sequences code for parts of three different polypeptides (Figure 23.1). Upon cell lysis hundreds of progeny phages are released.

Double-Stranded DNA Phages

Phages of this category abound. T-odd phages – T1, T3, T5, and T7 – have medium-sized, linear chromosomes (40 to 120 kb), mostly with small terminal redundant sequences. T-even phages – T2, T4, and T6 – have large chromosomes ($\approx$165 kb) with $\approx$5-kb terminal redundant sequences. All the T phages are virulent. The temperate phages of *E. coli* (λ, P22, P2, P4, P1, and Mu) alternate in reproductive mode between lytic and lysogenic; they range in size from 12 to 88 kb. dsDNA phages vary in size, complexity, suppression of host functions, and dependency on the host for transcription and replication machinery.

The 48-kb chromosome of phage λ has 12-nt single-stranded 5′ ends called cohesive sites (*cos*) because the base sequences of

Fig 23.1 A viral DNA sequence containing two open reading frames (ORFs) in one strand and one ORF in the other.

	amino acid sequences
	1 = MDPWPDYFR..
	2 = MARLFQ..
	3 = MGPSG..

the two ends are complementary. The capsid consists of an icosahedral head, which houses the chromosome, and a hollow tail, which is adsorbed by the cell envelope. The capsid is made of many different proteins. After the receptor (a protein of the outer membrane that transports maltose) adsorbs the tip of the phage's tail, the chromosome is injected into the host, perhaps along with a protein that facilitates injection. Shortly after the DNA enters the cell, the single-stranded 5' cohesive ends base pair and the ends are ligated, making the linear chromosome circular. The phage enters either the lysogenic or the lytic mode of reproduction, depending on many viral and host factors (Figure 23.2).

In the lysogenic state the genes for lytic reproduction are repressed, and the λ chromosome integrates into the host chromosome by site-specific recombination. The integrated λ chromosome, now called a **prophage**, is replicated by the host.

Mutagens or other inhibitors of DNA synthesis induce the lytic mode of reproduction. In lytic reproduction, a subset of phage genes are expressed, and viral proteins excise the λ chromosome as a closed circle of DNA. Early replication of the λ chromosome is by the bidirectional theta mechanism. Transcription is initiated near the origin of replication with host RNA polymerase; then two viral proteins bind to the origin. DNA synthesis uses host proteins – primosome, single-stranded DNA binding protein, DNA polymerase III holoenzyme, gyrase, and dnaB helicase. When several copies of the λ chromosome have been made, late

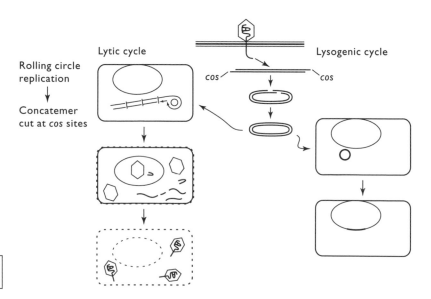

Rolling circle replication

Concatemer cut at cos sites

Lytic cycle

Lysogenic cycle

cos   cos

Fig 23.2 Life cycle of λ phage.

replication ensues by the rolling circle mechanism. The products of rolling circle replication are linear concatamers. Viral terminase enzyme cleaves the concatamers, making staggered cuts at cohesive end sites (*cos*). Host proteins IHF and THF help pack the linear chromosomes into the capsid. The cell lyses, releasing about 100 phages.

Viruses of Archaea

The viruses of archaea have dsDNA genomes. Both linear and circular chromosomes exist, 7 to 230 kb. The capsid of most archaeal viruses resembles that of λ phage. Both lytic and lysogenic life cycles are represented.

Viruses of Eukarya

Eukaryal viruses run the gamut in their genetic material – single-stranded or double-stranded, RNA or DNA, one or many chromosomes, small or big. All the examples given below are animal viruses, which represent much of the diversity in life cycles of eukaryal viruses. Much more is known about viruses that parasitize animals, plants, and fungi than those of early-branching eukarya (ancient lineages).

Viral genomes may consist of one or more chromosomes. A genome of one chromosome is **monopartite** (=**unsegmented**); a genome of several chromosomes is **multipartite** (=**segmented**). All viruses have a genome and a protein capsid. In addition, some viruses are **enveloped** in membrane stolen from the host. Viral proteins may be added to the envelope. Some viruses carry enzymes in the capsid.

Single-Stranded RNA Viruses That Replicate RNA→RNA

The entire life cycle takes place in the cytoplasm. Viruses of this kind fall into three categories, (+) **strand**, (−) **strand**, and **ambisense**. In ambisense viruses, both the parental and complementary strands contain mRNA sequences. Ambisense viruses (e.g.,

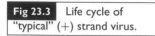

Fig 23.3 Life cycle of "typical" (+) strand virus.

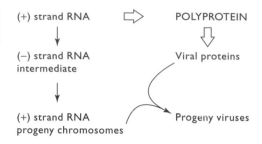

hantavirus, Lassa fever) have segmented genomes, and replicate like (−) strand RNA viruses.

(+) Strand RNA Viruses

Some (+) strand RNA viruses are enveloped (e.g., yellow fever, dengue, hepatitis C, rubella, and foot and mouth disease); others are not (e.g., poliovirus and hepatitis A); apparently, all are unsegmented. After entry into the cell, the parental RNA is translated. Viral proteins degrade proteins of the host's mRNA cap recognition factor, shutting off translation of host mRNA. Viral RNA contains a 5′-nontranslated region that binds to the 40S subunit to initiate translation. In some (+) strand viruses, parental RNA is translated into a single polyprotein, which is subsequently cleaved by viral protease into functional proteins that catalyze synthesis RNA from parental RNA and form the capsid (Figure 23.3).

(−) Strand RNA Viruses

The (−) strand RNA viruses are either unsegmented (e.g., mumps, measles, Ebola, and rabies) or segmented (e.g., influenza). After entry into the cell and uncoating, viral mRNAs are synthesized for each gene by a viral RNA polymerase/replicase enzyme. The polymerase/replicase of nonsegmented (−) strand RNA viruses also caps, methylates, and polyadenylates the mRNAs; mRNA of segmented (−) strand RNA viruses do not cap or polyadenylate mRNAs. When replicating the chromosome(s), the viral replicase synthesizes intermediate (+) strand template RNA from the parental (−) strand, and then synthesizes progeny (−) strand chromosomes from the (+) strand intermediates (Figure 23.4).

Double-Stranded RNA Viruses

dsRNA viruses commonly have segmented genomes. The entire virion enters the cytoplasm, where (+) strand mRNA is transcribed

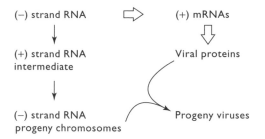

(−) strand RNA ⟹ (+) mRNAs

↓ ⇩

(+) strand RNA   Viral proteins
intermediate

↓

(−) strand RNA   Progeny viruses
progeny chromosomes

Fig 23.4 Life cycle of unsegmented (−) strand virus.

from each chromosome inside the partially opened capsid, using a viral transcriptase. Each species of mRNA is translated into a viral protein. The capsids assemble, viral mRNAs enter the capsids, and within the capsid a (−) strand of RNA is synthesized to yield progeny chromosomes. Thus, the life cycle resembles that of (−) strand RNA viruses.

Retroviruses

Retroviruses, including HIV and various oncogenic (cancer-causing) viruses infecting vertebrate animals, have small, unsegmented ssRNA genomes that code for dsDNA. They spend part of their life cycle in the nucleus, where the dsDNA copy of the genome integrates into the host's genome (Figure 23.5). Retroviruses are enveloped and therefore enter the cell by fusing the viral and cell membranes. The capsid contains two copies of the chromosome, reverse transcriptase (RNA-dependent DNA polymerase), tRNAs from the previous host, and ribonuclease H. A complex of tRNA and reverse transcriptase binds to the genomic RNA and synthesizes a complementary (−) strand of DNA, yielding an RNA-DNA duplex. Ribonuclease H, an enzyme that specifically attacks the RNA of RNA-DNA hybrids, digests the RNA strand, and reverse transcriptase directs the synthesis of a (+) strand of DNA.

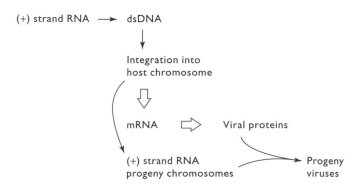

(+) strand RNA ⟶ dsDNA

↓

Integration into
host chromosome

⇩

mRNA ⟹ Viral proteins

(+) strand RNA   Progeny
progeny chromosomes   viruses

Fig 23.5 Life cycle of retroviruses.

The dsDNA copy of the chromosome enters the nucleus and integrates into the host genome by site-specific recombination, using host integrase enzyme. The provirus (integrated viral genome) may be quiescent for a time. At a later time, the viral genes are transcribed into genome-length RNA molecules and shorter mRNAs that are translated into viral proteins. Virions assemble in the cytoplasm and are extruded from the cell, enveloped in a bit of cytoplasmic membrane.

Single-Stranded DNA Viruses

ssDNA viruses that infect animals (parvoviruses) are uncommon. These viruses replicate in the nucleus, some independently and some with the aid of a second, helper virus (a dsDNA virus such as adenovirus). The telomeres of the viral chromosome contain self-paired hairpin loops that prime the synthesis of daughter chromosomes. Host RNA polymerase synthesizes mRNAs.

Double-Stranded DNA Viruses

This is a large, diverse group of viruses, including adenovirus (common cold), pox, hepatitis B (partly dsDNA and partly ssDNA) and herpes. Some reproduce in the nucleus, others in the cytoplasm, and a third group in both the nucleus and cytoplasm. The genomes are heterogeneous in size (2 to 280 kb) and monopartite, and some are enveloped while others are not. Two rather different life cycles are described here, those of herpes simplex and hepatitis B.

Herpesviruses are enveloped and have 120-kb to 240-kb linear chromosomes; they reproduce in the cell's nucleus and use the host's RNA-synthesizing machinery (Figure 23.6).

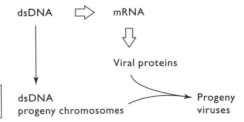

Fig 23.6 Life cycle of herpes simplex, a dsDNA virus.

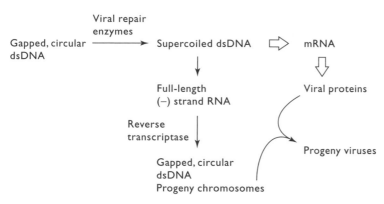

Gapped, circular dsDNA →(Viral repair enzymes)→ Supercoiled dsDNA ⇒ mRNA

Supercoiled dsDNA → Full-length (−) strand RNA

mRNA ⇓ Viral proteins

Full-length (−) strand RNA →(Reverse transcriptase)→ Gapped, circular dsDNA / Progeny chromosomes

→ Progeny viruses

Fig 23.7 Life cycle of hepatitis B (hepadnavirus).

Hepadnaviruses, including hepatitis B, have small (about 2 to 4 kb), circular, gapped genomes (Figure 23.7). After the virus enters the host's nucleus, viral repair enzymes fill in the gaps and make a supercoiled circle, which is transcribed into viral proteins. The genome is also copied as a full-length (+) strand of RNA. The viral (+) strand mRNA enters a new capsid and is replicated using viral reverse transcriptase, first into a DNA-RNA hybrid, then a DNA chromosome. Viral ribonuclease H digests the RNA in the DNA-RNA hybrid before the new (+) strand of DNA is synthesized.

Further Reading

Calender R (ed.). 1988. *The Bacteriophages*. Plenum Press, New York.

Chinchar VG. 1999. Replication of viruses. In: Webster RG, Granoff A (eds.). *Encyclopedia of Virology*, 2nd ed., vol. 3, p. 1471–1478, Academic Press, San Diego.

Roizman B. 1990. Multiplication of viruses. In Fields BN, *et al.* (eds.). *Virology*, 2nd ed., Raven Press, New York.

Chapter 24

Genetic Processes in Development

Overview

When a multicellular organism develops from a fertilized egg, four genetically regulated processes are at work: cell proliferation, programmed cell death, differentiation, and association of functionally related cells.

Genetic mechanisms of development are of two kinds, differential gene expression and changes in genome structure. Differential gene expression – spatial and temporal variation in the rates of synthesis of gene products – is by far the more prevalent. Gene expression, often triggered by signals from outside the cell, is regulated at the level of transcription, RNA processing, or protein synthesis.

Basic Developmental Processes

Cell Proliferation

Cell proliferation via cell cycling (Chapter 19) is universal. In plants and animals, cells proliferate when the organism grows, replaces dead cells, metamorphoses (remodels the body during development), or regenerates lost body parts. In the first 270 days or so of human development, cell number increases exponentially from 1 to $\sim 10^{14}$; the rate of cell proliferation far exceeds the rate of cell death. By conservative estimate, 99.9% of human cells die and are replaced, making a person's lifetime cell number $\sim 10^{17}$, although the true figure may be orders of magnitude greater than this.

The $G_1 \rightarrow S$ transition, blocked by RB-E2F, is critical to cell proliferation (Chapter 19); once a cell passes the R point it is committed to enter S phase, and passage through a full cycle normally occurs. External signaling molecules influence the $G_1 \rightarrow S$ transition: **mitogens** stimulate the transition and growth inhibitors block it.

Growth factors (e.g., epithelial growth factor [EGF], insulin-like growth factor-1 [IGF-1]) are protein mitogens that bind to receptors on the external cell surface, thereby activating a signal transduction pathway that ends in the transcription of genes encoding cell cycle regulators. These regulators cause increases in CDK4-cyclinD and CDK2-cyclinE, which together stimulate cell division by phosphorylating and thus inactivating RB. Some mitogenic hormones, such as estrogen (17β-estradiol) stimulate cell proliferation by entering the cell and binding to receptors in the cytoplasm; the receptor-hormone complex enters the nucleus and activates transcription of cell cycle regulators, either directly or indirectly. Growth inhibitors cause increases in proteins that block G_1 CDK-cyclins.

Programmed Cell Death

Cells can die by trauma, but usually they die by programmed cell death, called apoptosis [ăp·ō·tō·sis] in animals. Apoptosis rids the organism of damaged, infected, or cancerous cells and causes cells to die after their developmental role has been completed. During apoptosis, a cell destroys itself enzymatically. When other cells signal a cell to enter apoptosis, it may be viewed as "assisted suicide."

During apoptosis, endonucleases cut nuclear chromatin, which condenses in deposits near the nuclear envelope; following this the nucleus breaks down and the degraded chromatin is packaged in bits of membrane. Proteases cut proteins, molecular debris is packaged in vesicles, blebs appear on the cell membrane, and the cell shrinks and becomes a target for phagocytes (cell-devouring cells).

The instruments of cell death comprise two classes of enzyme: highly specific proteases called **caspases** (cysteine-aspartic acid proteases) and DNA endonucleases; mammals have 14 caspases and several endonucleases. The endonucleases nick nuclear chromosomal DNA at ≈200-nt intervals, preferentially in the

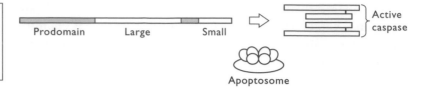

Fig 24.1 The apoptosome activates caspases by cleaving precursor polypeptides. Adapted from Adrain and Martin 2001.

linker region between core nucleosomes. Each caspase cuts target proteins at a specific sequence of amino acids. The expressed precursors of caspases are 30- to 50-kDa, inactive polypeptides with an N-terminal prodomain, a large-subunit domain, and a small-subunit domain (Figure 24.1). When a caspase is activated, the three domains are cut apart, the prodomain is degraded, the small and large subunits form a heterodimer, and two heterodimers form an active tetramer. Caspases are synthesized constitutively and are ever-present in the cell; hence, apoptosis begins by caspase activation and not by transcription of caspase genes.

There are three pathways to caspase activation: one initiated by cellular damage, one triggered by death receptors, and one induced by killer T cells (Figure 24.2). Proteins comprising the Bcl-2 family (16^+ members in mammals) are key players in the promotion and inhibition of apoptosis; 6 proteins of this family (e.g., Bcl-2) inhibit apoptosis, and 10 members of this family (e.g., Bax, Bid) promote apoptosis. Mitochondria participate in two of the three pathways.

Fig 24.2 Three pathways to caspase activation. Adapted from Adrain and Martin 2001.

Severe DNA damage triggers an internal pathway via p53, which activates transcription of *bax*. Bax causes the mitochondrion to leak cytochrome *c*. Apaf-1, cytochrome *c*, and caspase-9

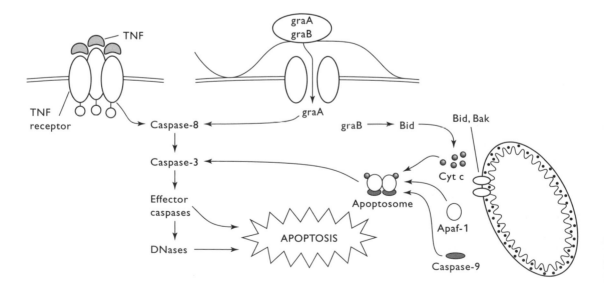

form a 700^+-kDa multimer, the **apoptosome**. Apoptosomes activate caspases in the cell by cleaving the inactive precursor polypeptides. Bcl-2 inhibits leakage of cytochrome c from the mitochondrion and also inhibits apoptosome formation; proapoptotic factors such as Bax inhibit Bcl-2.

Killer cells – cytotoxic T lymphocytes (CTLs) and natural killer (NK) cells – are the professional hit men of the immune system. CTLs and NK cells target tumor cells and cells infected by viruses or other parasites, and they can induce apoptosis by releasing death-inducing peptides called tumor necrosis factors (TNFs) or two proteases, granzymes A and B (graA and graB). TNFs bind to the extracellular part of TNF receptors (complex transmembrane multimers) thereby activating caspase 8, which begins apoptosis by activating other caspases. Granzyme B travels into the target cell's cytoplasm, where it activates caspases, triggering apoptosis.

Cell Differentiation

Cells range in degree of differentiation (specialization). **Stem cells** are unspecialized, and give rise to fully **differentiated** cells. A cell, whether it be specialized or unspecialized, is **totipotent** if it can develop into a reproductively competent multicellular organism, and it is **pluripotent** if it can develop into many specialized cells types, yet is not totipotent. Embryonic stem cells (**ES cells**) are totipotent cells from a very early animal embryo.

The zygote is the original stem cell; undifferentiated and totipotent, it gives rise to all the differentiated cell types of a multicellular organism. A differentiated cell has a specialized phenotype, which reflects the particular genes being expressed in that cell: each cell type expresses a subset of the genome needed to produce the phenotype of that cell type. The primary genetic cause of phenotypic specialization is regulated gene expression, often at the level of transcription or RNA processing. Some genes are expressed in all tissues, and other genes are expressed in a tissue-specific manner. Mammals have $\sim 10^4$ universally expressed genes. For each of the >100 different mammalian cell types, the average number of different gene products that are specific to cell type and to developmental stage exceeds 1000. Brain cells have the largest repertory of tissue-specific mRNAs, over 10,000.

Cells of a multicellular organism differentiate in response to asymmetries in cytoplasmic inheritance during cell division and to intercellular signals (hormones or proteins on the surfaces of adjacent cells).

In animals, cell differentiation is normally a one-way process; fully differentiated cells do not normally dedifferentiate or redifferentiate. For example, a hematopoietic stem cell gives rise to all the types of blood cells: mature B cells, T cells, macrophages, and red blood cells. However, a B cell never becomes a hematopoietic stem cell, nor does it become a macrophage.

Three dramatic exceptions to the "rule" against dedifferentiation and redifferentiation in plants and animals are regeneration, cancer, and clonal reproduction. When an annelid worm is cut transversely in two, the tail portion grows a new head portion and vice versa. When an adult salamander's leg is amputated, a new limb grows, perfectly replacing the missing part. Tumor cells often dedifferentiate and redifferentiate, by unknown mechanisms. Many species of cnidaria (e.g., sea anemones and hydras) can reproduce asexually by splitting down the middle. Flowering plant species, too, commonly reproduce by suckers, runners, and other asexual methods. The artificial asexual reproduction of useful plants from plant parts is an ancient practice, from which the term "clone" originated. A leaf or branch, cut from a plant, can easily be induced to form a new daughter plant, which is a genetic duplicate of the original, barring mutation and organelle heterogeneity. It is even possible to grow entire plants from single, differentiated, somatic cells; the rate of success is very high, often > 90%.

The artificial cloning of vertebrate animals has been done only recently and is more difficult than the artificial cloning of plants. Whole vertebrate animals have never been cloned from animal parts, say a foot, nor have they been cloned from a somatic cell. To clone a vertebrate animal, the nucleus from a somatic cell is transferred to an enucleated oocyte or ovum, which is then placed in an appropriate environment to develop. Frogs (*Xenopus laevis*) have been cloned by injecting nuclei from tadpole intestinal epithelia into irradiated eggs (irradiation destroys the nucleus but leaves the cytoplasm intact), decades before "Dolly," the celebrated cloned sheep. Sheep, cattle, and mice have also been cloned, by introducing nuclei from somatic cells of adults into enucleated eggs and then implanting the resulting embryos in the uteri of pregnant mothers. The sources of somatic cell nuclei have been mammary glands (sheep) and ovary (cattle, mice). It is not known whether these cells were fully differentiated.

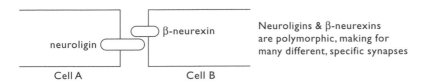

neuroligin

β-neurexin

Cell A   Cell B

Neuroligins & β-neurexins are polymorphic, making for many different, specific synapses

Fig 24.3 Junctions form between neurons when neuroligins bind to β-neurexins.

Cloned plants are phenotypically normal, while cloned animals have physiological and reproductive deficiencies. In animal cloning experiments to date, the success rate has been very low because fewer than 1% of the eggs with implanted nuclei have developed sufficiently even to be born.

Migration, Growth Pathways, and Cell Association

During development cells assemble into functional aggregates by migrating or by growing along guided pathways, and by forming junctions (Figure 24.3). Consider the following two examples.

Germ cells and soma of mammalian gonad originate in different parts of the embryo. The primordial germ cells (germ cell precursors) migrate from the allantois through the gut to the genital ridge, where they join somatic cells to form the gonad.

Neurons often become extremely elongated, forming nerves by cell growth. Neurons also make synaptic junctions with other neurons or with effectors (muscles or glands). As neurons grow, their path is guided by signals from adjacent cells. Junctions between cells depend on cell surface proteins, notably β-neurexin and neuroligin. Thousands of β-neurexins and hundreds of neuroligins make possible more than 10^5 different combinations, each specific to a theoretically separate class of synapse. How many genes encode this huge array of proteins? Merely six, three neurexin genes and three neuroligin genes. Each neurexin gene has two promoters and two primary transcripts. The six neurexin pre-mRNAs and the three neuroligin pre-mRNAs undergo alternative splicing. As a result, β-neurexins and neuroligins are highly polymorphic among cells. Junctions between cells in the brain form when the β-neurexin of one cell binds to the neuroligin of another cell.

Differential Gene Expression: Transcription

The Role of Transcription Factors

In animals and plants, most regulatory sequences – short DNA sequences to which transcription-activating proteins bind – are specific to tissues and to the stage of development. Transcription factors first bind to regulatory sequences and then stimulate the binding of RNA polymerase to the promoter. A protein-coding gene typically has several enhancers; a large genome therefore contains $\sim 10^5$ enhancers. Accordingly, many specific transcription factors (TFs) are required to recognize this multitude of enhancers. Over 1500 TF genes have been found in *Arabidopsis*, and to date the estimated number of TF genes in the human genome exceeds 2000. The potential number of unique TFs is likely to be higher than the number of TF genes, owing to the prevalence of alternative splicing of pre-mRNA.

TFs, being proteins, are themselves synthesized in a genetically regulated way. Signals external to the cell can trigger the activation or the synthesis of transcription factors. Such signals include hormones and proteins displayed on the surface of adjacent, touching cells.

Regulatory Sequences – Enhancers and Response Elements

Tissue-specific and development-stage-specific gene expression depends on the binding of transcription factors to enhancers and to response elements. A gene's enhancers are located within $\sim 10^1$ kb of its promoter and, by definition, are never located inside the promoter. Response elements, defined by coordinate regulation of functionally related genes, are binding sites for TFs that are similarly activated. For example, estrogen activates the TFs that bind to estrogen response elements (consensus sequence = AGGTCANNNTGACCT). Other examples are heat shock response elements (induced by high temperature), glucocorticoid response elements, and metal response elements (induced by specific metal ions). A gene's response elements are usually upstream of its promoter and within 1 kb upstream of the start site, but they may reside in the promoters themselves.

Facultative Heterochromatin

One way to regulate an entire chromosome is to render it transcriptionally silent by turning it into facultative heterochromatin (Chapter. 10). This happens to all but one X chromosome in somatic cells of female mammals. In somatic cells of male bugs in the family Leconoidae the entire paternal set of chromosomes becomes facultative heterochromatin.

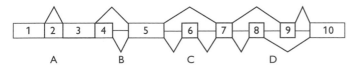

A 1-3 vs. 1-2-3 B 3-5 vs. 3-4-5 C 5-7 vs. 5-6-7 D 7-9-10 vs. 7-8-10

Fig 24.4 Some patterns of alternative splicing. Numbered boxes are exons, lines are introns.

Differential Gene Expression: Alternative Splicing of Pre-mRNA

Pre-mRNA may be alternatively spliced by the spliceosome. Splice variants may code for functionally distinct proteins, which expands the coding capacity of a single gene and increases the number and diversity of proteins encoded by a genome. Every intron has a 5′ splice site and a 3′ splice site, but various splice sites or combinations of splice sites may be included or excluded. A few of the possibilities are depicted in Figure 24.4.

In humans, more than half of protein-coding genes appear to have at least two splice variants. Splice variants may be produced at different times during development, in different tissues, or in different cells within a tissue. In each case, the different proteins have different functional qualities.

Activation of Gene Expression

How can specific cells be induced to proliferate, differentiate, or commit suicide? One way is by extrinsic signals, which trigger a response in the cell's protein-synthesizing machinery. Hormones, neurotransmitters, growth factors, growth inhibitors, and proteins in adjacent cells activate or repress gene expression. If the signaling molecule enters the cell, it binds to a receptor, which then becomes a TF. Alternatively, some signaling molecules bind to receptors in the cell membrane, thereby activating a signal transduction pathway whose end products are TFs.

A second way specific cells can be induced to respond differentially is via asymmetric inheritance of cytoplasmic factors – mRNAs and proteins that become concentrated in some daughter cells and not in others.

Activation of Transcription Factors
Peptide hormones, such as vasopressin, insulin, and growth factors, are genetically encoded extrinsic signals that can trigger

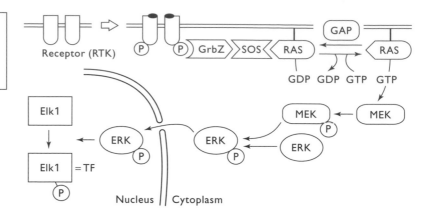

Fig 24.5 Ras signal transduction pathway. • = extracellular signal; phosphorylated Elk1 activates transcription of target genes.

differentiation or stimulate cell division. When the ligand (here, the signaling molecule) binds to the receptor, that receptor consequently triggers a signal transduction pathway that activates transcription. In such a signal transduction pathway, a **mitogen-activated protein kinase (MAPK)** activates a TF. MAPK pathways help regulate development in plants, animals, and fungi.

A typical MAPK pathway uses Ras, a membrane-bound G protein (guanine nucleotide binding protein) (Figure 24.5). Components of the Ras pathway are an external signal, a **receptor tyrosine kinase (RTK)**, an adapter protein such as Grb2, a guanine nucleotide release factor (GNRF) such as SOS (encoded by *Son of sevenless*), Ras, Raf (a MAPK kinase kinase, or MAPKKK), MEK (a MAPK kinase, or MAPKK), ERK (a MAPK), and a transcription factor. In many cases, the transcription factor initiates a cascade of transcription.

In response to ligand binding, the receptors dimerize, becoming an active, self-phosphorylating RTK. The phosphorylated RTK binds to the adapter protein, which binds to the GNRF, which activates Ras, which activates Raf. Raf phosphorylates MEK, which phosphorylates ERK, which enters the nucleus and there phosphorylates TFs, thereby activating them.

Asymmetrically Inherited Cytoplasmic Factors

A gene has a **maternal effect** if its phenotype depends on expression in the mother, usually oocytes and early embryos. The mRNAs and proteins encoded by some maternal-effect genes are distributed asymmetrically in the egg cytoplasm, which can result in a corresponding asymmetrical distribution of mRNAs and proteins in the early embryo. If one end of the embryo has a high concentration of an mRNA that encodes a transcription factor, and the other end has a low concentration of this mRNA, then target genes will be transcribed only at one end of the embryo, leading to early differentiation.

The *Drosophila* egg is an ellipsoid 0.5 mm long. The zygotic nucleus undergoes successive mitotic divisions without cell division, soon making a ball of nuclei enclosed in one cytoplasm. After nine divisions the 500+ nuclei migrate to the surface of the embryo, making the blastoderm. The multinucleate blastoderm is a layer of cytoplasm on the surface of the embryo, called the **syncitium**. Later, membranes grow between nuclei, producing a multicellular embryo. Very early, the *Drosophila* embryo gains anterior-posterior and dorsal-ventral polarities by generating concentration gradients of TF-encoding mRNAs within the continuous cytoplasm of the syncitium.

Before fertilization, each maternal mRNA has a characteristic distribution in the egg cytoplasm: bicoid (*bcd*) mRNA is concentrated in the anterior, nanos (*nos*) mRNA is concentrated in the posterior, while both hunchback (*hb*) and caudal (*cad*) mRNAs are evenly distributed. Both *hb* and *cad* are master regulatory genes. Consequently, Bcd and Nos proteins are concentrated in the anterior and posterior of the embryo, respectively. Bcd represses translation of *cad* mRNA, making the concentration of Cad protein higher in the posterior embryo. Nos protein represses translation of *hb* mRNA and Bcd activates transcription of the *hb* gene in anterior nuclei, making the concentration of Hb protein higher in the anterior embryo. Cad and Hb proteins are transcription factors that activate other genes. In this manner, the genes *bcd*, *nos*, *cad*, and *hb* participate in the beginnings of cell differentiation (Figure 24.6).

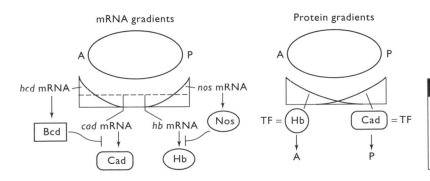

Fig 24.6   Molecular gradients determine polarity of *Drosophila* embryo. A = anterior, P = posterior, → = activate, ⊣ = inhibit. Vertical axes, amount of gene product.

Genomic Alterations during Development

Chromosome diminution, chromosome elimination, and polyploidy are kinds of chromosomal change that occur in the development of some organisms. Gene amplification and somatic recombination are associated with cell specialization.

Chromosome Diminution is the fragmentation of chromosomes during development, accompanied by massive loss of nuclear DNA. Chromosome diminution appears to be uncommon. It occurs in somatic cells of the worm *Ascaris*, the crustacean *Cyclops*, and hagfish. Chromosome diminution occurs in some ciliates, such as *Tetrahymena* and *Paramecium*. In the macronuclei of *Tetrahymena*, each of the $2N = 10$ chromosomes breaks into 50 to 200 microchromosomes about 600 kb long; more than 10% of the genome is eliminated. The microchromosomes acquire telomeres, do not acquire centromeres, and exist in multiple copies to compensate for the lack of regular mitotic segregation during nuclear division.

Chromosome Elimination, the wholesale loss of one copy of the genome from somatic cells, occurs in some flies and wasps early in development, leaving only the germ line cells with complete copies of the genome. The biological significance of chromosome elimination is unknown.

Many, if not most, species of plants and animals contain somatic tissues with polyploid cells, cells with more than two copies of the genome per nucleus – usually an even number. Polyploidy is achieved by modification of the cell cycle; for example, a diploid cell that omits karyokinesis from a cell cycle will become a tetraploid cell instead of giving rise to two diploid daughter cells. Larval salivary glands and brains in *Drosophila* contain polyploid cells, as do mammalian livers. In *Drosophila*, the chromosomes of polyploid cells are polytene, for the multiple copies of each homolog synapse during interphase. All other things being equal, a polyploid cell has a higher rate of gene expression, relative to a diploid cell, because there are more copies of each gene per nucleus.

Gene Amplification is the synthesis of multiple copies of a gene. Gene amplification can occur normally during development or in cancer cells. The oocytes of some vertebrates undergo gene amplification for rRNA genes. A strange and poorly understood kind of gene amplification in some animal embryos and in cells

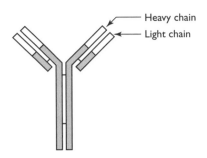

Heavy chain
Light chain

Fig 24.7 Structure of immunoglobin (1g) protein (antibody). Each 1g protein is a tetramer consisting of two identical light chains (polypeptides) and two identical heavy chains. Each chain is encoded by a single gene. Dark area in each chain is the constant region; light area in each chain is the variable region.

experiencing genetic instability (associated with cancer and aging) entails the manufacture of circular DNA: small, extrachromosomal circles of double-stranded DNA that are rich in multiple, tandem copies of repeated sequences.

Site-specific recombination can cause **allele-switching**, and can produce novel alleles or genes. In yeast, allele-switching at the mating type locus, causes the cell to change mating type. In trypanosomes, site-specific recombination generates novel alleles of the genes encoding proteins displayed on the cell surface, protecting cells from attack by the host's immune system. In cells of the vertebrate immune system, immunoglobulin (Ig) genes, which encode antibodies, are assembled by site-specific recombination from "prefab" DNA segments.

Immunoglobulins are proteins displayed on the outer surface of immature B cells and secreted by mature B cells (Figure 24.7). Every human possesses millions of distinct clones of B cells: each clone expresses a different Ig protein by virtue of having two unique, single-copy Ig genes (one allele per gene). Ig genes are manufactured in the precursors of B cells (pre-B cells); only the ("pre-gene") DNA segments used to make them, are inherited in the germ line. Ig proteins are tetramers consisting of two identical Ig **light chains** (about 220 aa/polypeptide) and two identical Ig **heavy chains** (about 440 aa). Disulfide bonds between cysteines join the four chains. Each chain has a **variable region** and a

Fig 24.8 Modular DNA segments.

Locus	Approximate number of:	V*	D	J	C	Combinations
Heavy (H)		200	12	4	9**	$\sim 10^4$
Kappa (κ)		300	0	5	1	1.5×10^3
Lambda (λ)		50	0	4	4	800

*The number of copies of V sequences per locus is not known precisely.
**The various C sequences are used in class switching, and do not contribute to the diversity of immunoglobins.

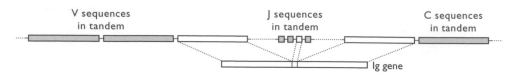

Fig 24.9 Site-specific recombination produces an Ig gene from V, J, and C sequences.

constant region. Three loci In the human genome contain segments used to make an *Ig* gene: heavy (H, chromosome 14q), kappa (κ, 2p), and lambda (λ, 22q).

In every pre-B cell, one copy of the heavy locus and one copy of either the κ or the λ locus (but not of both loci) undergo site-specific recombination to produce one heavy-chain gene and one light-chain gene. The restriction of the recombination event to a single allele encoding each polypeptide is called **allelic exclusion**: the possibility of forming a second allele is excluded. Each immunoglobin locus is about 2000 kb and contains modular DNA sequences, to encode the variable and constant portions of the Ig proteins. The types of modular DNA segments are **V** (**variable**), **D** (**diversity**, only at the heavy locus), **J** (**joining**), and **C** (**constant**) (Figure 24.8).

The potential number of distinct *Ig* genotypes is $\sim 10^7$. Site-specific recombination joins modular segments in a random combination in each pre-B cell (Figure 24.9). As the B cell matures, it undergoes **class switching**, in which the type of Ig changes, the types (or classes) being designated with letters: M, D, G, A, and E. Class switching from IgM to IgD is accomplished by alternative splicing of the Ig$_{heavy}$ pre-mRNA, while switching to the other classes of Ig is accomplished by further genomic rearrangement. Subsequently, *Ig* genes undergo a high rate of mutation ($\sim 10^3$ greater than normal), generating virtually infinite variation in *Ig* genotypes.

Further Reading

Adrain C, Martin SJ. 2001. The mitochondrial apoptosome: a killer unleashed by the cytochrome seas. *Trends Biochem. Sci.* 26:390–397.

Bashirullah A *et al*. 1998. RNA localization in development. *Annu. Rev. Biochem.* 67:335–394.

Coyne RS *et al*. 1996. Genome downsizing during ciliate development. *Annu. Rev. Genet.* 30:557–578.

Gilbert SF. 2000. *Developmental Biology*, 6th ed. Sinauer Associates, Sunderland, MA.

Gravely BR. 2001. Alternative splicing: increasing diversity in the proteomic world. *Trends Genet.* 17:100–106.

Lopez AJ. 1998. Alternative splicing of pre-mRNA. *Annu. Rev. Genet.* 32:279–305.

Missler M, Südhof TC. 1998. Neurexins: three genes and 1001 products. *Trends Genet.* 14:20–25.

Mizzen CA, Allis CD. 2000. New insights into an old modification. *Science* 289:2290–2291.

Pirrotta V. 1997. Chromatin-silencing mechanisms in *Drosophila* maintain patterns of gene expression. *Trends Genet.* 13:314–318.

Strasser A, *et al.* 2000. Apoptosis signaling. *Annu. Rev. Biochem.* 69:217–240.

Chapter 25

Sex Determination and Dosage Compensation

Overview

Most animal species and many plants are **dioecious**, meaning that there are two sexes, female and male. Sexual identity develops by genetically regulated pathways, but not by genes alone is any trait determined. Two principles are at work here. First, an environmental or genetic trigger channels development in a male or female direction. Second, the trigger initiates a cascade of gene-environment interactions that ends in the expression of mature sexual traits.

Species with XX females and XY males have a problem: both sexes have two copies of each autosomal gene, but females have twice as many copies of X-linked genes as do males. **Dosage compensation** solves the problem.

This chapter surveys mechanisms of sex determination and then compares both sex determination and dosage compensation in mammals and *Drosophila*.

Triggers for Sex Determination

Sex determination can be initiated by environmental factors, including temperature, chemical signals between individuals, and amount of light.

The developmental trigger in some fish and reptiles is incubation temperature. In some turtles, eggs incubated at above 30°C develop into females; eggs incubated at cooler temperatures develop into males. In other turtles, females develop at extreme incubation temperatures and males at intermediate incubation temperatures. Perhaps temperature-sensitive transcription factors bind to the enhancers or promoters of sex-determining genes.

Sex is determined by short-range chemical signals in some mollusks (e.g., the slipper limpet) and a marine worm *Bonellia*. In these animals, a larva that settles on a mature or maturing female becomes a male, while a larva that settles on the sea floor becomes a female.

Sex-determination in the horsetail *Equisetum* depends on light level. At light intensities above 24 μmol s^{-1} m^{-2} (equivalent to a bright reading light), most seedlings become females, while at light intensities below 6 μmol s^{-1} m^{-2}, most seedlings become males; the sex ratio varies continuously between those extremes of light intensity.

In many species of animals and plants, sex is determined by sex chromosomes, often called X and Y: females are XX and males are XY. Females are **homogametic**, for they produce only X-bearing eggs. Males are **heterogametic** sex, because the X and Y chromosomes pair and segregate 1:1 in meiosis, yielding equal numbers of X-bearing and Y-bearing sperm. In other species with chromosomal sex determination, females are heterogametic and males are homogametic; e.g., gallinaceous birds, moths and butterflies, and some fishes. To avoid confusion with the XX-XY system, females are ZW and males are ZZ. Grasshoppers, crickets, and roaches have an XX-XO system – females are XX and males have a single X.

Some dioecious species lack sex chromosomes; a single gene determines sex. Asparagus has a sex-determining gene, *M*, with two alleles (different versions), *M* and *m*. *M/M* and *M/m* individuals are male, while *m/m* individuals are female. *M/m* males sire females and males in a 1:1 ratio, but *M/M* males have only male progeny. The parasitic wasp *Habrobracon* has a sex-determining gene with many alleles. Individuals with two different alleles, such as A^1/A^2 are female, and individuals with two identical alleles, such as A^1/A^1 are male. Here, the sex ratio depends on the number of alleles and their relative population frequency; the more alleles, the more common are females, and the frequency of females is maximum when alleles are equally common (Chapter 37).

Sex Determination in Mammals

Primary Sex Determination

Primary sex determination is the development of ovary or testis from the indifferent gonad. The trigger for primary sex determination in placental mammals is the presence or absence of a Y chromosome. In humans, for example, XO persons are sterile females with rudimentary ovaries (Turner syndrome), and XXY persons are sterile males with nonfunctional testes (Klinefelter syndrome). In humans, then, the Y chromosome is necessary but not sufficient for full male development, and the absence of a Y chromosome is necessary but not sufficient for full female development.

The bipotential human gonad develops during week 5; it contains both somatic and germ line cells. The gonad develops into a testis under the control of **male-pathway genes** (**MPGs**). It develops into an ovary under the control of **female-pathway genes** (**FPGs**). Most MPGs and FPGs are autosomal. MPGs are activated and FPGs are repressed during testis development; FPGs are activated and MPGs are repressed during ovary development.

In week-7 males, testis determination is triggered by a regulatory protein encoded by the Y-linked **SRY** gene (SR = sex reversal), which is briefly expressed in the embryonic gonad. The 233-amino acid SRY protein is thought to be a transcription factor, as it has a DNA-binding domain. First to be determined are Sertoli cells, somatic testicular cells that send testis-determining signals to other gonadal cells.

In week-8 females, ovaries are determined. An early gene in the female pathway is **DAX1**, which inhibits expression of MPGs. SRY and DAX1 each inhibit expression of the other (Figure 25.1).

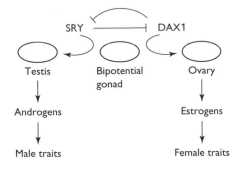

Fig 25.1 Determination of male and female traits.

Secondary Sex Determination

Once the sex of the gonad has been determined, it produces and secretes hormones that signal the embryonic development of the urogenital system, many of whose internal parts derive from the Wolffian ducts in males or from the Mullerian ducts in females.

In males, Sertoli cells secrete anti-Mullerian hormone (AMH), a 145-kDa glycoprotein, and Leydig cells of the testes secrete steroid androgens, notably testosterone. Testosterone stimulates development of the Wolffian ducts and AMH causes the Mullerian ducts to regress. Wolffian ducts become the vas deferens system of sperm ducts. Androgens signal development of other male fetal structures, such as the external genitalia. Androgens also influence brain development, resulting in male behavior.

Consequently, deficiency in an enzyme required for androgen synthesis can cause severe abnormalities in male development. Target cells take up **testosterone**, and most convert it to dihydroxytestosterone with the enzyme 5α-reductase. The active androgen, either dihydroxytestosterone or testosterone, binds to receptor proteins, which enter the nucleus. The receptor-hormone complexes stimulate transcription of male-determining genes. Mutations in the gene for 5α-reductase can dramatically affect the sexual development of males but not of females. Males deficient in 5α-reductase are born with ambiguous external genitalia and are sometimes raised as girls. At puberty, when testosterone production surges, external male genitalia suddenly grow. Mutations in genes that encode intracellular androgen receptors cause males to resemble females: they lack all male external genitalia and develop both female body form and female behavior despite high production of androgens in the undescended testes.

In females, the Wolffian ducts regress because androgen level is insufficient to maintain them, and the Mullerian ducts persist because AMH is absent (Figure 25.2). The Mullerian ducts develop into the oviducts, called Fallopian tubes in humans. A key gene in this process (studied in mice, also present in humans) is **Wnt-4,** which is essential for the development of the kidneys, formation and development of Mullerian ducts, steroidogenesis in the gonad, and development of oocytes. Recessive, null mutations of Wnt-4 masculinize females but have no effect on males. The ovaries later produce

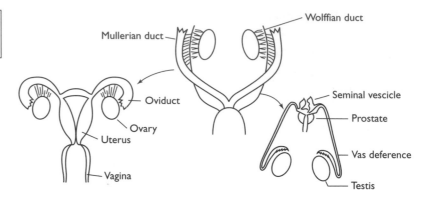

Fig 25.2 Development of female and male duct systems in mammals.

and secrete estrogens, notably estradiol-17β, which stimulate secondary sex characteristics, such as mammary gland development and female brain development (the basis of female sexual behavior).

Sex Determination in *Drosophila*

In the fly *Drosophila melanogaster*, females are XX and males are XY. The X chromosome carries about a fifth of the genome, while the Y chromosome carries very few genes. Having one X chromosome makes for male development, and having two X chromosomes makes for female development; the Y chromosome does not determine sex but is required for male fertility.

Somatic sex determination is a hierarchical process. At the top is the ratio of number of X chromosomes to number of sets of autosomes, the X:A ratio (1 in females, 1/2 in males), which is detected by X-linked numerator genes and autosomal denominator genes (sometimes gene is called element) (Figure 25.3). The X:A ratio determines whether a master regulatory gene, *Sex-lethal* (*Sxl*), is transcribed in early embryos. *Sxl* was discovered as

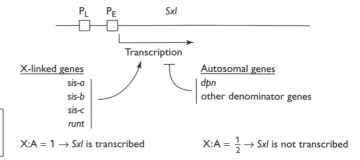

Fig 25.3 How *Drosophila* embryos count chromosomes (X:A ratio).

a mutation that kills females (the only surviving *Sxl* mutants are males).

After a few nuclear divisions post-fertilization, nuclei of the syncitium (a ball-shaped multinucleate layer of cytoplasm) "count" chromosomes to determine the X:A ratio. The X-linked numerator genes include runt and three genes called sisterless (*sis-a*, *sis-b*, and *sis-c*), which encode transcription factors that activate *Sxl*. The autosomal denominator genes include *deadpan* (*dpn*); they act in a negative way on *Sxl*. In the early female embryo, the numerator genes overpower the denominator genes and activate transcription of *Sxl*, but in the early male embryo, the denominator genes win out.

Sxl has two promoters, here designated early (E) and late (L): P_E is used in the early embryo, and P_L is used from mid-embryo to adulthood.

In females, the ratio of numerator genes to denominator elements is sufficiently high to activate transcription of *Sxl* at P_E. *Sxl* mRNA is translated into female-specific early SXL protein, a splicing factor. In the cellular blastula, *Sxl* is transcribed at P_L, and the transcript is spliced by early SXL protein to yield a female-specific late SXL protein, a splicing factor present in females from then on (Figure 25.4).

In male embryos, the denominator genes inactivate or repress the numerator genes, so that *Sxl* is not transcribed early. *Sxl* begins to be transcribed at P_L in all cells of the blastula, but since there is no SXL protein, the transcript is, by default, spliced in a male-specific pattern. The male-specific *Sxl* mRNA contains multiple stop codons in an upstream exon, which stops synthesis of late SXL protein (Figure 25.4).

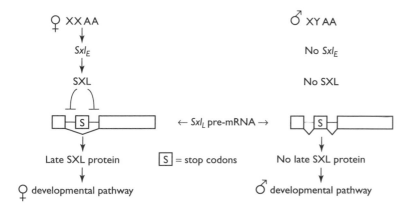

Fig 25.4 SXL protein in female and male *Drosophila*.

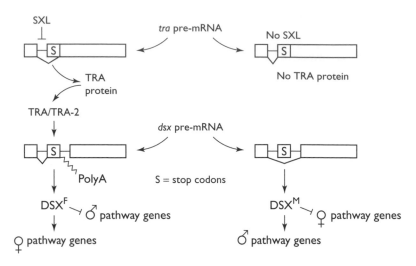

Fig 25.5 Some sex-determination genes in *Drosophila*. S = stop codons

Null mutations in two genes transform XX individuals into males, *transformer* (*tra*) and *transformer-2* (*tra-2*); mutations of these genes have no effect on XY individuals. SXL protein helps to splice *tra* transcripts into female-specific *tra* mRNA, which is translated into TRA protein. In the absence of SXL protein, *tra* RNA is spliced in a default male-specific pattern. The male-specific *tra* mRNA contains a stop codon that terminates translation early, and no TRA protein is synthesized.

Meanwhile, TRA-2 protein dimerizes with TRA. The TRA/TRA-2 dimer is a splicing factor that binds to the transcript of the next gene in the hierarchy, *doublesex* (*dsx*). Female-specific *dsx* mRNA is translated into a female-specific protein, DSX^F. Without TRA/TRA-2, the *dsx* transcript is spliced by default into a male-specific *dsx* mRNA, which is translated into a male-specific DSX^M protein (Figure 25.5).

Dosage Compensation

In XX-XY species, females have twice as many copies of each X-linked gene as males, but the two sexes both have two copies of each autosomal gene. The gene products of X-linked and autosomal genes work together, and for the most part their functions are unrelated to sex. To achieve the same balance of X-linked gene products and autosomal gene products in females and males, either expression of the female's X chromosomes must be halved or expression of the male's X chromosome must be doubled. The first of these happens in mammals and the second of these happens in *Drosophila*.

Mammals

To halve the rate of X-linked gene expression in female mammals, one X chromosome becomes inactivated. This makes the effective dosage of X-linked genes equal in the two sexes. The inactivation of an X chromosome is achieved by condensing it into heterochromatin, whose DNA is not transcribed. In the human XX embryo, one X chromosome is inactivated at the $\sim 10^3$-cell stage (2 weeks). In marsupials, the paternal X chromosome is inactivated, and the maternal X remains active. In placental mammals, one of the X chromosomes, either the maternal copy or the paternal copy, is randomly "chosen" for inactivation in each somatic cell. The inactive X of germ line cells soon becomes reactivated, but the inactivated X of somatic cells remains inactivated for the remainder of life. The heterochromatic X is a darkly staining blob in the nucleus of each somatic cell, called a **Barr body** after its discoverer. XO cells lack Barr bodies; an XXY cell has one Barr body; an XXX cell has two Barr bodies.

About 15% of genes on the X chromosome escape inactivation, although how this works is not known. On the short arm of the X chromosome, Xp, a higher proportion of genes (about 30%) escape inactivation than on the long arm, Xq (about 3%). For inactivated genes, female placental mammals that have genetically different X chromosomes are a random patchwork of cells, some expressing the maternal copy and others expressing the paternal copy. Female mammals are therefore effectively genetic **mosaics**, organisms in which a random subset of cells have one genotype and the other cells have another genotype. Mosaicism is apparent only for the few traits expressed in external parts. Calico cats, whose fur is a random pattern of black and orange colors, are females having different copies of an X-linked gene affecting coat color, the O gene. The O^+ allele makes for orange pigmentation, and the O^b allele makes for black pigmentation. O^+/O^b cats have patches of orange fur where the O^b allele is inactive and patches of black fur where the O^+ allele is inactive.

If X-linked mosaicism cannot be observed directly for a gene, then a female carrying one functional and one nonfunctional copy will appear completely normal. A human gene required for red color vision illustrates the point. The *RCP* gene on Xq encodes an opsin found in red-sensitive cone cells of the retina. A male with a defective copy of this gene (*RCP⁻*) has red-weak vision ("red color blindness"), while a heterozygous female (*RCP⁺/RCP⁻*) has normal color vision. The heterozygous female has the same ability to see color as the female with two normal copies of the gene, even though half of her "red" cones are nonfunctional. The functional red cones, millions of them in each eye, compensate for the nonfunctional ones, so that color vision is normal.

Inactivation of the X chromosome is signaled by transcription of *Xist*, which encodes a 17-kb, spliced, polyadenylated, nontranslated RNA (Figure 25.6). Inactivation begins in the X-inactivation center (*Xic*), a region containing the *Xist* locus. Each copy of the *Xist* gene is expressed. *Xist* RNA coats *Xic*. Meanwhile, an autosomal *Xist* blocking protein prevents one *Xic* from being coated by *Xist* RNA; the blocking protein works in an unknown way. On the blocked X chromosome, a DNA methyltransferase methylates CpG islands of the *Xist* gene, which silences its transcription. Thus, the presence of *Xist* RNA at the *Xist* locus ensures continuing transcription of the *Xist* gene, and the absence of *Xist* RNA leads to silencing of the *Xist* gene.

Only one *Xist* gene is silenced, though, for there is only enough blocking factor to block *Xist* RNA on one *Xic*. An X chromosome with a silenced *Xist* gene remains euchromatin, while an X chromosome with an active *Xist* gene is heterochromatized. The *Xist* RNA, which binds to chromosomal proteins, spreads and coats the entire X chromosome. Soon thereafter, each RNA-coated

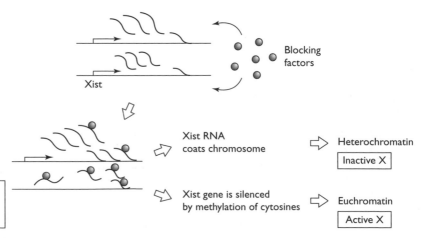

Fig 25.6 *Xist* and inactivation of the mammalian X chromosome.

X chromosome is inactivated. Once an inactive X chromosome turns into heterochromatin, transcription of the *Xist* gene stops.

XY cells have a single X chromosome, whose *Xist* gene is silenced, preserving its euchromatic status. In XX or XXY cells, one X becomes silenced and one does not (the choice of X^M or X^P for silencing is a random event). In XXX cells, two X's become Barr bodies.

Drosophila

The rate of X-linked gene expression in males is doubled, making the effective dosage of X-linked genes equal in the two sexes. Dosage compensation depends on five genes: *male-specific lethal-1, -2*, and *-3 (msl-1, msl-2,* and *msl-3*), *maleless (mle)*, and *males absent on the first (mof)* (Figure 25.7). The three MSL proteins form multimers that "paint" the male X chromosome, binding to hundreds of sites throughout the euchromatin, and attracting MOF, a histone acetyltransferase that acetylates histone H4. This loosens up chromatin structure and makes genes more accessible to transcription machinery. The MLE protein colocalizes with the MSL multimer on the X chromosome. MLE regulates X-linked gene expression. In one case, and maybe generally, MLE regulates gene expression by A-to-I RNA editing, which in turn affects the splicing of pre-mRNA (e.g., the transcript of the *para* gene, which encodes a sodium channel).

Females lack MSL, because the Sex-lethal protein (SXL), found only in females, binds to *msl-2* mRNA at 5'- and 3'-untranslated regions and blocks translation. This is an important example of

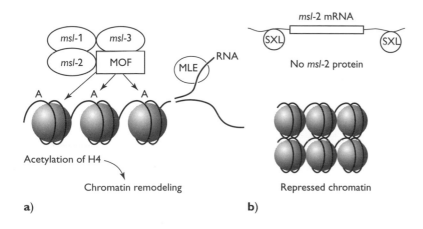

Acetylation of H4

Chromatin remodeling

a)

msl-2 mRNA

No *msl-2* protein

Repressed chromatin

b)

Fig 25.7 Dosage compensation in *Drosophila*. Activation in males (**a**). Blocking in females (**b**).

regulating gene expression by controlling translation rather than transcription or RNA processing.

Further Reading

Cline TW, Meyer BJ 1996. Vive la difference: males vs females in flies and worms. *Annu. Rev. Genet.* 30:637–702.

Kelley RL, *et al.* 1997. *Sex lethal* controls dosage compensation in *Drosophila* by a non-splicing mechanism. *Nature* 387:195–198.

Plath K, *et al.* 2002. *Xist* RNA and the mechanism of X chromosome inactivation. *Annu. Rev. Genet.* 36:233–278.

Stuckenholz C, *et al.* 2000. Guilt by association: non-coding RNAs, chromosome-specific proteins and dosage compensation in Drosophila. *Trends Genet.* 16:454–457.

Swain A, *et al.* 1999. *Dax1* antagonizes Sry action in mammalian sex determination. *Nature* 391:761–767.

Vaiman D, Pailhoux E. 2000. Mammalian sex reversal and intersexuality: deciphering the sex-determination cascade. *Trends Genet.* 16:488–494.

Chapter 26

Cancer

Overview

Cancer, a group of genetic diseases, is development gone wrong in a clone of somatic cells – a **tumor**. If a tumor destroys adjacent tissue it is **malignant**. Tumor cells:

- Accumulate mutations and become genetically unstable
- Grow in an unregulated manner
- Lose contact inhibition; i.e., growth is not inhibited by adjacent cells
- Lose the potential to undergo apoptosis
- May metastasize – migrate and establish subclones in other body locations

Characteristics of Cancer

The hallmark of tumors is uncontrolled cell proliferation. Cancer cells proliferate exponentially because they have gained the ability to self-stimulate cell cycling and have lost the ability to respond to extrinsic growth inhibitors. A tumor's potential to be lethal principally depends on its uncontrolled growth. The growth of normal cells is inhibited by contact with adjacent cells, whereas cancer cells have lost contact inhibition. The morphology of cancer cells changes and telomerase synthesis (not present in normal somatic cells) resumes. Cancer cells become immortal – they can go through an indefinite number of cell division cycles – and gain the ability to be cultured; a normal somatic clone can survive for a limited time, $\sim 10^2$ cell division cycles. Cancer cells often lose the ability to undergo apoptosis. Solid tumors may stimulate angiogenesis, the growth of blood vessels supplying the tumor. Many cancer cells metastasize – move into the blood and migrate to other locations in the body. Cancer cells may evolve the ability to evade the immune system.

All long-lived animals and many plants can develop tumors. In humans, over 100 clinically distinct kinds of cancer are recognized, which is not surprising, given the 250 or so human cell types and the diversity of genetic causes within and between types of cancer. A third to a half of humans living in the "developed world" have clinically recognized cancer at least once in a lifetime, and cancer accounts for a quarter of deaths. Every person who lives past the age of 70 is virtually guaranteed to develop several undetected tumors. Cancer rates vary dramatically by type of cancer. Factors that influence cancer rates by tissue type include cell division rate, exposure to mutagens, and hormonal responses of the tissue. Overall risk of cancer is considerably higher in developed countries (e.g., New Zealand or The Czech Republic) than in undeveloped countries (e.g., The Gambia or Ecuador).

Etiology

Cancer is a genetic disease, caused by multiple somatic mutations, and often by epigenetic change (noninherited modified gene functioning) as well. Both environment (e.g., exposure to mutagens) and genotype at conception can pose risks for cancer. Tens of genes may be mutant in any given tumor. Tumors are invariably mutant for genes whose products control the cell cycle, directly or indirectly. As **oncogenesis** (the development of a clone of cancer cells) proceeds, cells both accumulate mutations and become more mutable.

Oncogenesis is a multistep process, in which many genetic and epigenetic changes occur, producing the cancer phenotypes described above. The development of a single tumor can take decades. In any given cancer, genes whose products normally

Fig 26.1 Some critical parts of cell-cycle control. Circled letters = cyclins. Mutations causing any part of this control system to function promote cell proliferation and therefore tumorigenesis.

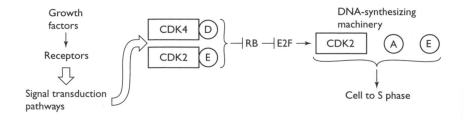

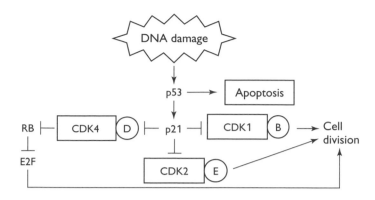

Fig 26.2 Some critical parts of DNA-damage response. Mutations affecting this control system cause damaged cells to proliferate, mutation rate to increase, or to block apoptosis.

brake progress through the cell cycle may either be mutated or underexpressed. Moreover, genes whose products stimulate cell-cycling may be overexpressed or may have gain-of-function mutations (Figure 26.1).

Several kinds of genetic changes occur in oncogenesis, including micromutations, chromosome rearrangements, gene amplification, aneuploidy, and genetic instability. Epigenetic change that promotes oncogenesis is gene silencing initiated by DNA hypermethylation of the cytosine of CpGs in regulatory regions of tumor suppressor genes (Figure 26.2).

There are many ways to disrupt the machinery that limits the growth of a somatic cell within normal bounds, and thus there are many ways to make a tumor. Every normal developmental pathway depends on the action of one specific set of genes, acting correctly. Oncogenesis is the opposite: diverse genetic and epigenetic abnormalities can destroy the normal controls of cell proliferation. Consequently, tumors are genetically heterogeneous. There appear to be $\sim 10^2$ genes, mutation of which contribute to oncogenesis. Any given tumor is mutant for a subset of these genes.

Genes That Feature Prominently in Cancer

Two broad classes of genes participate in oncogenesis, when they are mutant, **tumor suppressor genes** and **oncogenes** (Figure 26.3).

The proteins encoded by tumor suppressor genes (1) help block passage through the cell cycle unless appropriate positive signals are present, (2) participate in apoptosis, or (3) repair or regulate the repair of damaged DNA. Cancer-promoting mutations in

Gene	Product	Functions
Oncogenes		
SIS	growth factor	mitogen = growth factor
ERBB	receptor tyrosine kinase	mitogen receptor/signal transduction
KRAS	G protein	binds GTP/GDP/signal transduction
MYC	transcription factor (TF)	activates cell-cycle genes
Tumor suppressor genes		
RB1	inhibitor of TFs	master control, $G_1 \rightarrow$ S checkpoint
TP53[1]	TF	controls $G_1 \rightarrow$ S checkpoint, apoptosis
TP21	CDK inhibitor	controls $G_1 \rightarrow$ S checkpoint, R point
ATM	protein kinase	controls $G_1 \rightarrow$ S, $G_2 \rightarrow$ M checkpoints, activated by DNA damage
APC	inhibitor of TF; binds to cell adhesion proteins	roles in cell adhesion and cell migration[2]
Both[3]		
E2F1	TF	activates cell-cycle genes, signals apoptosis

[1] Over half of tumors lack functional p53 owing to mutations in the *TP53* gene that encodes this regulatory protein.
[2] Cells lacking APC tend to metastasize.
[3] E2F1 acts both as a tumor suppressor and as an oncogene.

Fig 26.3 Genes that are frequently mutant in tumors.

tumor suppressor genes are either null mutations or hypomorphs (they have reduced function), so that both copies of a tumor suppressor gene must suffer a loss-of-function mutation to promote tumorigenesis. Alternatively, tumor suppressor genes may be underexpressed because cytosines in promoters or enhancers become methylated.

The proteins encoded by oncogenes signal the cell to progress from G_1 to S phase, or else they inhibit apoptosis. Oncogene-encoded proteins include receptors for growth factors, proteins in signal transduction pathways, transcription factors, CDKs, cyclins, and apoptosis-inhibiting factors. Cancer-promoting mutations in oncogenes are the opposite of cancer-promoting mutations in tumor suppressor genes in that only one of the two copies of a gene in a diploid cell need be mutant. That is because oncogenes stimulate growth or inhibit apoptosis. If only one copy of an oncogene suffers a gain-of-function mutation (e.g., causing overly active gene product or overexpression of the gene), growth may be inappropriately rapid or stimulated, because the other copy of the gene does not exert a counterbalancing effect. Similarly, one gain-of-function mutation of an apoptosis-inhibiting gene suffices to inhibit apoptosis inappropriately or too strongly.

Somatic Mutations and Epigenetic Change

Species that develop tumors tend to be long-lived and to consist of a very large number of cells. By conservative estimate, a human's lifetime cell number is $\sim 10^{17}$ cells. If the rate of somatic mutation per gene per cell is 10^{-6}, then the probability that a cell becomes mutant for both copies of a gene is 10^{-12}. For oncogenes, only one copy need be overexpressed to promote oncogenesis, so a single mutation suffices. For tumor suppressor genes, both copies need to be mutated or silenced in a cell to promote oncogenesis. Because any given tumor is mutant for a total of 10 or so genes (oncogenes and tumor suppressor genes taken together), the probability that developing cancer by chance alone in the absence of exposure to mutagens is vanishingly small. For example, the probability that a somatic cell will become cancerous by virtue of independent mutations in nine oncogenes and one tumor suppressor gene is $\sim 10^{-66}$; under those conditions cancer would not develop. Normal mutation rates alone cannot account for tumorigenesis; other factors contribute.

Failure of DNA-repair systems accelerates the accumulation of mutations, and damage to the DNA-damage checkpoints allows cell proliferation in spite of mutation. Epigenetic silencing of tumor suppressor genes removes cell-cycle controls and disables the DNA-damage checkpoints. Selection is an invariable and essential process in tumorigenesis, too. Selective advantage is conferred to tumor cells over normal cells by mutations and epigenetic changes that promote uncontrolled growth, loss of ability to undergo apoptosis, ability to grow uninhibited by neighboring cells, and ability to metastasize.

Genetic Instability and Cancer

Both micromutations and chromosome mutations (rearrangements, aneuploidy) are observed in tumors. Micromutations can be nucleotide substitutions or small deletions or insertions within a single gene. About half of all tumors are mutant for the *p53* gene. Macromutations include aneuploidy, translocation, gene amplification, or replacement of one homolog by the other (chromosomal homozygosity). Multiple chromosomal mutations are very common in tumors and are not associated with any particular type of cancer. Simple chromosomal mutations – single, particular rearrangements – are less frequent overall in tumors but are consistently associated with certain cancers, such as leukemias and lymphomas. For example, the Philadelphia chromosome (a specific reciprocal translocation between 9q and 22q) characterizes chronic myelogenous leukemia. Most cancer cells have either two maternal copies or two paternal copies of several chromosomes, leading to an average chromosomal homozygosity of 25%. Advanced tumors sometimes contain an amplified chromosomal locus, containing many copies of an oncogene; N-myc is amplified in about 30% of advanced neuroblastomas.

Most tumors are genetically unstable as well, which is to say that they are prone to higher rates of mutation than normal cells. Genetic instabilities leading to micromutations are mutations in excision repair enzymes or mismatch repair enzymes. The main effect of deficient excision repair is increased rates of skin cancer. Deficient mismatch repair is observed in a small number of cancers, notably in 10% to 15% of colon cancers. There are many more ways to achieve chromosomal instability, and accordingly it is more common in tumors. It appears that mutations in genes that control the spindle checkpoint and that encode the mitotic machinery may be responsible for chromosomal instability, but this remains to be confirmed by direct empirical evidence. Epigenetic change (gene silencing by hypermethylation) presumably is a major cause of chromosome instability.

Risk Factors

Both environmental and genetic risk factors exist for cancer. Exposure to ultraviolet light and oxidative mutagens are the major environmental carcinogens. Tobacco products produce oxidative damage. Other environmental carcinogens are particulates in air pollution, industrial solvents (especially those containing aromatic compounds), asbestos, and nitrosamines (used as food additives). Estrogens and estrogen mimics are hypothesized to be carcinogenic.

Mutations in several genes increase risk for cancer – for example, *RB1*, *BRCA1*, and *BRCA2*.

Loss-of-function mutations in *RB1*, a tumor suppressor gene that encodes the Rb protein, causes retinoblastoma. In the retina of an $RB1^-/RB1^+$ heterozygote, the wild-type copy of $RB1$ becomes mutant or inactivated in one or more cells of the developing retina producing $RB1^-/RB1^-$ cells, promoting tumorigenesis. $RB1^-/RB1^+$ persons have increased risk for ovarian and colon cancer later in life; again, a second hit (mutation or inactivation) is required for tumorigenesis.

Around 80% to 90% of women who have inherited a null mutation of $BRCA1$ or $BRCA2$ will develop breast cancer by age 70, compared with about 10% of women who did not inherit one of these mutations, and the average age of onset is lowered 30 years or so. Heterozygotes for either gene are also at greatly increased risk for ovarian cancer – about 30-fold for $BRCA1$ and 15-fold for $BRCA2$. Proteins encoded by these two genes may participate in recombination and in DNA repair.

Further Reading

Laird PW, Janisch R. 1996. The role of DNA methylation in cancer genetics and epigenetics. *Annu. Rev. Genet.* 30:441–464.

Lengauer C *et al.* 1998. Genetic instabilities in human cancers. *Nature* 396:643–649.

Perera FP. 1997. Environment and cancer: who are susceptible. *Science* 278:1068–1073.

Welcsh PL *et al.* 2000. Insights into the functions of BRCA1 and BRCA2. *Trends Genet.* 16:69–79.

Chapter 27

Cutting, Sorting, and Copying DNA

Overview

Because the genes of all organisms consist of DNA, and because a gene's structure largely determines its functional properties, analysis of any particular gene demands the isolation and structural analysis of DNA. To manipulate DNA experimentally, or even to determine its base sequence, chromosomal DNA must be reduced to small fragments, which are then sorted, identified, copied, and stored for future use.

This chapter explains some basic principles of cutting DNA enzymatically into fragments, separating pieces of DNA or RNA by size, making DNA copies of RNA, identifying one particular piece of DNA or RNA, copying DNA, and establishing large collections of DNA fragments for storage and retrieval.

Cutting and Fractionating DNA

Restriction Enzymes

It is easy to isolate chromosomal DNA. However, chromosomes tend to be long, and, even after random breakage, the DNA isolated from chromosomes exists as intractably long pieces. It is convenient to cut DNA into appropriate-sized fragments with bacterial **restriction endonucleases** (often simply called restriction enzymes), which cut only at particular sites.

Bacteria protect themselves against viral infection by cutting foreign DNA with restriction enzymes, so called because they restrict foreign DNA's ability to invade the cell. The cell distinguishes its own DNA from foreign DNA by methylating certain bases shortly after replication, using **modification enzymes**. At least one strand of the bacterium's own DNA is modified and therefore resistant to restriction, whereas foreign DNA is not modified and therefore susceptible to restriction. One type of restriction enzyme, very useful to the experimentalist, binds to a short recognition sequence comprising 4 to 8 bp and cuts the backbone of both strands at that site. The recognition sequence has a peculiar symmetry: the base sequence of one strand, $5' \to 3'$, is identical to the base sequence of the complementary strand, also $5' \to 3'$. Biologists often call such a sequence a **palindrome** because it is reminiscent of a true palindrome, which is a meaningful sequence of letters, identical forward and backward, e.g., DOGEESESEEGOD. The names of restriction enzymes are based on the Latin name of the bacterium plus a strain designator (Figure 27.1). Most restriction enzymes make a staggered cut; some make a flush cut. There are thousands of restriction enzymes.

By choosing a particular restriction enzyme, the experimenter controls both the site being cut (each enzyme cuts only at a particular recognition sequence) and the mean size of the fragments produced from genomic DNA. How does that work? To a first approximation, any restriction enzyme's recognition site occurs randomly in target DNA. The frequency with which a random DNA sequence occurs depends on the base composition of the genome and on the length of the recognition site. Eukaryal genomes contain roughly equal amounts of the four main nucleotides. If the recognition site contains N nucleotide pairs, then its probability of occurrence is about 4^{-N}, and conversely the mean fragment length after complete digestion with a restriction enzyme is about 4^N. For example, in human nuclear DNA, the probability of the *Hae*III sequence, GGCC, is $(1/4)(1/4)(1/4)(1/4) = 1/256$; in other words, it occurs about every 256 nucleotides (nt) (4^4 nt).

Fig 27.1 Sample restriction enzymes and their recognition sites.

Enzyme size	Recognition site	Bacterial species	Mean fragment size
*Hpa*II	5'C▼CGG GGC▲C5'	*Haemophilus parainfluenzae*	0.25 kb
*Hae*III	5'GGC▼C C▲CGG5'	*Haemophilus aegyptius*	0.25 kb
*Eco*RI	5'G▼AATTC CTTAA▲G5'	*Escherichia coli*	4 kb
*Not*I	5'G▼CGGCCGC CGCCGGC▲G5'	*Nocardia ottidiscaviarum*	65 kb

The number of unique DNA fragments generated by complete digestion of genomic DNA is simply the size of the genome divided by the mean fragment size. The mean fragment size is larger if the genome is partially digested, and partial digestion is sometimes preferable. If only a fraction P of the sites are cut, then the mean size will be $4^N/P$. For example, if 10% of the sites are cut, then the mean fragment size will be ten times larger. Partial digestion of a genome yields a larger number of fragments; how much larger depends on the fraction of digested sites.

Solved Problem.   If DNA from human cells were completely digested with *Eco*RI, how many unique fragments would be produced? What if only half the sites were digested?

*Eco*RI is a "six-cutter" that recognizes a 6-nt sequence. The mean frequency of a 6-nt sequence $\approx 4^{-6} = 1/4096$, and the average fragment size ≈ 4.1 kb. The human genome consists of 3.3 million kb, so the number of unique fragments would be about $(3.3)(10^6)$ kb/genome $\div$ 4.1 kb/fragment $\approx$ 800,000 unique fragments/genome.

If only half the *Eco*RI sites were digested, then the average fragment size would be twice as large, $4^6 \div 1/2 = 8192 \approx 8.2$ kb.

Breaking DNA Physically

Cutting DNA with restriction enzymes is not always desirable or even feasible. Tracts of repeated sequences may contain too few or too many restriction sites to yield usable pieces. DNA can be broken into pieces by simple physical methods. Moving DNA rapidly through a constricted tube imposes shearing forces that break the sugar-phosphate backbone at random places. Sonication also breaks the sugar-phosphate backbone randomly. Any DNA, including physically broken pieces, can be ligated to short **linkers** that contain restriction sites.

Gel Electrophoresis

DNA, being an acid, ionizes in water and has a negative charge, mainly because of the phosphate groups of the backbone. In an electric field, DNA migrates to the anode (positive pole). To separate DNA fragments by size using gel electrophoresis, DNA is loaded into slots in a gel made of agarose or polyacrylamide, and the gel is submerged in a buffer solution (Figure 27.2). Gels are riddled with pores through which DNA can move. When a direct

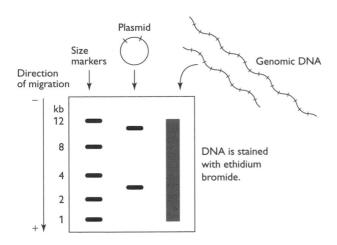

DNA is stained with ethidium bromide.

current electric field is applied across the ends of the gel, the DNA migrates through the gel toward the anode. Small pieces of DNA migrate fast, and large pieces migrate slowly; the speed of migration is inversely proportional to the logarithm of the fragment length. After a time, the electric field is turned off. DNA is detected by adding ethidium bromide, a UV-absorbing, fluorescent dye that binds to DNA; the gel is viewed and photographed under UV light. If a digest of genomic DNA is loaded into a slot, electrophoresis produces a smear of DNA, because a large number of different fragments are present, with short pieces near the anode and long pieces near the starting slot. If a digest of a small, purified piece of DNA (say a plasmid) is loaded into a slot, electrophoresis produces a few bands, each containing fragments of one length. Pieces of RNA can also be separated by size using gel electrophoresis.

cDNA

The genes transcribed in cells at one developmental stage or in one tissue comprise a subset of the genome. To obtain DNA from the protein-coding genes being expressed in particular cells, the experimenter makes **complementary DNA(cDNA)** – DNA copies of mRNA from these cells. cDNA lacks introns. To make cDNA *in vitro*, oligo(dT) primers, which bind to poly(A) tracts, are extended with a viral reverse transcriptase (an RNA-dependent DNA polymerase) (Figure 27.3). Following reverse transcription, the RNA is removed by digesting with ribonuclease. The resulting single-stranded DNA

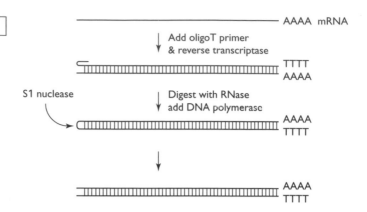

Fig 27.3 Making cDNA.

(ssDNA) tends to make a hairpin loop at the 5′ end, which primes synthesis of a cDNA sequence, with bacterial DNA polymerase. The hairpin loop is cut by S1 nuclease. The cDNA can then be copied and stored.

Copying DNA

DNA Cloning

To clone a fragment of DNA is to introduce that DNA into a life form (cell or virus) and then to grow a clone of the life form. A requisite tool is a **cloning vector**.

A cloning vector is a special, small, engineered chromosome made of double-stranded DNA (dsDNA) or ssDNA. It has an origin of replication and a **multiple cloning site** (**MCS**, a short sequence containing many different restriction sites in tandem), and it may have other features, such as a **selectable marker** (a gene whose presence can be detected in an organism). A fragment of DNA to be cloned is inserted into a cloning vector by *in vitro* recombination. It is sometimes desirable to cut a cloned fragment into smaller fragments and then to clone each of the smaller fragments. The smaller fragments are thus **subcloned**.

Of the many types of cloning vector, two are described here, circular plasmids and λ phage; then the properties of several types of cloning vector are summarized. This description is deliberately stripped of detail. Standard recipe books for molecular genetic experiments are many hundreds of pages long.

Circular Plasmids. Circular plasmids are vectors for cloning small fragments (usually <15 kb) of DNA (Figure 27.4). Typically,

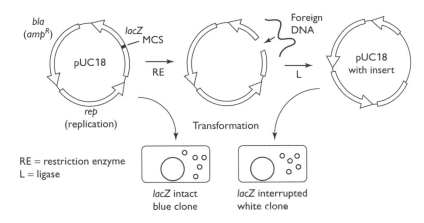

Fig 27.4 Cloning DNA in a plasmid vector.

RE = restriction enzyme
L = ligase

a plasmid cloning vector is about 3 kb of dsDNA having two se-lectable markers, a MCS inside one marker, and an origin of repli-cation. A widely used selectable marker is the *bla* gene, which encodes β-lactamase and confers resistance to the antibiotic ampi-cillin. Cells having a copy of *bla*$^+$ can grow in a medium contain-ing ampicillin, whereas ampicillin inhibits the growth of *bla*$^-$ cells.

A useful vector is pUC18, whose MCS is within a truncated copy of *lacZ*. To clone a gene in pUC18, both the DNA to be cloned and the plasmid are cut by the same restriction enzyme and then they are ligated. The resulting DNA mixture is used to transform competent, ampicillin-sensitive *E. coli* cells. Because transforma-tion is not a natural process in *E. coli*, cells must be treated with an agent that damages the cell envelope in order to render them competent (pervious to DNA). A small fraction of cells become transformed when the circular plasmid enters the cell, takes up residence, and replicates stably each cell cycle. The cells are plated on a medium containing both ampicillin and X-gal, a substrate of the β-galactosidase enzyme encoded by *lacZ*; β-galactosidase converts the white X-gal to a blue product. All clones growing on the plate are transformants, as only *bla*$^+$ transformants are ampicillin-resistant. Blue clones have functional β-galactosidase because DNA did not insert into the MCS within *lacZ*. White clones lack functional β-galactosidase – DNA inserted into the MCS in *lacZ* interrupts its coding sequence.

λ Phage Chromosomes. The λ phage chromosome is ≈46 kb of dsDNA. A 15-kb central region, not required to complete the lytic cycle, is removed by cutting with *Bam*HI, and the two distal

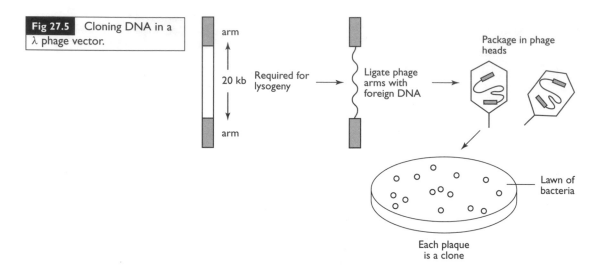

Fig 27.5 Cloning DNA in a λ phage vector.

fragments, or arms, are ligated to a heterogeneous sample of foreign, *Bam*HI-digested DNA fragments averaging 15 kb. The resulting recombinant DNA is packaged in purified phage heads, which can hold up to 50 kb of DNA; the maximum size of the insert is thus ≈20 kb. A suspension of these phages is poured on plates overgrown with *E. coli* cells. Phages infect cells with very high efficiency. Wherever a phage infects a cell, it lyses the adjacent cells within a small, circular area, or **plaque**. One large culture plate may contain ∼10^4 distinct plaques, each containing a different clone of recombinant λ phages (Figure 27.5).

DNA Libraries

A DNA library is a large collection of cloned fragments representative of the source of DNA. A **genomic library** comprises clones of genomic fragments, and a **cDNA library** comprises clones of cDNA. A library should contain at least one copy of each DNA sequence in the source and several copies of most sequences. A genomic library is made from a partial digest to ensure overlapping clones.

To reduce the probability that genomic sequences are missing from the library, the number of clones in the library must be quite a bit larger than the size of the genome divided by the average size of a cloned fragment. A simple formula estimates the required redundancy. Let N = the number of clones/library, P = the probability that a sequence will be included, S = the mean size of an insert, and G = the genome size; $N = \ln[1 - P]/\ln[1 - S/G]$.

Vector	Size of cloned fragment	Uses
Plasmid	<15 kb	Subcloning; cDNA library
λ phage	<20 kb	Genomic & cDNA libraries
Phagemid	<12 kb	„ „ „ „
Cosmid	30–45 kb	Genomic library
P1 phage	70–90 kb	„ „
BAC	100–500 kb	„ „
YAC	250–1000 kb	„ „

Fig 27.6 Characteristics of some cloning vectors.

Solved Problem. Suppose you plan to clone the human genome in a λ vector, and you want to be 99.99% sure that every sequence will be included.

The required library size is $N = \ln[0.0001]/\ln[1 - (1.5)(10^4)/(3)(10^9)] \approx 1.8$ million clones. A million clones seems unmanageable. Perhaps you should choose a vector that holds bigger inserts.

Other Cloning Vectors

Other types of cloning vector abound, and new ones are invented continually (Figure 27.6). A **phagemid** is part plasmid, part f1 phage; it facilitates sequencing of the cloned insert. A **cosmid** is a highly modified λ phage genome capable of carrying a large insert. Large pieces of DNA can be cloned in **BACs** (bacterial artificial chromosomes), and **YACs** (yeast artificial chromosomes).

Expression vectors promote the expression of the cloned genes cDNA is inserted into a plasmid that contains a bacterial promoter. Expression vectors are useful for screening libraries. Screening is accomplished with the help of a specific antibody tagged with a fluorescent label. The antibody binds specifically to the protein expressed in clones containing the gene of interest.

Amplifying a DNA Fragment with PCR

The **polymerase chain reaction** (**PCR**), like cloning, amplifies DNA (Figure 27.7). PCR is faster and easier than cloning, and only a tiny amount of DNA is needed to start – under ideal conditions as little as a single copy. One PCR amplifies a single, short sequence of DNA, smaller than 25 kb, and usually smaller than 1 kb.

In PCR, DNA is synthesized with a heat-resistant DNA polymerase (e.g., *Taq* polymerase, from the archaeon *Thermus aquaticus*) and short primers (≈ 20 nt long). Primers, chosen to flank the region to be amplified, are synthesized to order by commercial

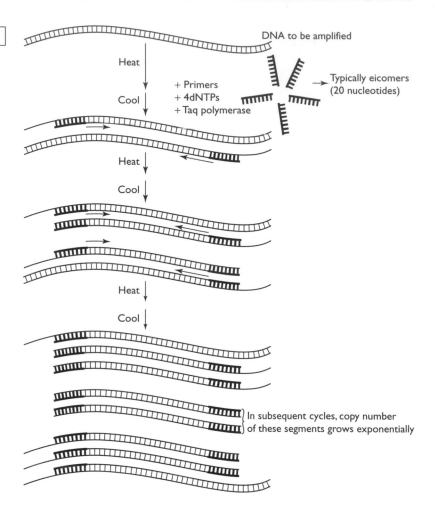

Fig 27.7 PCR scheme.

suppliers. In a small reaction tube, the four dNTPs, template DNA, primers, and *Taq* polymerase are heated to denature the DNA. Upon cooling, primers hybridize to template DNA and *Taq* polymerase extends the primers. The heating-cooling cycle lasts about 90 s and is repeated 25 to 45 times. Each cycle doubles the number of copies of the amplified sequence, or **amplicon**.

DNA Hybridization

It is possible to locate a specific piece of DNA by hybridizing (annealing) a cloned fragment that contains sequences complementary to part of the DNA to be located. One strand of the complementary fragment can base-pair with one strand of the searched-for DNA. Labeling such a complementary fragment, so that a small amount of it can be detected, converts it into a sharp tool for finding hidden DNA – a **probe**. To be useful as a probe, a DNA fragment must be sufficiently long and have a sufficiently complementary sequence to hybridize specifically to the hidden DNA. A probe must be labeled radioactively or by covalent attachment of a dye.

Labeled probe is mixed with target DNA, both are denatured to make ssDNA, and then the conditions are changed to allow renaturation. Probe fragments are in higher concentration than the target fragment. After hybridization, excess probe is washed off and the label is located; e.g., by autoradiography. DNA hybridization is useful for locating DNA fragments that have been fractionated by electrophoresis or clones within a DNA library.

Southern Blot Hybridization

After genomic DNA has been separated electrophoretically, fragments with a particular sequence can be located by hybridization. In preparation, the DNA is transferred from the gel to a membrane by blotting water out of the gel. This useful procedure, Southern blot hybridization, is named after its inventor (Figure 27.8). A similar procedure is **Northern blot hybridization**,

Fig 27.8 Southern blot procedure.

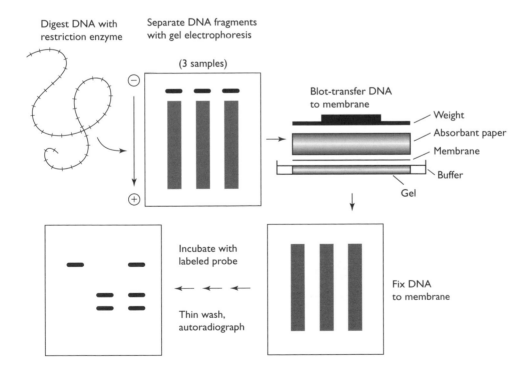

Digest DNA with restriction enzyme

Separate DNA fragments with gel electrophoresis

(3 samples)

Blot-transfer DNA to membrane

Weight

Absorbent paper

Membrane

Buffer

Gel

Incubate with labeled probe

Thin wash, autoradiograph

Fix DNA to membrane

in which RNA is separated by electrophoresis and transferred to a membrane before probing with labeled DNA.

Locating a Clone within a DNA Library

A clone within a DNA library can be located by displaying a copy of the library of clones on a surface and then hybridizing with a labeled probe. Following this, a copy of the clone can be sampled once its location on the surface is known.

Further Reading

The literature on this subject is vast and burgeoning. One consistently good source of reviews on molecular approaches to genetics is an annual series:

Setlow JK (ed.). 1979-present. *Genetic Engineering Principles and Methods.* Plenum, New York.

Chapter 28

Genotyping by DNA Analysis

Overview

An important task for geneticists is to determine individuals' **genotypes** – genetic make-up. Most hereditary analysis requires some knowledge about the genotypes of closely related individuals. When reproduction is purely sexual (excluding clones, twins, and the like), individuals tend to differ from each other in genomic DNA sequence. In sexually reproducing species, each individual is genetically unique. Two kinds of individual differences in DNA sequence are nucleotide substitution and variation in sequence length.

This chapter describes a few of the many approaches to genotyping. One valuable method is determining the nucleotide sequence of DNA (DNA sequencing). Other, easier, more rapid methods provide partial information about genotypes. Four such methods are described, all based on polymerase chain reaction (PCR).

Sequencing DNA

Most DNA sequencing entails synthesis of DNA by the **dideoxy method**, which is based on the fact that 2′,3′-dideoxynucleotides terminate DNA synthesis. The deoxyribose of DNA lacks a hydroxyl group on the 2′ sugar; dideoxyribose lacks the 3′ hydroxyl group as well. Recall that DNA chain growth is 5′→3′; the α phosphate of the incoming nucleotide makes a covalent bond with the oxygen of the 3′-OH group of the growing polymer (Figure 28.1). If the 3′ nucleotide of the polymer lacks a 3′-OH group, then a nucleotide cannot be added and chain growth stops. A dideoxynucleotide can add to a growing polynucleotide chain, but,

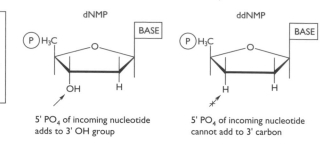

Fig 28.1 Normal deoxynucleotide (dNMP) versus dideoxynucleotide (ddNMP); incoming nucleotide can add to 3'-OH but not to 3'-H.

after it is incorporated into the polymer, it blocks further DNA synthesis.

When DNA is sequenced, one of the two strands of DNA is synthesized and the products of the synthesis reaction are separated electrophoretically. The DNA to be sequenced is a fragment cloned in a plasmid vector (Figure 28.2). In a low-tech version of the dideoxy method, the plasmid is added to four reaction tubes containing DNA polymerase, radioactively labeled primer,

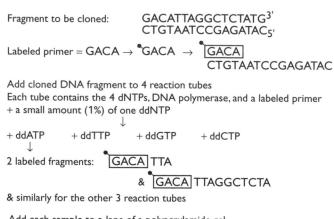

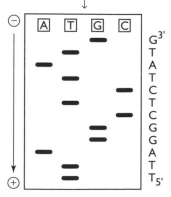

Fig 28.2 DNA sequencing: primer extension method.

and the four dNTPs. Additionally, each reaction tube contains a small amount of one dideoxynucleotide: ddTTP, ddCTP, ddATP, or ddGTP. In the tube with ddTTP, a fraction of each "T" position is filled by ddT instead of dT; the tube contains all fragments ending in "T" plus complete chains. Similarly, in the other reaction tubes, incomplete fragments end in the dideoxy version of one nucleotide. The reaction products are separated by electrophoresis in a very long polyacrylamide gel, which is then autoradiographed to reveal bands in each lane, corresponding to the ddN-terminated fragments.

Automated sequencing also uses the dideoxy method, but in place of a radioactive primer, the four ddNTPs are uniquely labeled with fluorescent dyes. Therefore, each ddNTP is color coded. A single reaction tube suffices, and the synthesized fragments are size-fractionated in a single lane of a gel. The gel is scanned by laser, which excites the four dyes and reveals the color of each band, corresponding to the nucleotide at that position.

Individual Genetic Differences Detectable by PCR

Genomes of eukarya are rich in **single nucleotide polymorphisms**, or **SNPs**. Polymorphism is a frequently encountered term referring to DNA sequence variants that occur in a population. The human genome contains $\sim 10^7$ SNP sites.

A second, common kind of genetic variation is associated with microsatellites – tandem arrays (10 to 100 repeats) of 1- to 5-nt sequence (Chapter 6). At each microsatellite locus, the array varies in length from individual to individual. Microsatellites are scattered throughout many genomes; the human genome contains $\sim 10^5$ microsatellite loci. Because of the great amount of individual variation in microsatellites, they are excellent forensic tools – **DNA fingerprinting**. Microsatellite loci are small and therefore can be analyzed by PCR.

Another kind of DNA fingerprinting uses minisatellites, tandem arrays (10 to 1000 repeats) of 10 to 40 nt (Chapter 6). Minisatellites must be analyzed by Southern blotting (a slow, tedious procedure) because minisatellite loci are too large for PCR.

Four PCR-Based Methods

Restriction Fragment Length Polymorphism (RFLP)

This method is used to detect SNPs (Figure 28.3). Suppose gene Z has two alleles (variants of a gene), differing by a single base substitution, A→G, within an *Eco*RI site: allele Z^1 has 5′GAATTC (recognized and cut by *Eco*RI), and allele Z^2 has 5′GGATTC (not recognized by *Eco*RI). To distinguish between Z^1Z^1, Z^1Z^2, and Z^2Z^2 individuals, the geneticist amplifies DNA containing the *Eco*RI site by PCR, digests the amplicons with *Eco*RI, and by electrophoresis separates the fragments, which are stained with ethidium bromide. There is a good chance a given genetic difference can be detected by some restriction enzyme, but this approach can be used efficiently only with prior knowledge about the target sequences.

Single-Strand Conformation Polymorphism

Another method for detecting SNPs is based on the tendency of 1-nt substitutions to cause large conformational changes in single-stranded DNA (ssDNA). The single-strand conformation polymorphism (SSCP) method is simple in concept (Figure 28.4). Genomic DNA is amplified by PCR. The amplicons are then denatured and subjected to electrophoresis. Many, but not all, single substitutions can be detected by the SSCP approach.

Suppose gene Z has two alleles, differing by a single transition, AT→GC: allele Z^1 has AT, and allele Z^2 has GC. To distinguish

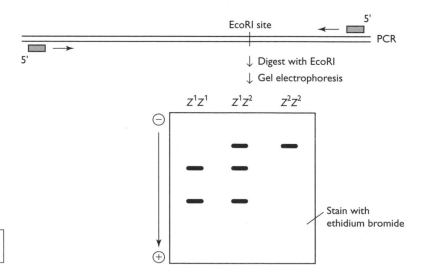

Fig 28.3 PCR-based RFLP analysis.

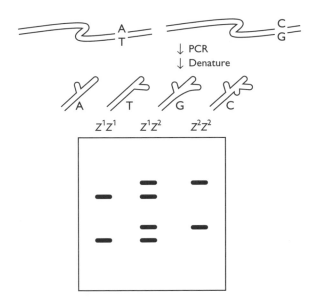

Fig 28.4   SSCP analysis.

between Z^1Z^1, Z^1Z^2, and Z^2Z^2 individuals, the region containing this site is amplified by PCR, the DNA is denatured, and the ssDNA is size-fractionated by polyacrylamide gel electrophoresis.

Microsatellites

Each microsatellite can have several alleles, and each allele has a different length. The key to analyzing a microsatellite is to develop primers for sequences that flank it. When flanking primers are available, the procedure is to amplify a microsatellite from the genomic DNA of an individual and size-fractionate the products by electrophoresis. The example illustrated here (Figure 28.5) is for a trinucleotide repeat in DNA from three persons, Di, Kim, and Liz. The number of repeats is written for each allele of each person.

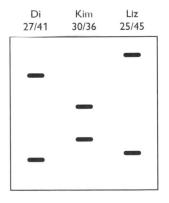

Fig 28.5   PCR of microsatellites. The number of trinucleotide repeats is given for both alleles of each person.

Amplified Fragment Length Polymorphism (AFLP)

The AFLP method is a powerful tool for comparative genome analysis, for a large number of DNA fingerprints are displayed at once in AFLP.

AFLP analysis is a four-step process (Figure 28.6). First, genomic DNA is digested with two restriction enzymes; a subset of all restriction products will be cut by enzyme 1 at one end and by enzyme 2 at the other end. Second, double-stranded oligonucleotide **adapters** are ligated to the fragments' ends. Third, the products

Fig 28.6 AFLP analysis.

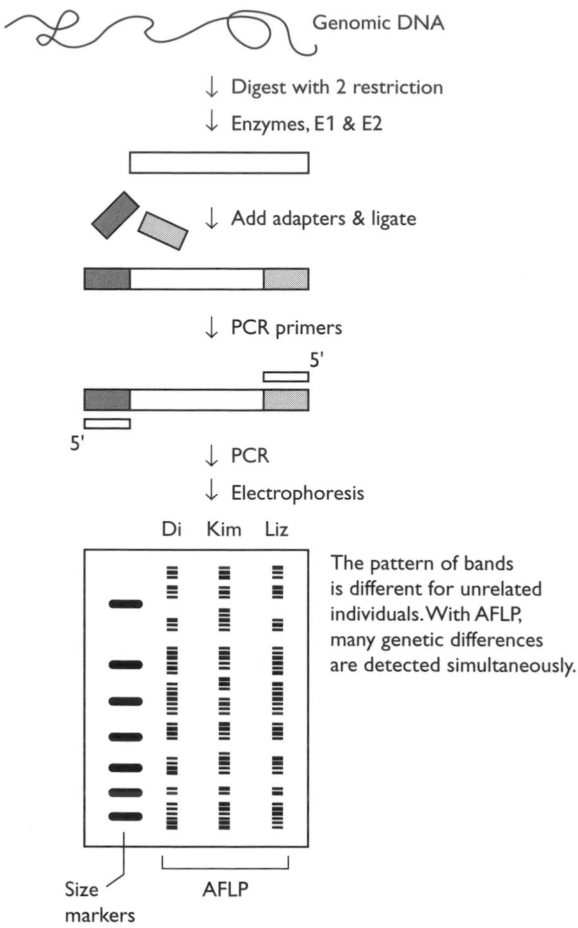

Genomic DNA

↓ Digest with 2 restriction
↓ Enzymes, E1 & E2

↓ Add adapters & ligate

↓ PCR primers

5′

5′

↓ PCR
↓ Electrophoresis

Di   Kim   Liz

The pattern of bands is different for unrelated individuals. With AFLP, many genetic differences are detected simultaneously.

Size markers   AFLP

are amplified by PCR. The primers are homologous to the adapter and the restriction site and extend a few bases on the 3′ ends. Fourth, the products are size-fractionated by acrylamide gel electrophoresis. AFLP gels have a large number of bands, and thus a large amount of information.

Further Reading

Jeffreys AJ *et al.* 1993. Minisatellite variant repeat mapping: application to DNA typing and mutation analysis. In: Pena SDJ, *et al.* (eds.). *DNA Fingerprinting: State of the Science.* Birkhäuser Verlag, Basel.

McPherson MJ, Møller SG. 2000. *PCR. The Basics.* Springer-Verlag, New York.

Vos P *et al.* 1995. AFLP: a new technique for DNA fingerprinting. *Nucleic Acids Res.* 23:4407–4414.

Chapter 29

Genetically Engineered Organisms

Overview

A plant or animal that is genetically altered by the artificial introduction of foreign genes is **transgenic**, for genes have been transferred from one genome to another. Donor DNA is usually recombined *in vitro* with a vector before being introduced into recipient cells. The donor DNA may reside temporarily in the transgenic recipient, or it may become a permanent part of the recipient's genome, where it is stably inherited. Any species can be a recipient when workable methods of gene transfer are developed. It is also possible to inject mRNA into a recipient for transient expression there.

Transgenic organisms are useful for studying gene expression, phenotypic effects of mutations, and gene interactions as well as for discovering novel genes. Agricultural crop plants are modified genetically with the aim of improving the yield and quality of food and fiber. **Gene therapy** is the genetic modification of human cells to treat genetic disease; its enormous potential has yet to be realized.

Transient Gene Expression

Experiments with transient gene expression are useful for finding out what a protein does and determining the function of parts of the protein by inducing mutations in the gene that encodes it. The transgenic organism is a testing ground.

It often suffices to express a transgene or mRNA *in vivo* for a short time. This works if the foreign protein functions in transgenic cells and can be assayed quickly. DNA or mRNA is introduced into cultured cells, where the transgenic protein is typically expressed for more than a day. DNA for transfer is first cloned into a plasmid or virus vector that carries a selectable marker. The DNA is either injected by micropipette into individual cells or else introduced into many cells by another method. For example, DNA can be delivered in **liposomes** (tiny water-filled lipid spheres) that fuse with the cell membrane, or by **electroporation** (brief, high-voltage shocks that makes hole in the cell membrane). The foreign DNA gets into the nucleus. mRNA can be injected into the cytoplasm of single cells.

Several inducible transgene systems are available, based on simultaneous introduction of a **regulatory vector** and an **expression vector**. One such system is based on *lacZ* from *Escherichia coli*. The regulatory vector carries a copy of the *lacI* gene, which encodes *lac* repressor, and the expression vector contains a *lac operator* upstream of *YFG* (your favorite gene – the one to be expressed in transfected cells) (Figure 29.1). Isopropyl β-D-thiogalactopyranoside (IPTG), an analog of lactose, induces *YFG*; IPTG is simply added to the growth medium. One inducible transgene system for mammalian cells is based on ecdysone receptors, in which 20-OH ecdysone is the inducer.

A widely used model organism for expression of mRNA is the oocyte or egg of *Xenopus laevis*; this is a large (d $\approx$ 1 mm), stable, easily manipulated cell. For example, to study the electrophysiological properties of a mammalian sodium channel, mRNA is synthesized *in vitro* from cloned cDNA and injected into the oocyte's cytoplasm, where it is translated. The channel polypeptides insert into the plasma membrane, where they become functional. All the electrophysiological characteristics of the channel can easily be measured by using either the whole cell or bits of membrane that contain the channel. By mutating the gene systematically

Fig 29.1   Two-vector transfection with the *lac* repressor, induced by IPTG. *lacI* = *lac* repressor gene, P = promoter, O = operator, *YFG* = your favorite gene.

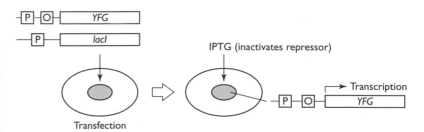

in vitro and using this assay, the functional significance of each part of the channel protein can be ascertained in as much detail as the experimenter desires.

Transgenic Organisms

A cell's genome can be modified permanently if the transgenes insert into the nuclear genome. The aim may be to put functional copies of a gene into mutant somatic cells, as in gene therapy. To modify a strain of sexually reproducing plants or animals genetically, the transgene must insert into the germ line.

Transgenic Somatic Cells: Attempts at Human Gene Therapy

Theoretically, there are three ways to modify somatic cells genetically: *in vivo*, *in situ*, and *ex vivo*. Delivering transgenic DNA via blood would be an *in vivo* method, which awaits the development of an injectable vector. The *in situ* approach is to place transgenic DNA on or in the affected tissue. Cystic fibrosis, a genetic disease due to mutations in a chloride channel gene, may someday be treated by *in situ* transfection with an adenovirus carrying a copy of the chloride channel gene, as adenoviruses, which cause the common cold, readily infect bronchial tissues. There is also hope of killing cancer cells by injecting a tumor with a vector carrying a toxin-encoding gene; a key problem will be to target tumor cells specifically. *Ex vivo* transfection is based on removing cells from the body, incubating them with retrovirus vectors, selecting transgenic cells, and returning those cells to the body.

The *ex vivo* method has been used in mice. For example, hematopoietic stem cells from bone marrow were cultured with retroviruses carrying both the green fluorescent protein (GFP) gene and a copy of the human dihydrofolate reductase gene and then injected into recipient mice. One day later, about half the bone marrow cells of recipient mice fluoresced green, showing that they were transgenic.

Germ Line Transformation of Mice

Three ways to make transgenic mice are to introduce DNA into: (1) a haploid pronucleus of a newly fertilized egg; (2) cultured

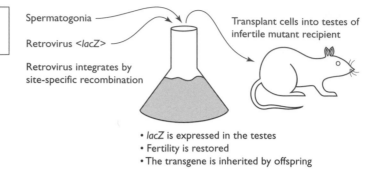

Fig 29.2 Transgenic mice from transfected spermatogonial stem cells.

Spermatogonia

Retrovirus *<lacZ>*

Retrovirus integrates by site-specific recombination

Transplant cells into testes of infertile mutant recipient

• *lacZ* is expressed in the testes
• Fertility is restored
• The transgene is inherited by offspring

spermatogonial stem cells, which are transplanted into recipient testes (Figure 29.2); and (3) cultured embryonic stem cells (ES cells), which are added to an early embryo. In methods (1) and (3), the resulting embryo is implanted in the uterus of a foster mother, with a yield of about 1% transgenic animals. In method (2), the recipients of transfected spermatogonia produce offspring by normal sexual reproduction, and the yield is higher. With method (3), the experimental mouse is a **chimera**, an organism derived from embryonic cells of two genetically distinct individuals. The transgenic DNA integrates into recipient chromosomes by recombination.

The transgenic DNA in a plasmid vector may integrate by homologous recombination, in which case the transgene replaces the genomic version; this is **targeted gene replacement**. When transgenic DNA interrupts a gene and renders it nonfunctional, the interrupted gene is "knocked out"; **gene knockout** is a widely used tool in mouse genetics. The experimenter may alter a transgene before it is put in a vector by directed *in vitro* mutagenesis. When transgenic DNA is integrated by nonhomologous recombination, DNA is added to the genome at an ectopic site unrelated to the normal locus of the host's homologous gene; in other words, the transgene may insert into various loci of any chromosome.

In the following example of targeted gene replacement, the human *ApoE3* allele replaced the mouse *ApoE* gene (Figure 29.3). *ApoE*, a gene on chromosome 19, encodes apolipoprotein E, which transports cholesterol and triglycerides. The *ApoE3* allele, the most common allele in humans, encodes a "normal" form of the enzyme. *ApoE3* knockout mice developed hypercholesterolemia and atherosclerosis when they are fed a diet high in cholesterol and total fat, unlike nontransgenic mice.

Integration of the transgenes is shown as a double crossover event here, but the actual mechanism is not known. For example, it may occur by a gene-conversion-like event.

Fig 29.3 Transgenic mice with human *ApoE3* replacing mouse *ApoE*. neo = neomycin-resistance gene, a selectable marker.

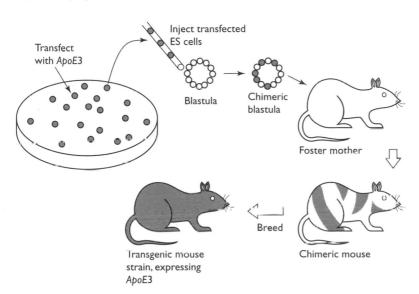

Germ Line Transformation of *Drosophila*

Transgenic strains of *Drosophila* are made by injecting plasmid vectors into the pole cells at the caudal end of the early embryo. Pole cells develop into germ cells. The vectors are based on the P element, a 3-kb transposon abundant in wild strains but absent from many laboratory strains. Two plasmid vectors are coinjected into the embryo, one containing the transgene(s) flanked by the inverted repeats of a P element, and the other containing a copy of the P-element's transposase gene (Figure 29.4). Transposase catalyzes recombination between the P-element's inverted repeats and genomic DNA in the recipient chromosome. A selectable marker is included in the transfecting sequence. The recipient embryo has only mutant, nonfunctional copies of the selectable marker. For example, the selectable marker w^+ (making for red eyes) is used for w/w (white-eyed) recipients. The recipient grows into an adult, which is crossed to another white-eyed w/w

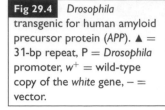

Fig 29.4 *Drosophila* transgenic for human amyloid precursor protein (*APP*). ▲ = 31-bp repeat, P = *Drosophila* promoter, w^+ = wild-type copy of the *white* gene, − = vector.

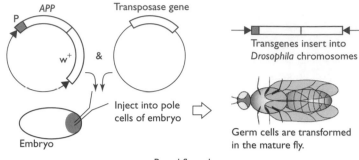

fly. Transgenic red-eyed offspring are selected and bred to make transgenic strains. There are many experimental uses of transgenic *Drosophila*.

Suppose a point mutation produces a clear-cut phenotype and maps to a chromosomal locus, but several candidate segments of DNA (putative genes) map to that same chromosomal locus, which comprises many kilobases of DNA. Each candidate gene can be checked for its ability to produce a wild-type phenotype by transfecting a mutant strain. If the transfecting DNA "rescues" the mutant, then it must code for the trait of interest. Phenotype rescue by transfection is proof that a piece of DNA *is* a particular gene. An example is the *period* gene of *Drosophila*; this gene is essential for normal circadian rhythms. The biological clock of an arrhythmic null mutant was restored by transformation with a functional copy of the gene, per^+ cDNA.

It is possible to make transgenic strains that express a gene conditionally, or in a specific tissue, allowing study of the developmental and spatial effects of that gene. A gene's promoter may control gene expression conditionally, as in genes coding for heat shock proteins; such genes are transcribed when the temperature is extremely high. Tissue-specific expression depends on the proximity of tissue-specific enhancers.

Transgenic Plants

Foreign genes can be introduced into plants. Transgenic plants can be made with relative ease because somatic cells are totipotent, leading to the direct cloning of plants from transfected, cultured cells. Cloned transgenic plants can reproduce sexually; the germ line is, of course, fully transgenic.

Vectors for making transgenic plants are derived from an ≈200-kb, conjugative, **tumor-inducing plasmid (Ti plasmid)**,

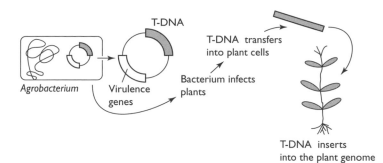

Fig 29.5 Tumor genes (T-DNA) in the Ti plasmid of *Agrobacterium tumefaciens* transfect plant cells; T-DNA inserts into plant chromosomes.

present in a soil bacterium, *Agrobacterium tumefaciens*, which can infect more than 600 plant species (Figure 29.5). *Agrobacterium* can transfer a 23-kb segment of the Ti plasmid, called **T-DNA**, to eukarya (plants and fungi) or to distantly related bacteria. Proteins encoded by five *vir* (*virulence*) operons (*virA-E*) of the Ti plasmid effect the transfer of a single strand of T-DNA to a plant cell and its integration into host chromosomes. The T-DNA inserts into host nuclear DNA, aided by the 25-bp direct repeats of the T-DNA.

The Ti plasmid is made into a vector by inserting foreign DNA (often, *YFG* plus one or two antibiotic-resistance genes – selectable markers) between the ends of the T-DNA (Figure 29.6). Cultured somatic cells from a plant are exposed to the engineered Ti plasmid, transfected cells are selected on medium containing antibiotics, and the cells grow into whole plants. The cloned plants reproduce sexually to make useable plant varieties that express the transgene.

Enhancer Traps

An **enhancer-trap element** is a transposon that contains a selectable marker and, in addition, either a reporter gene or a gene that encodes a transcription factor. The reporter gene has a weak promoter, so that it is expressed only if the enhancer-trap element

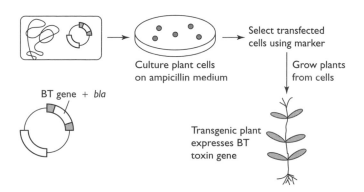

Fig 29.6 In a Ti plasmid vector, transgenes replace most of the T-DNA. BT toxin, from *Bacillus thuringiensis*, is insecticidal; somatic plant cells are totipotent. In this example, the selectable marker is *bla* (ampicillin resistance).

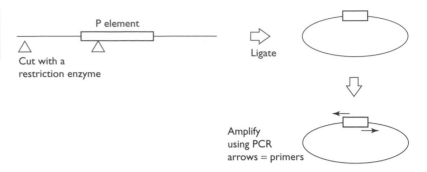

Using inverse PCR to amplify DNA sequences flanking an inserted P element.

inserts near enhancers. This is why enhancer-trap elements can be used to find enhancers.

Enhancer-trap elements are tools for insertional mutagenesis. Insertion of an enhancer-trap element is mutagenic by definition, and, if the inserted element disrupts a gene, a mutant phenotype likely results. By inducing the transposition of enhancer-trap elements, one can make a very large number of new insertion mutations. The reporter gene is expressed in specific tissues and at specific developmental stages. An often-used reporter gene is *lacZ*; wherever and whenever *lacZ* is expressed can be detected with X-gal, a substrate that β-galactosidase converts into a blue product.

Furthermore, a gene mutated by insertion of an enhancer-trap element can easily be cloned by inverse PCR. In this procedure, DNA is extracted from P-element-mutated flies and digested with a restriction enzyme that cuts at one end of the P element (there is only one P element per haploid genome). The restriction fragments are ligated to make circles, and PCR is performed. In this way, the DNA sequences on either side of the P element can be amplified (Figure 29.7). The amplified fragments are then cloned in a plasmid.

Enhancer-trap mutants are useful for studying the effects of directed misexpression of a gene. The methods are too lengthy to describe here; consult references below for further information (Kaiser 1993 is an excellent introduction). Enhancer-trap collections have been made in several species, including *Drosophila*, *Arabidopsis*, maize, rice, and a moss.

Further Reading

Abbud R, Nilson JH. 1999. Recombinant protein expression in transgenic mice. In: Fernandez JM and Hoeffler JP (eds.). *Gene Expression Systems*. Academic Press, New York.

An X et al. 2001. The effects of ectopic *white* and *transformer* expression on *Drosophila* courtship behavior. *J. Neurogenet.* 14:227–243.

Fossgreen A et al. 1998. Transgenic *Drosophila* expressing human Amyloid Precursor Protein shows g-secretase activity and a blistered wing phenotype. *Proc. Na. Acad. Sci. USA* 95:13703–13708.

Kado CI. 1998. *Agrobacterium*-mediated horizontal gene transfer. *Genet. Eng.* 20:1–24.

Kaiser K. 1993. Second generation enhancer traps. *Curr. Biol.* 3:560–562.

Kay MA, High K. 1999. Gene therapy for the hemophilias. *Proc. Na. Acad. Sci. USA* 96:9973–10075.

Nagano M et al. 2001. Transgenic mice produced by retroviral transduction of male germ-line stem cells. *Proc. Na. Acad. Sci. USA* 98:13090–13095.

Peterson K. 1997. Production and analysis of transgenic mice containing yeast artificial chromosomes. *Genet. Eng.* 19:235–255.

Rossant J, Spence A. 1998. chimeras and mosaics in mouse mutant analysis. *Trends Genet.* 14:358–363.

Sullivan PM et al. 1997. Targeted replacement of the mouse apolipoprotein E gene with the common human APOE3 allele enhances diet-induced hypercholesterolemia and atherosclerosis. *J. Biol. Chem.* 272:17972–17980.

Chapter 30

Genomics

Overview

Genomics, the newest branch of genetics, is the study of genome structure and function: massive genome-wide mapping, determination of primary nucleotide sequence for whole genomes, analysis of spatial relationships of various sequences or classes of sequence within and between chromosomes, genomic inventory by the sequence or gene class, and global analysis of gene expression. Genomics emphasizes genes over nontranscribed, nonregulatory sequences. A major challenge in genomics is the analysis of very large amounts of information.

Genome Cloning

The first step in genomic analysis is construction of a fully representative, high-quality genomic library. A large, pure sample of the life form of interest is collected and treated physically to separate genomic DNA from other components of the life form. The DNA is extracted chemically, purified, and cleaved. The fragments are cloned in a suitable vector, commonly cosmid, bacteriophage P1, BAC, or YAC (Chapter 27). To ensure that the library contains overlapping clones that span the entire genome, the DNA is digested partially, and the cloned segments comprise a large random sample, typically an average of 10 to 30 copies per sequence. A set of overlapping, cloned, sequenced DNA segments is called a **contig**, because the sequence of the region spanned by the segments has no gaps (Figure 30.1). In genome sequencing, it is ideal to render each chromosome a **contig** – an array of fragments covering the chromosome's entire DNA molecule.

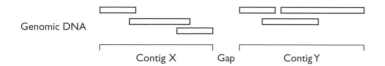

Some DNA sequences cannot be maintained stably in a plasmid or may even be toxic to *Escherichia coli*; "unclonable" genes include the metastasis-suppressor gene *KA11* and the mouse *mucin* gene *MUC2*. Highly repeated sequences in eukaryal genomes are difficult to isolate and to clone and have been left out of genomic sequencing, as the payoff is small relative to the effort required. However, it is often possible to isolate highly repeated sequences by physical methods.

Physical and Chromosomal Mapping

A high-density, physical map of the genome is a helpful tool in full genome sequencing. The principle of physical mapping is easy, but the work is tedious and expensive. Sequence-tagged sites (STSs) are used to make a physical map (Figure 30.2). An STS is a site within the genome that contains a unique sequence (200 to 500 nt); the short nucleotide sequence tags that site but no other. To discover and locate an STS, a genomic library is probed with a labeled fragment. Especially useful are STSs near the ends of cloned fragments, because such STSs "hit" two or more overlapping clones. This procedure was successfully used to make a physical map of autosomal genes of *Drosophila*, using 1923 STS oligonucleotides to probe a BAC library of 17,540 clones.

A detailed chromosomal map of cloned sequences in eukarya can be made by *in situ* annealing of labeled probes to condensed, stained chromosomes, usually from mitotic metaphase, the method of choice being fluorescence *in situ* hybridization

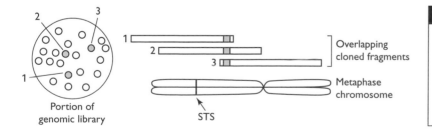

Portion of genomic library

Overlapping cloned fragments

Metaphase chromosome

STS

Fig 30.2 Using a STS to map cloned DNA sequences. One STS is cartooned here. A series of STSs would be used to tag chromosomal sites for all cloned DNA fragments making up the whole-chromosome contig.

(FISH) (Chapter 34). A short segment is subcloned from each cloned sequence to be mapped, and copies of the segment are labeled with a fluorescent dye. FISH can resolve two probed loci on mitotic chromosomes separated by as little as 1000 kb. When FISH is applied to interphase nuclei; resolution ranges from 50 to 2000 kb. In the polytene chromosomes of *Drosophila*, this method has worked extremely well.

A promising method for large-scale fine mapping is **dynamic molecular combing**. To perform dynamic molecular combing, genomic DNA is prepared in a way that minimizes shearing, so that it is in very long pieces; the DNA is embedded in agarose. A glass coverslip coated in trichlorosilane is dipped in the agarose-embedded DNA solution and then pulled out precisely. The DNA ($\sim 10^2$ genomes) adheres to the treated glass surface and is pulled and stretched in parallel lines (combed). When the coverslip dries, the stretched DNA adheres tightly to the surface. It is probed with dye-labeled, cloned DNA; if two close sequences are to be mapped, two dyes are used, say red and green. The physical distance between adjacent red and green spots on the cover slip measures the DNA distance between the sequences, in kilobases; the method is useful in the range 7 to 150 kb.

Sequencing Genomes

Determining the nucleotide sequence of an entire genome requires accurate, redundant sequencing of overlapping cloned segments (≈ 0.5 kb), and assembly of the sequences.

Accurate sequencing is performed on many overlapping segments, so that each segment is sequenced several times. In the case of *Drosophila*, 550-bp segments were sequenced with 15-fold coverage of sequences. More than 97% of cloned and sequenced segments were in euchromatin; highly repeated sequences in centric heterochromatin (about 60 Mb) were not sequenced. The final accuracy of sequencing was said to be at least 99.99% for sequences located in euchromatin and 99.5% for sequences in heterochromatin.

Two alternative strategies are **systematic** and **shotgun sequencing** (Figure 30.3). In the systematic approach, overlapping segments within every cloned piece of DNA are sequenced in series: the nth primer is based on the sequence of the $(n - 1)$th

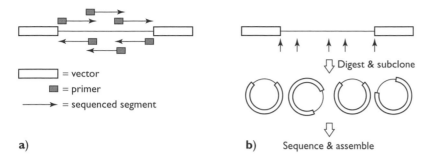

a)

= vector

= primer

= sequenced segment

b)

Digest & subclone

Sequence & assemble

segment. In the shotgun approach, overlapping segments within every cloned piece of DNA are sequenced in parallel: a cloned piece of DNA is cut or sheared randomly into small segments, which are subcloned and sequenced. The order of sequences is determined afterward as the first step of sequence assembly. In practice, the shotgun approach has proved to be faster.

The first entire genome to be sequenced was that of bacteriophage ϕX174 (5.4 kb) in 1977; 24 years later, sequencing of the euchromatic part of the human genome ($\approx$3000 Mb) was announced to be nearly complete. At this writing, about a thousand organismal genome projects are underway. Sequencing of the genomes of *Saccharomyces cerevisiae*, *E. coli*, *Caenorhabditis elegans*, and *Drosophila melanogaster*, all workhorses of experimental genetics, were completed, to a first approximation, in 1996, 1997, 1998, and 2000, respectively. Genome sequencing has been completed for 40+ bacteria, 11 archaea, 150+ viruses, and 50+ mitochondria and chloroplasts; the euchromatic portions of 9 eukaryal genomes have been sequenced. These sequences are available in public databases – i.e., highly organized repositories of information – in this case, primary DNA sequences. Anyone can access the gigantic database maintained by the National Institutes of Health (U.S.A.) via the Internet at http://www.ncbi.nlm.nih.gov/.

The completeness and accuracy claimed for genome sequencing should be viewed with healthy skepticism for the next few years, as should claims of the form, "The genome of species X contains N genes."

Fig 30.3 Systematic sequencing (**a**). Shotgun sequencing (**b**).

Analyzing Genome Structure

The nucleotide sequence of an organism's entire genome contains a huge amount of information. The body of methodology used to mine information from data bases of genomic DNA sequences is **bioinformatics**. The main goals of bioinformatics are to establish, maintain, and manage genomic databases as well as to analyze the structural and functional aspects of genes and proteins on a genome-wide basis. Most genomic and protein databases and the basic computer software to access and analyze them is freely and publicly accessible. Bioinformatics has extensive and growing commercial applications in health sciences and agriculture.

The levels of information range from the frequencies and distributions of short strings of nucleotides (1 to 10) to the classification, numbers, sizes, structures, physical and chromosomal distributions, and functions of large sequences, notably genes and their parts (codons, exons, introns, and regulatory sequences). Another task of genomic analysis is to discover how the genome is partitioned into chromosomes; in bacteria, plasmids must be distinguished from essential chromosomes. A few examples of genomic structural analysis follow.

Each nucleotide within one strand of DNA varies in frequency throughout a genome. For example, near origins of replication in circular bacterial chromosomes one strand is G-rich, making it possible to predict where origins of replication are in newly analyzed chromosomes. Genes acquired by horizontal transfer from other species may differ from the genome average in GC content.

Some oligonucleotide sequences are distributed nonrandomly. An example in *E. coli* is chi, the octamer (GCTGGTGG) to which recA protein binds to initiate recombination (Chapter 14). Chi is evenly distributed on the chromosome, but it is concentrated manyfold in the leading strand of replication. Chi's even distribution may reflect a selective advantage of the capacity for recombination repair everywhere in the genome. Chi's concentration on the leading strand may be related to its role in recombination repair at replication forks. The ability of primase to bind to part of the complementary sequence may also contribute to both deviations from randomness.

To inventory the coding portion of a genome, the database is scanned for sequences similar to known gene sequences, for open reading frames (ORFs), and, in eukarya, for exon-intron junctions. In the first approach, the genome is scanned for target sequences that match a search template (sequence of a known gene or functional portion of a gene). **Orthologs** are homologous genes in different species (e.g., the mouse and human TFIIA gene);

paralogs are homologous genes within one species (e.g., human β-globin genes). Inevitably, searches based on homology fail to find all genes in a genome. A gene may be missed because it is novel and has no homolog, or because it is too dissimilar from the search template, despite homology. To locate an ORF, one scans a sequence (both strands, both directions) for stop codons in the three possible reading frames; then one looks in the 5′ direction for ORFs, an initiator codon, and a promoter. Looking for genes based on ORFs is tricky in eukarya, in part because a large fraction of protein-coding genes have introns, which contain stop codons.

The larger a genome, the harder it is to inventory its genes. By some estimates, the 12.1-Mb genome of *Saccharomyces cerevisiae* contains 6340 genes, and the 4.6-Mb genome of *E. coli* contains about 2600 genes (transcription units), but yeast has 5885 different known proteins, only about 30% more than the 4288 different proteins of *E. coli*. It seems likely that more genes will be discovered in yeast in the next few years. The completely sequenced genome of *Drosophila* has about 14,000 genes, which may be an underestimate. There is wild disagreement about the number of genes in the human genome; since the year 2000, estimates between 26,000 and 154,000 have been proposed. Estimates have been based on ORFs in published sequences and on the number of unique cDNAs detected. An inventory of transcription units, splice variants of pre-mRNAs, variant peptides and polypeptides, pseudogenes, and potentially transcribed transposons has yet to be done in human genomics.

Functional Genomics

Genomics, like traditional genetics, can be applied to questions of gene expression and to the relationship between genes and phenotypes. Genomics can be used to identify genes that influence a common set of phenotypes, or genes whose expression is spatially, temporally, or developmentally restricted, or genes encoding a biochemical activity, or genes whose products bind to each other. Functional genomics is also helpful in analyzing alternative splicing, a prominent feature of animals. Three methods of functional genomics are described here: probing cDNA libraries with ESTs, RNA-DNA hybridization using DNA "chips," and the yeast two-hybrid system.

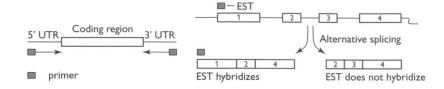

| **Fig 30.4** | Deriving ESTs by PCR. ESTs can distinguish splice-variant mRNAs. |

Expressed Sequence Tags

An **expressed sequence tag** (**EST**) is a segment of cDNA that identifies that cDNA. A few hundred base pairs long, an EST is synthesized by PCR from the 5′ or 3′ end of cDNA (Figure 30.4). ESTs can identify novel genes and alternative splice variants. Because splice variants appear at different times of development and in different tissues, ESTs are a tool to study tissue-specific or temporally patterned gene expression.

To discover alternative splice variants, a cDNA library is probed with ESTs. An EST hybridizes to clones containing different cDNAs from alternatively spliced mRNAs if the cDNAs share a 5′ exon or a 3′ exon from which the EST was synthesized.

DNA on Chips: Probing Microarrays to Study Genome Expression

Tissue-specific patterns of gene expression can be studied for large numbers of genes by probing a large collection of ESTs with labeled RNA. DNA sequences to be probed are attached to tiny spots on a silicon chip, arranged in a rectangular array, making a **DNA microarray** or **chip** (Figure 30.5). Each EST is assigned to a position in the microarray, which can be extraordinarily dense: $\sim 10^5$ different sequences can be fixed to one chip, 1.3 cm^2. The microarray is probed with labeled RNA (obtained from a specific tissue or from a specific developmental stage or specific time of day), which RNA hybridizes only to the DNA sharing sequences with

| **Fig 30.5** | Dye-labeled mRNA hybridized to a DNA microarray. |

Single mRNA   Mixture of mRNAs

the genes that encoded the RNA. A DNA chip can be reprobed a large number of times. DNA chips are used in basic science as well as in medical studies to establish a link between genes and a class of phenotypes (e.g., schizophrenia, circadian rhythms).

Two-Hybrid Systems That Aid the Genetic Dissection of Pathways

Proteins often work by binding to each other. For example, polypeptides bind to form a functioning multimer; proteins send signals by binding to other proteins; regulatory proteins bind co-operatively to DNA; protein-modifying enzymes bind to substrate proteins. The biologist may wish to search for novel proteins that interact in a pathway or process. If proteins X, Y, and Z bind to each other and act in a pathway but only X is known, Y and Z can be discovered by the two-hybrid method, a powerful tool to search for proteins that interact.

There are many versions of the two-hybrid method. A yeast two-hybrid system designed to study non-yeast genes is described here. This system is an *in vivo* assay for interacting proteins, using a DNA library cloned in yeast, and expressed by the host yeast cells.

The assay uses two reporter genes, *lacZ* and *Leu2*. The β-galactosidase encoded by *lacZ* converts a white substrate, X-gal, into a blue pigment, so that a clone expressing *lacZ* and grown on X-gal is blue. *Leu2* encodes an enzyme required for leucine synthesis. Cells with the null mutation $Leu2^0$ are auxotrophic for leucine, but $Leu2^0$ cells containing a $Leu2^+$ transgene are prototrophs.

Suppose one is interested in human protein X and wants to identify all the human proteins that work with X by binding to it. A series of proteins are tested, one at a time. To assay for binding between proteins X and Y, one makes a transgenic strain of yeast that synthesizes both human proteins and that has reporter genes whose transcription requires X and Y to bind. To make the expression of reporter genes contingent on X-Y binding, they are engineered so that their transcription requires an artificial gene activator, and the activator is constructed so that it works only when X and Y bind. How is that done?

Activators have both a DNA-binding domain (D) and an activation domain (A) that binds to RNA polymerase holoenzyme. Two hybrid polypeptides are present in the cells used in this assay, one polypeptide having the DNA-binding domain fused to protein Y

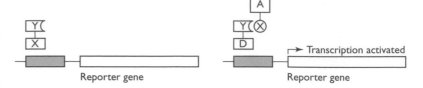

Fig 30.6 Bait and prey proteins bind to activate transcription in a two-hybrid system. D = DNA-binding domain, A = activation domain. Adapted from Brent and Finley 1997.

(D-Y) and the other having the activation domain fused to protein X (A-X). If X binds to Y, then the dimer activates transcription of the reporter genes. In this experiment, the investigator is fishing for proteins that interact with protein X; D-Y is the **bait**, while A-X is the fish or **prey** (Figure 30.6).

The assay requires a library of bait strains and a single prey strain (Figure 30.7). Each bait strain is chromosomally $Leu2^0$ and has a plasmid containing a fusion gene that encodes the DNA-binding domain connected to protein Y; the plasmid also contains $Leu2^+$ and $lacZ$, each with an upstream sequence to which the bait binds. The prey strain is also $Leu2^0$ and is transformed by a fusion gene encoding the activation domain and the prey protein X. To perform the assay, haploid bait strains are crossed to the haploid prey strain. The resulting diploid cells are cultured on two replicate plates, one lacking leucine and the other containing leucine and X-gal.

A positive clone could indicate a meaningful interaction between proteins, or it could be a false positive. A negative clone could indicate a genuine lack of interaction between X and Y, or it could be a false negative. False positives happen if proteins X and Y happen to stick to each other even though they do not in nature, or if they stick together in the assay and in nature, but their interaction has nothing to do with the process being

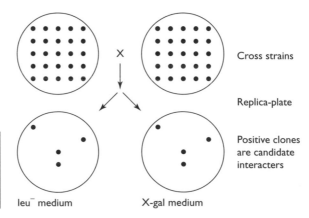

Fig 30.7 Searching for prey in a cross between a prey strain and many bait strains. Adapted from Brent and Finley 1997.

studied. False negatives happen if the proteins' secondary or tertiary structures are different in nature and in the assay, or if their interaction is too weak to measure in the assay. Candidate interactors are tested further.

Further Reading

Adams MD *et al.* 2000. The genome sequence of *Drosophila melanogaster. Science* 287:2185–2195.

Ainscough R *et al.* 1998. Genome sequence of the nematode *Caenorhabditis elegans:* a platform for investigating biology. *Science* 282:2012–2018.

Blattner FR *et al.* 1997. The complete genome sequence of *Escherichia coli* K-12. *Science* 277:1453–1462.

Brent R, Finley RL. 1997. Understanding gene and allele function with two-hybrid methods. *Annu. Rev. Genet.* 31:663–704.

Duggan D *et al.* 1999. Expression profiling using cDNA microarrays. *Nature Genet.* 21:10–14.

Gad S *et al.* 2001. Color bar coding the *BRCA1* gene on combed DNA: A useful strategy for detecting large gene rearrangements. *Genes Chromosomes Cancer* 31:75–84.

Goffean A *et al.* 1996. Life with 6000 genes. *Science* 274:546–567.

Hoskins RA *et al.* 2000. A BAC-based physical map of the major autosomes of *Drosophila melanogaster. Science* 287:2271–2274.

Morse R. 2000. The awesome power of yeast biochemical genomics. *Trends Genet.* 16:49–51.

Pandey A, Lewitter F. 1999. Nucleotide sequence databases: a gold mine for biologists. *Trends Biochem. Sci.* 24:276–280.

Zhu H *et al.* 2001. Analysis of expressed sequence tags in time-of-day-specific libraries of *Neurospora crassa* reveals novel clock-controlled genes. *Genetics* 157:1057–1065.

Chapter 31

Behavior of Genes and Alleles

Overview

To learn anything meaningful about a gene – its identity, its effects on an organism's traits, its interaction with other genes – one must study its behavior and inheritance *in vivo*. It is alarming how common are the patently false beliefs that a gene is a piece of DNA divorced from the organism and that a gene's existence, characteristics, functioning, and genetic interactions can be known from molecular analysis alone.

The inheritance of genes in complex organisms has two hallmarks: (1) it is inextricably tied to sex, in which chromosomes segregate, assort, and recombine, and (2) it works according to quantitative rules. This chapter introduces the rules of inheritance and ties them to the behavior of chromosomes in sexual reproduction.

Genetic Terminology

A diploid organism's genetic makeup is its **genotype**; depending on the context, this can refer to a single gene or to many genes (Figure 31.1). An organism's traits are its **phenotype(s)**. Genotype and environment both contribute to the phenotype. Variant forms of a gene (variant DNA sequences) are **alleles**. Functional alleles that occur in nature and make for a normal phenotype are **wild type** (this term is not applied to humans); abnormal alleles, especially those induced experimentally, are **mutant**. A gene's name usually relates to its phenotypic effects, and that name is abbreviated with one or a few letters; alleles are ideally designated by

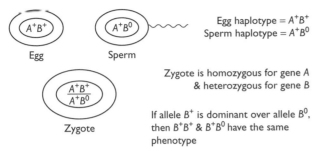

Egg haplotype = A^+B^+
Sperm haplotype = A^+B^0

Zygote is homozygous for gene A
& heterozygous for gene B

If allele B^+ is dominant over allele B^0,
then B^+B^+ & B^+B^0 have the same
phenotype

Fig 31.1 Some genetic terminology for gametes and zygote, for two genes A and B. Lines and slashes may be written to remind us that the genes are on chromosomes. $+$ = wild type; o = null allele.

Write A^+/A^+ B^+/B^0 if A & B are on different chromosomes.
Write A^+ B^+/A^+ B^0 if A & B are on the same chromosome.

meaningful superscripts. A genotype of two identical alleles is **homozygous** and a genotype of two different alleles is **heterozygous**; the corresponding nouns are **homozygote** and **heterozygote**. The combination of alleles in a haploid cell, or within a single chromosome, is a **haplotype**. A gene's name is normally italicized.

The haploid cells produced by meiosis receive a copy of each chromosome, except for sex chromosomes. During meiosis, paired homologous chromosomes undergo recombination, segregate, and assort independently. Gametes unite at random to produce diploid individuals, whose genotypes are combinations of genes present in the gametes. The resulting patterns of inheritance are the **segregation of alleles** and the **assortment of genes**.

Simple dominance is the masking of one allele by another in a heterozygote. If A^1/A^2 resembles A^1/A^1 rather than A^2/A^2, then A^1 is **dominant** to A^2, and A^2 is **recessive** to A^1. Simple dominance is dichotomous. In contrast, **quantitative dominance** admits of degrees: one allele may be partially dominant over another (Chapter 33). This chapter considers only simple dominance.

A **null allele** (sometimes called an **amorph**) does not code for a functional gene product either because the gene is not expressed or because the gene product is so aberrant as to be functionally equivalent to no gene product. A null allele is typically symbolized with the superscript 0. An allele that makes for a partially functional gene product is a **hypomorph**, **weak allele**, or **leaky allele**.

Underlying Assumptions of Genetic Analysis

An assumption underlying the analysis of sexual genetics is the lack of correlation between the haplotypes of uniting gametes. If two parents are A^1/A^2, it is assumed that the A^1 and A^2 gametes unite at random. There are very few exceptions to this assumption. The rules of segregation and assortment are often stated in a way that assumes equal survival of the genotypes and haplotypes. In many cases, this assumption is false.

A word of advice: when working on the genetics of sexual organisms, "think gametes." Focus on the haplotypes of gametes and their frequencies. The zygotic genotypes and their frequencies follow easily, based on random fertilization.

Transmission of Genes: Diploid Parent → Gametes

Genes Located on Different Chromosomes

Consider two genes in a diploid heterozygote, genotype A^1/A^2 B^1/B^2, where genes A and B are on different chromosomes.

In a heterozygote, half the haploid cells made in meiosis carry one allele, and half carry the other allele. This is a fundamental rule for sexually reproducing organisms.

> **Alleles segregate 1:1 in meiosis.**

Alleles A^1 and A^2 move into the four daughter cells made in meiosis independently of alleles B^1 and B^2 because the two chromosomes are physically independent and their arrangement on the spindle apparatus is random with respect to the alleles of any gene. Consequently, the four haplotypes, A^1B^1, A^2B^2, A^1B^2, and A^2B^1, are equally frequent in the haploid products of meiosis (Figures 31.2, 31.3).

> **Two genes on different chromosomes assort independently in meiosis.**

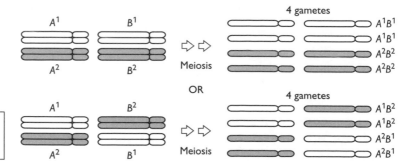

Fig 31.2 Two equally likely patterns of assortment for two chromosomes.

Chromosome 1

	$\frac{1}{2}A^1$	$\frac{1}{2}A^2$
$\frac{1}{2}B^1$	$\frac{1}{4}A^1B^1$	$\frac{1}{4}A^2B^1$
$\frac{1}{2}B^2$	$\frac{1}{4}A^1B^2$	$\frac{1}{4}A^2B^2$

Chromosome 2 *(label to the left of the table)*

Fig 31.3 Assortment table.

Genes Located on the Same Chromosome

Consider a diploid organism heterozygous for two genes, A and B, located on the same chromosome. Alleles segregate 1:1, of course, but with this arrangement, some pairs of genes do not assort independently. In Figure 31.4, an exchange occurs between genes A and B, yielding recombinants.

If the organisms' parents are $A^1B^1//A^1B^1$ and $A^2B^2//A^2B^2$, then the heterozygous offspring is $A^1B^1//A^2B^2$. The two genes may recombine during pachynema of meiosis.

If genes A and B occupy distant loci on a chromosome, they well may recombine at a maximal rate, behaving as if they were located on different chromosomes; in that case, A and B assort independently. Genes that assort independently are said to be genetically unlinked, even if they reside on the same chromosome.

If genes A and B tend to assort together, they are genetically linked. Genetic linkage is a restriction of recombination, usually caused by simple proximity of genes on a chromosome. Two genes that never recombine are completely linked.

Recombination between two genes is a random event, and chromosomes that are recombinant for two genes are produced with probability **r**. A heterozygote is predicted to produce r recombinant gametes and 1 − r nonrecombinant gametes.

> **Definition: For two genes, r is the probability that a gamete is recombinant.**

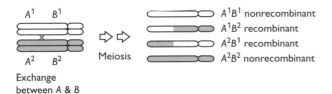

A^1 B^1

A^2 B^2

Exchange
between A & B

Meiosis

A^1B^1 nonrecombinant
A^1B^2 recombinant
A^2B^1 recombinant
A^2B^2 nonrecombinant

Fig 31.4 Exchange between A and B.

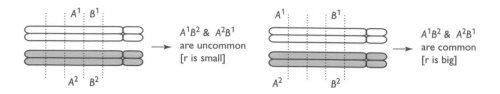

Fig 31.5 Relation of r to distance between genes A and B.

Not only does the average frequency of recombinant gametes produced by a heterozygote $A^1B^1//A^2B^2$ equal r, but the two kinds of recombinant, A^1B^2 and A^2B^1, are equally common on average; the frequency of each recombinant is $\frac{r}{2}$. Similarly, the two kinds of nonrecombinant, A^1B^1 and A^2B^2, are equally common on average. Their total frequency is $1 - r$ and the frequency of each one is $\frac{1}{2}(1 - r) = \frac{1}{2} - \frac{r}{2}$. Notice that these frequencies apply no matter what r is. The value of r can be discovered only empirically.

Figure 31.5 shows that r is related to the distance between genes A and B on the chromosome. Vertical lines represent points of exchange between paired chromosomes. Exchanges are chance events.

The range of r is from zero (no recombination) to $\frac{1}{2}$ (independent assortment). Why is $\frac{1}{2}$ the maximum value of r? If two genes are located on different chromosomes, they behave independently, making for equal frequencies of recombinant and nonrecombinant haplotypes in the gametes. The frequency of recombinant haplotypes, r, is $\frac{1}{2}$ when genes assort independently, and the frequency of recombinants is maximal with independent assortment.

Complete linkage and independent assortment are special cases of the general rule for gene assortment (Figure 31.6). If $r = 0$ then A and B exhibit complete linkage, and the frequency of each type of nonrecombinant gamete is $\frac{1}{2}$. If $r = \frac{1}{2}$ then A and B assort independently, and the frequency of each of the four haplotypes is $\frac{1}{4}$; $r = \frac{1}{2}$ if A and B are on different chromosomes or if they are far apart on one chromosome.

Fig 31.6 $A^1B^1//A^2B^2 \rightarrow$ meiosis $\rightarrow$ gametes. Gamete frequencies for assorting genes depend on r.

	$0 < r < \frac{1}{2}$	$r = 0$	$r = \frac{1}{2}$
Gamete	General Rule	Complete linkage	Independent assortment
A^1B^1	$\frac{1}{2} - \frac{1}{2}r$	$\frac{1}{2} - \frac{1}{2}r = \frac{1}{2}$	$\frac{1}{2} - \frac{1}{2}r = \frac{1}{4}$
A^2B^2	$\frac{1}{2} - \frac{1}{2}r$	$\frac{1}{2} - \frac{1}{2}r = \frac{1}{2}$	$\frac{1}{2} - \frac{1}{2}r = \frac{1}{4}$
A^1B^2	$\frac{1}{2}r$	$\frac{1}{2}r = 0$	$\frac{1}{2}r = \frac{1}{4}$
A^2B^1	$\frac{1}{2}r$	$\frac{1}{2}r = 0$	$\frac{1}{2}r = \frac{1}{4}$

Transmission of Genes:
Diploid Parents → Diploid Offspring

Predicting the Results of Testcrosses

A simple experiment to detect the haplotypes of the gametes is a **testcross** between the heterozygote $A^1B^1//A^0B^0$ and the recessive homozygote $A^0B^0//A^0B^0$ (Figure 31.7). The aim of a testcross is to observe and count the gametic haplotypes made by the heterozygous parent. In preparation for a testcross, homozygous parents (**P generation**) are crossed to yield heterozygous $A^1B^1//A^0B^0$ offspring (**F$_1$ [first filial] generation**). Assume equal survival of all genotypes.

All the male gametes are A^0B^0. The frequency of eggs with recombinant chromosomes is r, and the frequency of eggs with nonrecombinant chromosomes is $1 - r$. Therefore the frequency of each recombinant type in the offspring is $\frac{r}{2}$, and the frequency of each nonrecombinant type in the offspring is $\frac{1}{2}$ $\frac{r}{2}$.

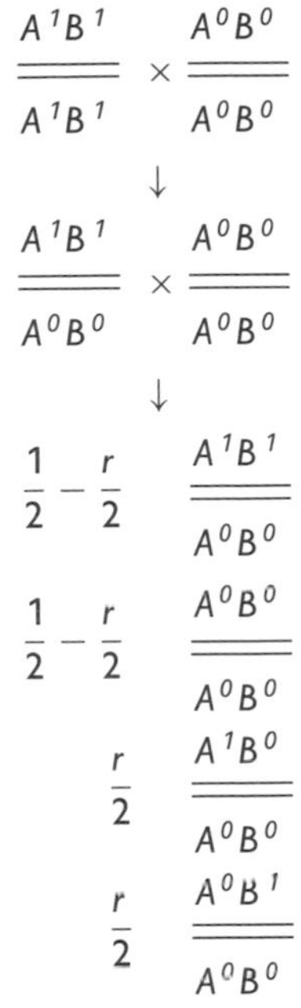

$$\frac{A^1B^1}{A^1B^1} \times \frac{A^0B^0}{A^0B^0}$$

$$\downarrow$$

$$\frac{A^1B^1}{A^0B^0} \times \frac{A^0B^0}{A^0B^0}$$

$$\downarrow$$

$$\frac{1}{2} - \frac{r}{2} \quad \frac{A^1B^1}{A^0B^0}$$

$$\frac{1}{2} - \frac{r}{2} \quad \frac{A^0B^0}{A^0B^0}$$

$$\frac{r}{2} \quad \frac{A^1B^0}{A^0B^0}$$

$$\frac{r}{2} \quad \frac{A^0B^1}{A^0B^0}$$

Fig 31.7 Segregation of alleles and assortment of genes in a testcross.

In a testcross with two genes, $A^1B^1//A^0B^0 \times A^0B^0//A^0B^0$, if all genotypes survive equally then the offspring are expected to occur in the following ratios:

- Segregation. Half the offspring are expected to be A^1/A^0 and the other half A^0/A^0; similarly, half are expected to be B^1/B^0 and the other half B^0/B^0.
- Assortment. $1 - r$ of the offspring are expected to have non-recombinant genotypes and r to have recombinant genotypes. The expected ratio of the four genotypes is: $\frac{1}{2} - \frac{r}{2} : \frac{1}{2} - \frac{r}{2} : \frac{r}{2} : \frac{r}{2}$. If two genes assort independently, $r = \frac{1}{2}$, and the testcross ratio is expected to be 1:1:1:1.

The expected ratios from genetic crosses are like expected ratios from tossing coins. The more times a coin is tossed, the closer the ratio of heads to tails is to 1. Similarly, large numbers of offspring must be observed to reduce the effects of random sampling. An important difference is, the probability of heads, $\frac{1}{2}$, is known beforehand, whereas the probability of getting a recombinant chromosome, r, is not known and must be estimated from experiment.

Sometimes testcrosses are not feasible, and more complicated crosses are performed instead. For example, if the recessive allele causes lethality or sterility, one may analyze the F$_2$ generation;

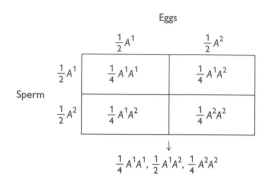

Fig 31.8 Fertilization table for a cross between two heterozygotes, $A^1A^2 \times A^1A^2$.

$F_1 \times F_1 \to F_2$. This cross is shown in Figure 31.8 for one gene; it can easily be extended to two or more genes. If there is partial linkage, then r can be calculated from the F_2 ratios but with difficulty.

If an experiment leads to ratios that differ from those expected under the segregation and assortment rules, then one or more assumptions about the genetic hypotheses or about the test are wrong; Chapter 33 considers some of these issues.

More Tests for Segregation and Assortment

Segregation and assortment were discovered by Mendel, who studied homozygous varieties of garden pea, *Pisum sativum*, in controlled, reciprocal crosses – each variety was used as female parents and as male parents in paired sets of crosses. He identified seven clear-cut genetic factors, which now we understand to be seven genes, and observed the segregation of alleles for each gene. He also observed the assortment of genes in crosses involving four pairwise combinations. In each of the four combinations, assortment was independent.

Mendel's Experiments, Example 1. Consider a testcross involving two genes of *P. sativum*, the r (rugosa) gene, which encodes starch-branching enzyme (SBEI) and the i gene, which influences endosperm color. Seeds homozygous for the recessive r^0 allele become wrinkled when they dry, owing to a deficiency in starch production, which makes the pea taste sweet and shrink when dried. r^+/r^+ ("+" = wild type) and r^+/r^0 peas retain a round shape when they dry, because they are full of starch. Seeds homozygous for the recessive i^0 allele have green endosperms, while i^+/i^+ and i^+/i^0 peas have yellow endosperms. The r gene is located on chromosome 5, and the i gene is located on chromosome 1, facts of which Mendel was unaware.

	Eggs		Observed	
Pollen	all $r^0 i^0$			
$\frac{1}{4} r^+ i^+$	$r^+/r^0 i^+/i^0$	round, yellow	55	r^+ & r^0 segregated $\sim$ 1:1
$\frac{1}{4} r^+ i^0$	$r^+/r^0 i^0/i^0$	round, green	51	i^+ & i^0 segregated $\sim$ 1:1
$\frac{1}{4} r^0 i^+$	$r^0/r^0 i^+/i^0$	wrinkled, yellow	49	The genes assorted independently.
$\frac{1}{4} r^0 i^0$	$r^0/r^0 i^0/i^0$	wrinkled, green	<u>53</u>	
		Total	208	

A cross between two varieties, $r^+/r^+ \ i^+/i^+$ and $r^0/r^0 \ i^0/i^0$, yielded round and yellow ($r^+/r^0 \ i^+/i^0$) F_1 seeds. Testcrosses, $r^0/r^0 \ i^0/i^0 \times r^+/r^0 \ i^+/i^0$, are expected to yield the four phenotypes in a 1:1:1:1 ratio (Figure 31.9). Each seed is an individual offspring, and as such represents independent meiosis and fertilization events.

Fig 31.9 Testcross between $r^0/r^0 i^0/i^0$ female and $r^+/r^0 i^+/i^0$ male.

Mendel's Experiments, Example 2. In another experiment, $r^+/r^0 \ i^+/i^0$ plants were self-fertilized to yield an F_2 generation (by definition, $F_1 \times F_1 = F_2$). Because r^+ and i^1 assort independently, figure out the expected phenotypic ratios for each gene and then combine the results. From the previous page, we know: $r^+/r^0 \times r^+/r^0 \rightarrow \frac{1}{4} r^+/r^+$, $\frac{1}{2} r^+/r^0$, and $\frac{1}{4} r^0/r^0$; because of dominance, any pea with an r^+ allele is round. Using a dash to indicate the second allele, the frequencies are $\frac{3}{4} r^+/-$ and $\frac{1}{4} r^0/r^0$. The same holds for the i gene, $i^+/i^0 \times i^+/i^0 \rightarrow \frac{3}{4} i^+/-$ and $\frac{1}{4} i^0/i^0$ (Figure 31.10).

Each predicted number is the product of the predicted frequency and 556, the total number of progeny observed in the experiment. The observed numbers of round versus wrinkled and yellow versus green are very close to those predicted by 1:1 segregation of alleles, and the observed numbers of the four phenotypes are very close to those predicted by independent assortment.

Fig 31.10 Table combining r and i genotypes (predicted frequencies).

	$\frac{3}{4} r^+/-$	$\frac{1}{4} r^0/r^0$
$\frac{3}{4} i^+/-$	9/16 $r^+/- i^+/-$	3/16 $r^0/r^0 i^+/-$
$\frac{1}{4} i^0/i^0$	3/16 $r^+/- i^0/i^0$	1/16 $r^0/r^0 i^0/i^0$

Phenotype	Genotypes	Observed	Predicted frequency (number)
round, yellow	$r^+/- i^+/-$	315	9/16 (312.75)
round, green	$r^+/- i^0/i^0$	101	3/16 (104.25)
wrinkled, yellow	$r^0/r^0 i^+/-$	108	3/16 (104.25)
wrinkled, green	$r^0/r^0 i^0/i^0$	32	1/16 (34.75)
Total		556	16/16 (556)

The F_2 is less informative than the testcross, because dominant phenotypes may be either homozygous or heterozygous; it is also harder to analyze. For these reasons, avoid the cross $F_1 \times F_1 \to F_2$, whenever a testcross is possible.

A Critical Cross Mendel did not Perform, Example 3. Here is a thought experiment. What if Mendel had performed a cross involving two genes, v and le, that he checked for segregation but not for assortment? These genes reside on chromosome 4, rather close together. The v allele is recessive and makes for constricted pea pods, and the dominant allele v^+ makes for inflated pea pods. The le gene affects the length of internode, between leaves on the stem; the recessive le allele makes for short internode length and dwarfing (plant height $\approx$30 cm) and the dominant le^+ allele makes for long internode length (plant height $\approx$200 cm). In the thought experiment, we cross $v/v\ le^+/le^+$ plants to $v/v\ le/le$ plants and testcross the $v^+/v\ le^+/le$ F_1 progeny the double recessive type. The exact recombination frequency is not known, but it is approximately 10%. Assuming that $r = 0.1$, the expected results would be:

$$v^+\ le^+ /\!/\ v\ le \times v\ le /\!/\ v\ le$$
$$\downarrow$$

$\frac{1}{2} - \frac{r}{2}\ v^+le^+ /\!/\ v\ le$	or 45% inflated pod, tall plant
$\frac{1}{2} - \frac{r}{2}\ v\ le /\!/\ v\ le$	or 45% constricted pod, dwarf plant
$\frac{r}{2}\quad v^+\ le /\!/\ v\ le$	or 5% inflated pod, dwarf plant
$\frac{r}{2}\quad v\ le^+ /\!/\ v\ le$	or 5% constricted pod, tall plant

There are 21 possible pairwise combinations of seven genes, two alleles/gene. The combination of v and le is one of the 17 possible combinations of traits Mendel did not study, and the only one that would clearly have failed to show independent assortment. Mendel wrongly concluded that independent assortment was a rule. Now we know that independent assortment is a special case and not a rule, even though it occurs commonly.

Sex Linkage

Some animals have **sex chromosomes**, often designated X and Y, which are essential in primary sex determination. All the other chromosomes are called **autosomes**. Sex-linked inheritance and autosomal inheritance follow different patterns. XY males, having a single copy of the X chromosome, are **hemizygous** for all genes on that chromosome, while females are either homozygous or heterozygous.

Solved Problem. In humans, red and green color vision depends on adjacent, X-linked genes that encode opsins, light-absorbing proteins. The opsin for green reception has an absorption peak at 450 nm, and the opsin for red reception has an absorption peak at 550 nm. Mutations in one or both genes are rather common in European populations. Males having a mutation in the gene for green-absorbing opsin have green-weak vision, because they lack a wild-type allele. A male inherits his X chromosome from his mother and his Y chromosome from his father. A female inherits an X chromosome from each parent. Therefore, males inherit red-weak and green-weak color vision from their mothers, while red-weak or green-weak females inherit defective alleles from both parents.

Suppose a woman's father has green-weak vision. She has normal color vision and she marries a man with normal color vision. What is the probability that their son will be green-weak?

The woman must be heterozygous for green-weakness.

Therefore P[son will be green-weak] $= \frac{1}{2}$.

What is the probability that the woman will have a green-weak daughter by this marriage? Zero, because each daughter receives a wild-type allele from her father.

Testing for Allelism between Recessive Variants

There are about 30 mutant strains of *P. sativum* with the wrinkled phenotype. These represent single mutations in five separate genes: *r*, *rb*, *rug3*, *rug4*, and *rug5* (*rug3*, *rug4*, and *rug5* are different genes, not alleles of the same gene) Each of these five genes codes for a different enzyme required for starch synthesis. The *r* gene studied by Mendel is on chromosome 5, *rb* is on chromosome 3, and the locations of the other three genes are not known.

However, determining whether two recessive mutations are alleles does not require knowing their chromosomal locations. A quick experiment, the **complementation test**, usually suffices. This test in animals and plants consists of crossing the two homozygous (recessive) mutant strains with identical or similar phenotypes to make a heterozygote. For example, all the homozygotes may be sterile, or may have light-colored eyes. If the offspring's phenotype is mutant like the parents' phenotype, then the mutants fail to complement, whereas if the offspring's phenotype is wild type, then the mutants complement. Conversely, the simplest explanation for failure to complement is that the two mutations are alleles, while the simplest explanation for complementation is that the two mutations are in different genes.

Solved Problem. Suppose you are asked to test five wrinkled-pea strains, each independently derived, and each carrying a mutation in a single gene. You are asked to find out which strains have allelic mutations, if any. For now, call the strains A, B, C, D, and E. Cross the strains in all possible combinations, A × A, A × B, . . . D × E. Excluding reciprocal crosses (e.g., not doing B × A if you do A × B), that makes 5!/2!3! = 10 crosses. ("!" = "factorial"; e.g., 3! = (3)(2)(1) = 6.)

Each wrinkled strain crossed to itself yields wrinkled offspring. A × B, A × E, and B × E also have wrinkled offspring, so A, B, and E must be homozygous for alleles of a single gene. C and D give wild-type offspring when crossed to any other strain, so they must be homozygous for recessive mutations in two more genes. In this example, the complementation test has identified three genes. The results of the 10 crosses, plus 5 control crosses (self-fertilization), can be summarized in a table, where the entry is w for wrinkled offspring and + for round offspring (Figure 31.11).

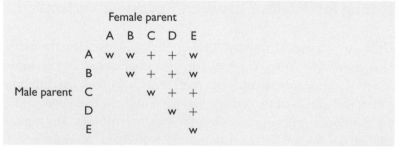

Fig 31.11 Results of 10 crosses plus 5 control cross. w = wrinkled, + = round.

		Female parent				
		A	B	C	D	E
	A	w	w	+	+	w
	B		w	+	+	w
Male parent	C			w	+	+
	D				w	+
	E					w

Strains A, B, and E have mutations in the *r* gene, strain C is homozygous for a mutation in the *rug4* gene, and strain D is mutant in the *rb* gene, facts you did not know when you performed the complementation test. Check your understanding by writing out the genotypes of the two parents and their offspring, for representative crosses. In the following, only "+" alleles are wild type and dominant. For example, strain A × strain B

$$r^1/r^1 \; rb^+/rb^+ \; rug4^+/rug4^+ \times r^2/r^2 \; rb^+/rb^+ \; rug4^+/rug4^+$$

$$\downarrow$$

$r^1/r^2 \; rb^+/rb^+ \; rug4^+/rug4^+$ wrinkled because it has only mutant alleles of r

strain A × strain C

$$r^1/r^1 \; rb^+/rb^+ \; rug4^+/rug4^+ \times r^+/r^+ \; rb^+/rb^+ \; rug4/rug4$$

$$\downarrow$$

$r^1/r^+ \; rb^+/rb^+ \; rug4^+/rug4$ round because it has a wild-type allele of each gene.

Further Reading

McKusick VA *et al.* 1998. *Mendelian Inheritance in Man.* Johns Hopkins University Press, Baltimore.

Mendel GJ. 1865. Versuche uber Planzen-Hybriden, Verh. Natur. Ver. Brunn 4 [English translation], In: Peters (ed.) Classic Papers, Prentice-Hall, Eastwood Cliffs NJ.

Sturtevant AH, Beadle GW. 1962. *An Introduction to Genetics.* Dover Publications, New York.

Chapter 32

Probability and Statistics Toolkit

Overview

Genetics is a quantitative science. Many genetic processes (mutation, recombination, assortment of genes) are not deterministic but are based on random events. The outcome of a random process cannot be known; its frequency of occurrence is predicted by its **probability**, or expected frequency. Thus, one needs statistics, the mathematical analysis of random variables. To design an experiment in genetics and to interpret the results correctly – or at all – requires knowledge of probability and statistics.

This chapter introduces a few rules of probability and gives the bare minimum of statistics for describing genetic phenomena and for testing genetic hypotheses. It is just a little toolkit, to be hauled out when you need it.

Rules of Probability

A probability is a number between 0 and 1 that predicts the frequency of a random event; the sum of the probabilities of all possible outcomes is 1. During meiosis in a heterozygote A^1/A^2, a gamete is equally likely to receive A^1 or A^2, and the probability of each is $\frac{1}{2}$. In general:

If event E happens in m of n equally likely events, then the probability of event E is P[E] = m/n.

Solved Problem. In a cross between two heterozygotes, $A^1/A^2 \times A^1/A^2$, what is the probability of a heterozygous offspring? Because alleles segregate 1:1, half the eggs and sperm are A^1 and half are A^2. Fertilization is random with respect to the gametes' haplotypes, so there are four equally likely outcomes for each progeny: A^1/A^1, A^1/A^2, A^2/A^1, or A^2/A^2. There are two ways to get a heterozygous offspring (egg = A^1 and sperm = A^2, or vice versa), and m/n = 2/4. Thus, the probability that one of the progeny will be heterozygous is $\frac{1}{2}$. This can be visualized in a fertilization table (Figure 32.1).

		Eggs		
		$\frac{1}{2}A^1$	$\frac{1}{2}A^2$	
Sperm or pollen	$\frac{1}{2}A^1$	$\frac{1}{4}A^1/A^1$	$\frac{1}{4}A^2/A^1$	$\rightarrow \frac{1}{2}A^1/A^2$
	$\frac{1}{2}A^2$	$\frac{1}{4}A^1/A^2$	$\frac{1}{4}A^2/A^2$	

Fig 32.1 Fertilization table for $A^1A^2 \times A^1A^2$.

Multiplication Rule. The probability that two independent events X and Y will both happen is the product of their probabilities.

$$\mathbf{P[X\ and\ Y] = P[X]\ P[Y].}$$

Solved Problem. A couple plan to have three children. They want to know the probability that all three will be boys.

$$P[3\ boys] = \{P[boy]\}^3 = [\tfrac{1}{2}]^3 = 1/8.$$

Solved Problem. Two genes are linked, and the recombination fraction r = 0.2 (This means that 20% of the chromosomes produced by meiosis are recombinant for the two genes). In a marriage between two heterozygous individuals ($A^1B^1//A^2B^2$), what is the expected frequency of $A^1B^2//A^2B^1$ children?

Half the recombinant chromosomes are A^1B^2, and half are A^2B^1, and the frequency of each is $\frac{r}{2} = 0.1$. $P[A^1B^2//A^2B^1] = (0.1)(0.1) = 0.01$, or 1%.

Addition Rule. If X and Y are independent events, the probability that X or Y will happen is the probability of X plus the probability of Y minus the probability of their joint occurrence. "X or Y" includes the co-occurrence of X and Y. If X and Y are mutually exclusive, then P[X and Y] = 0 and the addition rule still applies.

$$\mathbf{P[X\ or\ Y] = P[X] + P[Y] - P[X\ and\ Y].}$$

Solved Problem. Two genes, A and B, are genetically unlinked. A woman and man, both with genotype $A^1/A^0\ B^1/B^0$, plan to have children. They want to know the probability of having a child with genotype A^0/A^0 or $//B^0/B^0$, including $A^0/A^0B^0/B^0$.

$P[A^0/A^0 \text{ or } B^0/B^0] = P[A^0/A^0] + P[B^0/B^0] - P[A^0/A^0 \text{ and } B^0/B^0] = \frac{1}{4} + \frac{1}{4} - 1/16 = 7/16.$

Conditional Probability. The probability of event X, given that event Y has occurred, is $P[X|Y]$. Prior knowledge of event Y makes $P[X|Y]$ larger than $P[X]$:

$$P[X \mid Y] = P[X \text{ and } Y]/P[Y],\ P[Y] > 0.$$

Solved Problem. Lesch-Nyan disease is due to a recessive null allele (H^0) of the gene that encodes the enzyme hypoxanthine-guanine phosphoribosyl transferase (HGPRT). Homozygous H^0/H^0 individuals suffer a tragically painful life that invariably ends in early childhood. Both parents of a family are heterozygous, H^+/H^0. What is the probability that a surviving child is H^+/H^0?

The probability that a zygote will be H^0/H^0 is $\frac{1}{4}$, and the probability that a zygote will be either H^+/H^+ or H^+/H^0 is $\frac{3}{4}$.
$P[H^+/H^0|(H^+/H^+ \text{ or } H^+/H^0)] = P[H^+/H^0]/P[H^+/H^+ \text{ or } H^+/H^0] = \frac{1}{2}/(1 - \frac{1}{4}) = \frac{1}{2}/\frac{3}{4} = \frac{2}{3}.$

Using the Binomial Distribution

Suppose a random process has two possible outcomes, A and $\bar{A}$ (A and not-A), whose probabilities are p and q; $p + q = 1$. These events have a binomial distribution. Binomial probabilities apply to transmission genetics. Consider a random sample of n binomial events. The probability that outcome $\bar{A}$ occurs k times and outcome $\bar{A}$ occurs $n - k$ times is

$$P[A \text{ occurs k times}] = n!/[k!(n - k)!]\ p^k q^{n-k}.$$

In practice, one usually estimates p and q from observation or experiment, based on counts of A and $\bar{A}$ outcomes in a large number of cases, sampled randomly. The frequency of outcome A in a random sample is an estimate of the probability p. In n cases, the frequency of A is a $f_1 = k/n$ and the frequency of $\bar{A}$ is $f_2 = 1 - f_1 = (n - k)/n$. The **sample mean**, $\bar{X}$, is f_1. The **sample variance**, s^2, is a measure of variation around the mean, and the square root of the sample variance (s) is the **standard deviation**.

$$s^2 = f_1 f_2 / (n - 1)..........\text{formula for binomial sample variance.}$$

It is useful to compute the **95% confidence interval** for p: it is 95% certain that the true value of p lies in this interval. The exact formula for the 95% confidence interval of a binomial probability is too complicated to explain here; see Zar (1999). However, if $np \geq 100$, where p is the smaller proportion, the following approximation is good.

The 95% confidence interval of $p \approx [f_1 \pm 1.96s] + \frac{1}{2n}$, $np \geq 100$.

Solved Problem. What is the probability that a family of six children will contain three boys and three girls?

P[3 boys and 3 girls] $- [6!/3!3!] (1/2)^3 (1/2)^3 = [20](1/64) = 5/16 \approx 0.31$.

Solved Problem. Both parents of a family have genotype A^+/A^0. What is the probability that exactly three of six of their children will have genotype A^0/A^0?

P[3 of 6 children are A^0/A^0] $= [6!/3!3!](1/4)^3(3/4)^3 = (20)(27/4096)$
$$= 540/4096 = 135/1024 \approx 0.13.$$

Solved Problem. Genes A and B are linked. A testcross, $A^+B^+// A^0B^0 \times A^0B^0//A^0B^0$ is replicated until 3000 progeny are counted. Exactly 300 of the progeny are recombinant (A^+B^0 or A^0B^+). What is the 95% confidence interval for r?

Since $nr = 300$, use the formula

$$r = f_r \pm 1.96s + \frac{1}{2n}$$
$$f_r = 300/3000 = 0.100, s = \sqrt{[(f_1 f_2)/(n - 1)]}$$
$$= \sqrt{[(0.1)(0.9)/2999]}$$
$$= 0.0055$$
$$r = 0.10 \pm (1.96)(0.0055) + 1/6000, r = 0.10 \pm 0.01.$$

If the only error is due to random sampling, one is 95% sure that r lies between 0.09 and 0.11.

Using the Poisson Distribution

Imagine a discrete random process with a rare outcome, such as encountering a dead skunk on the side of a 1-km stretch of country road. The random variable X is the number of dead skunks in any 1-km stretch of road. X has a Poisson probability distribution whose mean and variance are λ. The probability that X is an integer k is:

$$P[X=k]=\frac{e^{-\lambda}\lambda^k}{k!}.$$

$k! = k(k - 1)(k - 2) \ldots (1)$, and $0! = 1$. Several random processes with a rare outcome turn up in genetics, including mutation of a gene and recombination within a small chromosomal interval. Poisson probabilities are useful to predict the frequency of rare, random events. Use a hand calculator to get $e^{-\lambda}$.

Example. A *his⁻* strain of *Salmonella* bacteria (requiring histidine to grow) mutates back to *his⁺* spontaneously. When the bacteria from a *his⁻* culture are added to Petri plates containing minimal medium (no histidine in the medium), each *his⁺* colony represents an independent back mutation. Each of 200 Petri plates was inoculated with 10^7 *his⁻* cells. From the results summarized below, calculate the mutation rate and the expected number of plates for each value of k (k $\geq$ 4 lumped).

k	N of plates	N of colonies	P[X = k]	Expected N of plates (200P)
0	60	0	0.273	54.5
1	68	68	0.354	70.9
2	43	86	0.230	46.1
3	20	60	0.100	20.0
4+	9	46	0.043	8.5
Total	200	260	1.000	200.0

$\lambda = 260/200 = 1.3$ colonies/plate; $e^{-\lambda} = e^{-1.3} = 0.273$

Calculate the probability and expected number of each result using the Poisson formula.

$$P[X = 0] = (e^{-1.3})(1.3)^0/0! = (0.273)(1)/1 = 0.273,$$

and the expected number of plates = $(0.273)(200) = 54.5$.

$$P[X = 1] = (e^{-1.3})(1.3)^1/1! = (0.273)(1.3)/1 = 0.354,$$

and the expected number of plates $= (0.354)(200) = 70.9$

... and so on for the remaining numbers of colonies per plate.

The mutation rate was 260 mutants/(200 plates)(10^7 cells/plate) $=$ 1.3×10^{-7} mutations per cell.

Using the Normal and t Distributions

Now consider phenotypes measured on a continuous scale – e.g., mass, length, and fertility. The frequency distribution of most continuously varying phenotypes is normal, which is a bell-shaped curve. When a random variable, X, is the sum of many other random variables, then X is distributed normally. The normal distribution has a mean of μ and a variance of σ^2. The normal curve is ubiquitous in biology because phenotypes are usually the result of many factors, each with a small effect that adds its influence. Usually, the geneticist is interested in a phenotype of a population, such as the fertility of humans or the fetal growth rate of mice, but has available only a random sample from the population of interest. In that case, the sample mean, $\bar{X}$, is used to estimate μ, and the sample variance, s^2, is used to estimate σ^2. The population mean and population variance are parameters of the population, while the **sample mean** and **sample variance** are **statistics**. It is best to perform most statistical calculations by computer, but the computational formulas for the sample mean and sample variance are easy.

$$\bar{X} = \frac{\Sigma x_i}{n}$$

$$s^2 = \frac{[\Sigma x_2{}^i - (\Sigma x_i)^2/n]}{(n-1)}.$$

The symbol Σ tells you to sum the numbers to the right of it; x_i is the ith observation in the sample, and $x_i{}^2$ is the square of the ith observation.

If a population has been studied intensively, then the mean and variance are known, for practical purposes. If a phenotype is normally distributed, and the population mean and variance are known, then any individual in that population can be compared with others in a quantitative way. First, convert every individual's phenotype, x_i, into a **standard normal score**, $z = (x_i - \mu)/\sigma$. If

Table 32.1. Areas under the normal curve

z	P	z	P	z	P	z	P
0.0	0.5000	1.0	0.1587	2.0	0.0228	3.0	0.0013
0.1	0.4602	1.1	0.1357	2.1	0.0179	3.1	0.0010
0.2	0.4207	1.2	0.1151	2.2	0.0139	3.2	0.0007
0.3	0.3821	1.3	0.0968	2.3	0.0107	3.3	0.0005
0.4	0.3446	1.4	0.0808	2.4	0.0082	3.4	0.0003
0.5	0.3085	1.5	0.0668	2.5	0.0062	3.5	0.00023
0.6	0.2743	1.6	0.0548	2.6	0.0047	3.6	0.00016
0.7	0.2420	1.7	0.0446	2.7	0.0035	3.7	0.00011
0.8	0.2119	1.8	0.0359	2.8	0.0026	3.8	0.00007
0.9	0.1841	1.9	0.0287	2.9	0.0019	3.9	0.00005

Area under the standard normal curve from z to ∞ (which is the probability P that a randomly chosen value from this distribution is as big as or bigger than z) tabled for values of z in tenths, up to 3.9. Interpolate to get an approximate area for a value between two tabled values of z.

the x_i's follow a normal distribution, then the z values follow a normal distribution with $\mu = 0$ and $\sigma = 1$. The tail lengths of mice or of elephants, and the longevity of nematodes or of turtles can be analyzed with the same standard curve. About two-thirds of a population have z scores between -1 and $+1$, and about 95% of the population have z scores between -2 and $+2$. Table 32.1. gives some values of z and their corresponding probabilities (i.e., $P[Z \geq z]$.

Solved Problem. The mean adult height of males born in the United States during the 1940s is 1.75 m, $\sigma = 0.05$ m. Calculate z scores for Bill Clinton (1.85 m) and George W. Bush (1.76 m), the 42nd and 43rd presidents of the United States.

Bill Clinton: $z = (1.85 - 1.75)/0.05 = 2.00$ (in the top 2.3%)

George W. Bush: $z = (1.76 - 1.75)/0.05 = 0.20$ (in the top 42%)

The experimentalist often wants to know how good the estimate of μ is for a normally distributed phenotype. To calculate the 95% confidence interval for μ, it is necessary to use a distribution called t, which is a bell-shaped curve, like the normal distribution, but with wider limits. t is a family of curves, and each member of the family has a parameter called **degrees of freedom** (**df**, or ν). To get the appropriate value of t in order to calculate a confidence interval, we need two numbers, ν and α; $\nu = n - 1$, where n is the sample size, and α is the probability of making the

Table 32.2. Critical values of the t distribution

ν	$\alpha = 0.10$	$\alpha = 0.05$	ν	$\alpha = 0.10$	$\alpha = 0.05$
1	6.314	12.706	15	1.753	2.131
2	2.920	4.303	20	1.725	2.086
3	2.353	3.182	25	1.708	2.060
4	2.132	2.776	30	1.687	2.042
5	2.015	2.571	40	1.684	2.021
6	1.943	2.447	60	1.671	2.000
7	1.895	2.365	120	1.658	1.980
8	1.860	2.306	∞	1.645	1.960
9	1.833	2.262			
10	1.812	2.228			

A few values of t are given for two probabilities, $\alpha = 0.10$ and $\alpha = 0.05$; α is the area under the curve for all $|t| > t_{\alpha/2}$; ν = degrees of freedom (df).

experimentalist's most dreaded mistake: rejecting a true hypothesis. Table 32.2 gives t for several values of ν and for two values of α. One is 95% sure that the population mean μ lies in the range $\bar{X} \pm t_{\nu;0.05} s/\sqrt{n}$.

Solved Problem. The heights (cm) of 10 pea plants of the dwarf variety (*le/le*) at first flowering were: 26, 26, 27, 28, 28, 29, 30, 31, 33, 34. [These data are not real.] What is the 95% confidence interval for the mean μ?

When we make a 95% confidence interval, we accept a 5% error rate (α). The sample size n = 10, so ν = 9. In Table 32.2, find t for ν = 9 and α = 0.05; $t_{9;0.05} = 2.262$. Calculate the mean and standard deviation: $\bar{X} = 29.2$, s = 2.8. The 95% confidence interval is: $\bar{X} \pm t_{\nu;0.05} s/\sqrt{n}$. $(t_9)(2.8)/\sqrt{9} = 2.1$, $29.2 - 2.1 = 27.1$ and $29.2 + 2.1 = 31.3$. Rounding to the nearest centimeter, we are 95% sure that the true mean of *le/le* peas grown under these conditions lies between 27 and 31 cm.

Solved Problem. In the section on the Poisson distribution, the rate of *his* back mutation was 1.3×10^{-7}. What is the approximate 95% confidence interval of this rate?

We observed that the expected and observed numbers of colonies per plate were pretty close, confirming what we had strong theoretical reason to believe, that mutations occur by a Poisson process. For this distribution, the variance equals the mean. We work with the raw number of colonies, 260. The total for 200 Petri plates was 260 mutant colonies, with $\sigma = \sqrt{260} = 16.1$. In this example, the sample size is so large that we use $\nu = \infty$; $t_{\infty;0.05} = 1.96$. The 95% confidence interval is $\mu \pm 1.96\sigma$, or $260 \pm (1.96)(16.1) = 260 \pm 31.6$. The total number of cells used to inoculate the plates was $(200)(10^7)$, so we divide (260 ± 32) by 2×10^9. We are 95% sure that the true mutation rate was between 1.1×10^{-7} and 1.5×10^{-7} per cell.

Correlation, Covariance, and Linear Regression

Covariance measures how two random variables, X and Y, vary together for pairs of observations (e.g., X = height and Y = mass of each object being measured). The sample covariance of X and Y is cov(X,Y); s_X and s_Y are the standard deviations of X and Y. Covariance can be negative, unlike variance, which is never negative. A correlation coefficient measures how two random variables, X and Y, covary, and ranges between -1 and $+1$. The product-moment correlation coefficient **r**, has these properties.

$$\mathbf{cov(X, Y) = [\Sigma X_i Y_i - \Sigma X_i \Sigma Y_i / n]/(n - 1)}$$

$$\mathbf{r = cov(X, Y)/s_X s_Y.}$$

Sometimes paired random variables, X and Y, are causally related. If X is a cause of Y, and if the quantitative relationship between them is a straight line, then linear regression is a procedure that estimates that relationship. Linear regression applies to the following situation: X and Y are paired, normally distributed variables observed in randomly and independently sampled items or individuals (there are other assumptions that need not concern us here). For example, the milk yield of cows (X) and of the cows' daughters (Y) could qualify for regression analysis. Linear regression produces an equation describing the relationship between X and Y or regression of X on Y. The method of least squares is a common way to get a regression equation. The line's intercept, a, is estimated from the means, $\bar{Y}$ and $\bar{X}$.

$$\mathbf{Y = a + bX}$$

$$\mathbf{b = cov(X, Y)/s^2_X.}$$

Solved Problem. Flower length, in millimeter, is measured for parents (X, the average of the two parents) and for offspring (Y). In the following sample, calculate parent–offspring regression.

Y 12 15 14 15 13 13 16 15 19 15
X 10 12 13 13 14 14 14 16 17 18
$\bar{Y} = 14.7$, $s^2_Y = 3.79$, $\bar{X} = 14.1$, $s^2_X = 5.66$,
$cov(X,Y) = [(10)(12) + \ldots + (18)(15) - (141)(147)/10]/9 = 2.81$
$r = 2.81/\sqrt{3.79}\ \sqrt{5.66} = 0.61$
$b = 2.81/5.66 = 0.50$
The regression equation is true for the X and Y means, so
$14.7 = a + (0.50)(14.1)$, $a = 7.7$.
The regression equation is: $Y = 7.7 + 0.50X$.

Testing Genetic Hypotheses Using Chi-Square

A goal of many genetic experiments is to check whether genotype
frequencies conform to the segregation and assortment rules. The
geneticist needs a rigorous statistical test to compare experimental results with the results expected under some genetic hypothesis, such as 1:1 segregation of alleles. The best test is based on a
family of curves called **chi-square** (χ^2). Table 32.3 gives some critical values of χ^2. Like the t curves, there is a χ^2 curve for every
integer called degree(s) of freedom (df). For simple genetics experiments, **but not in general**, df = the number of outcomes (e.g.,
phenotypes) minus 1. The χ^2 test can be applied to any experiment with two or more outcomes, where some theory predicts
the number of occurrences of each outcome. Call the number of
occurrences for the ith outcome the observed number (O_i) and
call the predicted number of occurrences for the ith outcome the
expected number (E_i). From these two sets of numbers, calculate

**Table 32.3. Critical values of the
chi-square distribution**

ν	$\chi^2{}_{0.05}$
1	3.841
2	5.991
3	7.815
4	9.488
5	11.070
6	12.592
7	14.067
8	15.507

Values of $\chi^2{}_{0.05,\nu}$ are given for $\nu = 1, \ldots, 8$; $\nu =$
degrees of freedom (df).

a statistic X^2 (X-squared) based on their differences:

$$X^2 = \Sigma(O_i - E_i)^2/E_i$$

To perform a χ^2 test, calculate X^2 based on the hypothesis. Choose the critical χ^2, with the appropriate degrees of freedom, to make the probability of rejecting a true hypothesis equal to 5%. If $X^2 \leq \chi^2$, then accept the hypothesis; otherwise reject it.

Thought Experiments. To get a feel for the test, perform a couple of thought experiments. Suppose you toss a coin 10 times and you get 4 heads and 6 tails. If the coin is fair, the probability of heads (p) is 1/2, the probability of tails (q) is 1/2, the expected number of heads is np = (10)(1/2) = 5, and the expected number of tails is nq = (10)(1/2) = 5. The results obviously fit what we expect to get with a fair coin, but do the χ^2 test anyway.

Outcome	O	E	D	D^2/E
heads	4	5	1	0.2
tails	6	5	−1	0.2
				0.4 = X^2

There are n = 2 outcomes and n − 1 = 1 degree of freedom. The critical value of χ_1^2 = 3.84. 0.4 < 3.84, so we accept the hypothesis that the coin is fair. This value of χ_1^2 is associated with a probability of 0.05, which is to say that 5% of the values of χ_1^2 exceed 3.84. Since X^2 is distributed as χ_1^2, then 5% of the time *by chance alone* one encounters a value of X^2 bigger than 3.84. Does the experiment prove that the coin is fair? No. Proof would require exhaustive tests of the coin's behavior.

Suppose you toss the same coin 100 times and get 40 heads and 60 tails, exactly the same proportions as in the first experiment. Now the coin seems biased. Obviously, sample size matters.

Outcome	O	E	D	D^2/E
heads	40	50	10	2.0
tails	60	50	−10	2.0
				4.0 = X^2

There is still 1 degree of freedom. $X^2 > \chi_1^2$, so reject the hypothesis that the coin is fair and accept the hypothesis that the coin is biased toward tails. We do this knowing that 1 time in 20 the χ^2 test will falsely tell us the coin is biased when it is fair. Inferences about random processes are probabilistic.

Solved Problem. Mendel performed a testcross with peas, r^+/r^0 i^+/i^0 × r^0/r^0 i^0/i^0. If the genes are unlinked, the female plant is expected to produce the four types of egg (r^+i^+, r^+i^0, r^0i^+, and r^0i^0) in a ratio of 1:1:1:1. He asked if the alleles segregated 1:1 and if the two genes sorted independently. Use chi-square to test these hypotheses.

Genotype	N	Phenotype	
r^+/r^0 i^+/i^0	55	round, yellow	106 r^+/r^0 and 101 r^0/r^0
r^+/r^0 i^0/i^0	51	round, green	103 i^+/i^0 and 104 i^0/i^0
r^0/r^0 i^+/i^0	48	wrinkled, yellow	
r^0/r^0 i^0/i^0	53	wrinkled, green	
	207		

Did r^+ and r^0 segregate 1:1? The expected numbers of r^+/r^0 and r^0/r^0 are (1/2)(207) = 103.5 of each. $X^2 = (103.5 - 106)^2/103.5 + (103.5 - 101)^2/103.5 = 0.39$ There is one degree of freedom, for $2 - 1 = 1$; $\chi_1^2 = 3.84$. Since $X^2 < \chi_1^2$, we accept the hypothesis that r^+ and r^0 segregated 1:1.

Did i^+ and i^0 segregate 1:1? Again, the expected numbers are 103.5 and 103.5. X^2 is $(103.5 - 103)^2/103.5 + (103.5 - 104)^2/103.5 = 0.005$, which is less than χ_1^2, so we accept the hypothesis that the i^+ and i^0 alleles segregated 1:1.

Did the two genes assort independently? The expected numbers of the four genotypes are (1/4)(207) = 51.75 each. There are 4 $- 1 = 3$ degrees of freedom. $X^2 = [(51.75 - 55)^2 + (51.75 - 51)^2 + (51.75 - 48)^2 + (51.75 - 53)^2]/51.75 = 0.52$ $X^2 < \chi_3^2 = 7.815$, so we accept the hypothesis that the two genes assort independently. Do Mendel's results prove 1:1 segregation and independent assortment in this instance? No, the true ratios may differ from the ones hypothesized by Mendel. Proof requires further experiments.

Solved Problem. Segregation and assortment was studied in *Drosophila* for two genes. The recessive *vestigial* (*vg*) mutation makes for no wings and the recessive *brown* (*bw*) mutation makes for brown eyes; *vg*$^+$ flies have wings and *bw*$^+$ have red eyes. A wingless *vg/vg* female was crossed to a brown-eyed *bw/bw* male, and female offspring were testcrossed to wingless brown-eyed males (*vg/vg bw/bw*). As above use chi-square for the tests.

The testcross vg/vg^+ bw/bw^+ × vg/vg bw/bw gave the following results:

Genotype	Phenotype	O	E	D	D^2/E
vg/vg^+ bw/bw^+	winged, brown-eyed	72	55	17	5.25
vg/vg^+ bw/bw^+	winged, red-eyed	44	55	−11	2.20
vg/vg^+ bw/bw^+	wingless, brown-eyed	40	55	−15	4.09
vg/vg^+ bw/bw^+	wingless, red-eyed	64	55	9	1.47
		220	220		$X^2 = 13.01$

Did vg^+ and vg segregate 1:1? The expected number of each is $(1/2)(220) = 110$, and there were 116 winged flies and 104 wingless flies. With two categories, df = 1. $X^2 = (116 - 110)^2/110 + (104 - 110)^2/110 = 0.65$, which is smaller than $\chi_1^2 = 3.84$. We accept the hypothesis that two alleles segregated 1:1. Similarly, bw and bw^+ segregated 1:1.

Did the two genes assort independently? The expected number of each genotype, with independent assortment, is 55. There are four categories, so df = 3. Because $X^2 = 13.01 > \chi_3^2 = 7.815$, we conclude that the two genes did not assort independently. They appear to be linked, with a map distance of $(40 + 44)(100)/220 = 38$ map units.

Further Reading

Kempthorne O. 1973. *An Introduction to Genetic Statistics.* John Wiley and Sons, New York.

Weir BS. 1996. *Genetic Data Analysis II.* Sinauer Associates. Sunderland, MA.

Zar, JH. 1999. *Biostatistical Analysis,* 4th ed. Prentice-Hall, Upper Saddle River, NJ.

Chapter 33

Genes, Environment, and Interactions

Overview

There is not a 1:1 relationship between a gene and a trait. Genes are **pleiotropic**: they affect many traits. Conversely, traits are **multigenic** – influenced by many genes. The environment, too, plays a role in determining phenotype. Further complexity arises from the quantitative nature of many traits – for example, size, fertility, or the probability of developing a tumor. Genetic effects considered in this chapter include **quantitative dominance**, **epistasis** (interactions between genes), **penetrance and expressivity** (variation in phenotype exhibited by one genotype), and **genotype-environment interactions**.

Multigenic Determination of Phenotype

Only the nucleic acid sequence of the primary transcript is surely and invariably determined by a single gene. All other classes of phenotype require the participation of two or more genes. However, from a practical standpoint, a single gene often can be considered to code for the amino acid sequence of a polypeptide, as other determinants of amino acid sequence usually do not act in a gene product-specific way. At high levels of biological organization, all traits are multigenic. Nearly all individuals of a species may be phenotypically alike for some trait, no matter how many genes influence it, because there may be no genetic variation for the trait, or else such variation may be hidden.

Example of Multigenic Influence. When peas dry they shrink. Excessive shrinkage, due to a shortage of starch, causes the seed coat to wrinkle. A mutation in any of five genes that encode enzymes essential for starch synthesis (*r*, *rb*, *rug3*, *rug4*, and *rug5*) can cause peas to wrinkle. Because the biosynthetic pathway from simple sugars to starch entails more than five enzymes (not to mention regulatory proteins), many more than five genes could, if mutant, cause peas to wrinkle.

Pleiotropy

Multiple phenotypic influences of a gene or genotype is pleiotropy. Most genes are pleiotropic; i.e., one gene typically influences many phenotypes. An apparent lack of pleiotropy in a gene likely reflects our ignorance of that gene.

Example of Pleiotropy: pku. The enzyme phenylalanine hydroxylase (PAH) converts phenylalanine to tyrosine. More than 380 recessive mutations in the human *PAH* gene (on chromosome 12) are known, and many alleles cause a deficit in PAH as well as many pleiotropic effects. Persons deficient in PAH use an alternative catabolic pathway to degrade phenylalanine. Phenylketones, which are by-products of the alternative pathway, are excreted in urine. Thus, the term **phenylketonuria** (**pku**) is applied to PAH^-/PAH^- homozygotes. PAH deficiency also causes phenylalanine to accumulate, because the alternative catabolic pathway is inefficient. The high level of phenylalanine during infancy causes brain damage. Over 90% of persons with untreated pku have an IQ below 50. The PAH deficiency causes low melanin production, so persons with pku have fair skin, blue eyes, and light hair.

Example of Pleiotropy: *dunce*. A gene of *Drosophila*, *dunce* (*dnc*), encodes cAMP phosphodiesterase, which converts cyclic adenosine monophosphate (cyclic AMP) into AMP. Mutations that reduce or abolish the enzyme's activity are pleiotropic, affecting behavior, fertility, and survival. The gene is called *dunce* because several mutant alleles of the gene result in a learning disability; *dnc* mutants do poorly in learning to avoid a novel odor paired with electric shock or to approach a novel odor paired with sugar. In addition, female *dnc* mutants mate more frequently than wild-type females, and their life span is reduced by half. Female *dnc* mutants are sterile; most eggs are not laid, and the few eggs that are laid are fragile, leading to death of the embryo.

Gene Interactions

For all multigenic phenotypes, there is scope for interaction between genes. Genetic interaction can take many forms. One gene may mask the effect of another gene; this is **epistasis**. A **modifier** gene can modify the expression of another gene in a positive or negative way. An **enhancer gene**, not to be confused with a gene's enhancer (a position-independent regulatory sequence to which transcription factors bind), increases the effect of another gene, while a **suppressor** gene decreases the effect of another gene.

Enhancer and suppressor genes are powerful tools for studying gene action and for discovering novel genes. An enhancer or suppressor gene product may work in many ways – e.g., by regulating transcription, regulating RNA processing, modifying translation, or acting in a biosynthetic pathway.

Example of Epistasis: Nonsense Suppressor. A chain-terminating mutation (e.g., UAG [≡stop]) can be suppressed by a mutation in a tRNA gene. For example, a codon for glutamine is 5′CAG; the tRNA for this codon has the anticodon 5′CUG. If the gene for this tRNA mutates to give the anticodon CUA, then the UAG in the mutated mRNA will be translated as glutamine, and the CAA→CUA mutation in the gln-tRNA gene is a suppressor mutation. In a strain of *Escherichia coli* that contains the CAA→CUA suppressor mutation, how can stop codons be read during translations of all the normal mRNAs, and how can glutamine be incorporated into proteins, using the normal UUG codon of mRNA? The first part is easy: most terminator sequences of mRNAs contain two or three stop codons in a row. The second part is harder. In this example, *E. coli* has two copies of the gene encoding glutamine tRNA with a CUG anticodon; the normal copy works for CAG codons. In other cases of stop-codon-suppressors, wobble-pairing allows for translation of the normal codon.

Example of Epistasis: Blood Antigens. ABO blood group antigens are on the surface of blood cells. Type A blood has A antigen, type B blood has B antigen, and type O blood scores negative on a test for A or B antigen. The *Secretor* (*Se*) gene controls secretion of these antigens into other body fluids, such as saliva. A recessive allele, Se^0, makes for no secretion of ABO antigens.

ABO Genotype	Secretor Genotype	Blood Type	Saliva Type
I^A/I^A or I^A/I^0	Se^+/Se^+ or Se^+/Se^0	A	A
I^B/I^B or I^B/I^0		B	B
I^0/I^0		O	O
I^A/I^A or I^A/I^0	Se^0/Se^0	A	O
I^B/I^B or I^B/I^0		B	O
I^0		O	O

Example of Epistasis: Beadex, a Regulatory Gene. Directing the development of a fly's wing or a baby's arm takes the concerted work of many genes. Investigation of these complicated processes is at an early stage. Studying mutations that produce birth defects in *Drosophila* has led to most of our current knowledge of developmental genetics. The first such mutation was *Beadex*, which causes scalloped wings. *Beadex* is a regulatory gene, a fact discovered by looking for second-site mutations (in other genes) that either enhance *Beadex* (make the birth defects more severe) or suppress *Beadex* (ameliorate the birth defects). Two genes of this kind are *apterous* and *Chip*, each of which is essential to development, and each of which was independently discovered in searches for mutations affecting wing morphogenesis or embryonic survival. Knowing that these genes interact in wing development led to the question, "How do they interact at the molecular level?"

The Beadex protein has two LIM domains; LIM domains bind to other proteins. Other regulatory genes in *Drosophila*, worms, and vertebrates code for LIM domains. Apterous also has two LIM domains, and Chip has a LIM-binding domain. Beadex, Apterous, and Chip form a multimeric complex regulating transcription during wing morphogenesis. In vertebrates, LIM-domain proteins form comparable complexes to regulate development of limbs and blood cells. Malfunction of LIM-coding genes contributes to development of T-cell leukemias.

In animals, some genes affect the sexes differently. Gene expression may be different between the sexes, or even limited to one sex, or the affected tissue may be limited to one sex.

Sometimes a distinction is made between sex-limited genetic effects (one sex is affected, the other is not) and sex-influenced genetic effects (one sex is affected more frequently or more strongly than the other sex).

Example of Sex-Limited Gene Expression. *BRCA1*. *BRCA1* is an autosomal gene that protects against cancer (it is a tumor suppressor gene). Rare null alleles of this gene are dominant in causing increased risk for several types of cancer. The lifetime risk of breast cancer for women in the United States is about 12%; for men, about 0.6%. The lifetime risk of breast cancer for a woman carrying one null mutation in *BRCA1* is about 85%, a 7-fold increase in risk. In males, a mutation in *BRCA1* does not appreciably increase the risk of breast cancer, but it does increase the risk of prostate cancer. *BRCA1* mutations are sex-limited for the risk of prostate cancer because only males have prostate glands; they are sex-limited for the risk of breast cancer for unknown reasons.

Penetrance and Expressivity

If a trait varies in a continuous way from individual to individual, it is a **quantitative trait**. A single genotype produces a range of phenotypes, which are often distributed normally. When a single genotype shows variability for a quantitative phenotype, then the genotype shows **variable expressivity**. The strength of expression of a genotype is its expressivity. When some individuals with a mutant genotype have a normal phenotype, the mutant genotype shows **incomplete penetrance**. The fraction of individuals with a mutant genotype that have a mutant phenotype is the penetrance of that genotype.

Example of Incomplete Penetrance and Variable Expressivity: Retinoblastoma. Retinoblastoma is a rare disease (300 cases/yr in the United States), tumors of the retina in early childhood. Forty percent of the cases are due to an inherited mutation of the *RB* gene and 60% are sporadic, due to non-inherited, spontaneous mutation of the *RB* gene. About 90% of children who inherit a mutant *RB* allele develop a tumor of the eye, meaning that the penetrance of *RB* mutations for eye tumors is 90%. A person heterozygous for an RB mutation is virtually certain to have a second cancer later in life. By age 30, more than 90% of RB^0/RB^+ persons develop at least one tumor outside the eye, often tumors of bone. Because the number of tumors that develop in an RB^0 heterozygote varies, RB mutations show variable expressivity as well as incomplete penetrance.

Simple Dominance

Simple dominance, which applies only to discrete traits, is an all-or-nothing phenomenon. For example in *Pisum sativum*, both le^+/le^+ and le^+/le plants are tall (mean $\approx$ 2 m), and le/le plants are dwarf (mean $\approx$ 30 cm). Mendel did not record the heights of plants in analyzing the le gene, and it appears no one else has either; a plant is scored as tall or dwarf, a clear-cut difference. Simple dominance usually has a simple explanation: the dominant allele codes for a functional gene product, while the recessive allele codes for a gene product with little or no function. The presence of a gene product from one allele of the heterozygote thus masks the absence of gene product from the other allele. When both alleles of a heterozygote make for distinct, co-expressed phenotypes, they are **codominant**. Differences in amino acid sequence of proteins are codominant, by definition.

Example of Simple Dominance: ABO Blood Types. ABO blood group alleles in humans illustrate simple dominance further. The H antigen is displayed on the surface of cells in nearly all humans. H antigen can be modified to make either A antigen or B antigen. The I gene codes for an enzyme that glycosylates the H antigen. The I^A allele codes for an enzyme that adds N-acetylglucosamine to H; the I^B allele codes for an enzyme that adds galactose to H; and the I^0 allele is a null allele (no H-modifying enzyme is encoded). I^A and I^B are dominant over I^0, but I^A and I^B do not mask each other; the alleles are codominant.

Genotype	Blood type	Dominance
I^0/I^0	O	I^0 is recessive
I^A/I^A, I^A/I^0	A	I^A is dominant to I^0
I^B/I^B, I^B/I^0	B	I^B is dominant to I^0
I^A/I^B	AB	I^A and I^B are codominant

Quantitative Dominance

In stark contrast to simple dominance, quantitative dominance admits of degrees, and can be quantified by a ratio. To establish the degree of quantitative dominance between two alleles A^1 and A^2, it is necessary to measure the phenotypes of all three genotypes, A^1/A^1, A^1/A^2, and A^2/A^2. Furthermore, many individuals of each genotype, independent of each other but sharing a similar environment, are sampled randomly and measured. Following this, the mean values are compared by a rigorous statistical procedure. These elaborate procedures are required because both environmental and genetic factors subject the phenotypes to random variation, producing uncertainty in their measurement. The genetic variation for the phenotype of interest and the genetic variation for the degree of dominance come from the **genetic background** – all the genome, other than the A gene.

When the phenotypic mean of A^1/A^2 is halfway between the means of A^1/A^1 and A^2/A^2, then neither A^1 nor A^2 is dominant: there is **no dominance**. When the mean of A^1/A^2 is statistically indistinguishable from that of A^1/A^1, then A^1 is **completely dominant** over A^2. When the mean of A^1/A^2 is statistically indistinguishable from that of A^2/A^2, then A^2 is completely dominant over A^1. Finally, when the mean of A^1/A^2 falls between the midparent value and the phenotype of A^1/A^1, then A^1 is **partially dominant** over A^2; similarly, A^2 may be partially dominant to A^1. The difference between the heterozygote and the nearest homozygote, divided by the difference between the homozygote and the midparent value, is **d**, the degree of dominance. Degree of dominance depends on environment and genetic background as well as the relationship between alleles:

$$\mathbf{d} = |\mathbf{H} - \mathbf{M}|/|\mathbf{P} - \mathbf{M}|.$$

H = the mean of the heterozygote, M = the midparent value, and P is the mean of one parent. The absolute values are used to make d positive. If the heterozygote is identical to the midparent value, then d = 0 (no dominance); if the heterozygote is identical to A^1/A^1, then d = 1 (complete dominance); if $0 < d < 1$, then dominance is incomplete. If the heterozygote has a more extreme value than one homozygote, then there is **overdominance**.

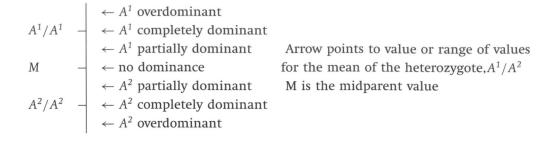

A^1/A^1 $\leftarrow A^1$ overdominant
$\leftarrow A^1$ completely dominant
$\leftarrow A^1$ partially dominant   Arrow points to value or range of values
M $\leftarrow$ no dominance   for the mean of the heterozygote, A^1/A^2
$\leftarrow A^2$ partially dominant   M is the midparent value
A^2/A^2 $\leftarrow A^2$ completely dominant
$\leftarrow A^2$ overdominant

Example of Quantitative Dominance: Dwarf Mice. The $Pit1^{dw}$ mutation of *Mus* causes dwarfing, owing to low levels of growth hormone. Adult $Pit1^{dw}/Pit1^{dw}$ mice have about 30% the mass of normal mice. In a controlled experiment, mice were reared in similar environments, and their mass was determined at 6 weeks of age.

Genotype	Mean mass
$Pit1^{dw}/Pit1^{dw}$	5.4
$Pit1^{dw}/+$	12.2
$+/+$	14.8

The three means were statistically different, and the heterozygote was statistically greater than the midparent value of 10.1. The degree of dominance is d = $(12.2 - 10.1)/(14.8 - 10.1) = 0.45$. Conclusion: the wild-type allele $Pit1^+$ is partially dominant over $Pit1^{dw}$, with d = 0.45.

Example of Quantitative Dominance: Eye Shape in *Drosophila*. The *Bar* (B) mutation of *Drosophila* makes for a narrow eye, and **always** exhibits simple dominance over B^+ (heterozygotes invariably have narrowing of the eye). Quantitative dominance varies from strain to strain, as it depends on genetic background. Environment was the same for both strains in the experiment.

Genotype	Mean width of eye, μm	
	Strain 1	Strain 2
$+/+$	400	420
$B/+$	190	390
B/B	110	120
midparent value	255	270

For strain 1, B is partially dominant over B^+. The degree of dominance is d = $(255 - 190)/(255 - 110) = 0.45$. For strain 2, B^+ is partially dominant over B. The degree of dominance is d = $(390 - 270)/(420 - 270) = 0.73$. This example illustrates the point that quantitative dominance is **not** a direct relationship between alleles but rather depends on interactions with other genes.

Genotype–Environment Interaction

Genotype–environment interaction is differential response of genotypes to different environments. When an environmental condition mimics the effect of a genotype, the resulting phenotype is called a **phenocopy**. Conditional alleles, such as temperature-sensitive alleles, exemplify genotype-environment interactions.

Example of Phenocopy: a Temperature-Sensitive Mutant. There are temperature-sensitive (ts) mutant alleles of the X-linked *paralytic (para)* gene in the fruit fly *Drosophila*. *Drosophila* is cold-blooded and assumes the temperature of its environment. The *para* gene encodes a voltage-dependent sodium channel found in nerve cells. Females homozygous for *para^{ts}* and males hemizygous for this mutation pass out at temperatures above 29°C. Flies recover completely and rapidly when the temperature is lowered to 25°C. The temperature-sensitive ion channel changes shape at restrictive temperatures and upon doing so becomes nonfunctional, blocking nerve conduction. The *para^{ts}* mutants are phenocopies of wild type at 25°C.

Example of Phenocopy: Curing pku. Phenylketonuria (pku), described in the section on pleiotropy, is caused in most instances by null mutations in the *PAH* gene. If a PAH^0/PAH^0 infant breast-feeds, then IQ in most instances is very low. On the other hand, if the infant with pku is given a diet low in phenylalanine, then IQ is usually normal. The normal mental development induced by dietary control is phenocopy of the PAH^+ genotype. For this reason, virtually all infants born in hospitals in the United States are tested for pku at birth. When a PAH^+ infant has high levels of blood phenylalanine from some other metabolic failure, IQ is also severely reduced; here, development of low intelligence is a phenocopy of pku.

Analysis of Human Pedigrees

To study human genetics is more difficult than to study genetics in experimental organisms. An important method is the analysis of pedigrees, where the genotypes of individuals can be either ascertained or inferred. In a pedigree analyzing one trait, a circle denotes a female, a square denotes a male, a darkened symbol denotes an affected person, and an open symbol denotes an unaffected person. Lines connect mates to each other and to their offspring. Children are shown below parents, and the birth order of children is left to right. An arrow points to the **proband**, the affected person whose condition led to the pedigree. Potential complications in reading a human pedigree are penetrance, environmental effects including maternal effects, sex-limited (versus sex-linked) traits, and misreported paternity. The following pedigrees illustrate single-gene traits. Approach each pedigree with specific hypotheses, and write down each hypothesis. For example, an X-linked mutation would be X^a; affected males would be X^aY and female carriers would be X^aX^+. An autosomal recessive mutation would be a, carriers would be a^+/a, and affected persons would be a/a. In the three example pedigrees, a mutation in a single gene is associated with the phenotype of interest.

Pedigree Analysis: Huntington Disease. Huntington disease (HD) is a late-onset, lethal neurodegenerative disease due to dominant mutations in an autosomal gene (Figure 33.1). The arrow points to the proband. HD is transmitted from father to son in one case, ruling out sex-linkage. It is found in females, ruling out Y-linkage (which anyway is nearly too rare in humans to consider as a reasonable possibility). About half the offspring of persons with HD develop HD, which is to be expected for an autosomal dominant.

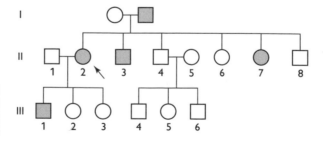

Fig 33.1 Pedigree of a family in which Huntington disease is inherited. II 2 = proband.

Pedigree Analysis: Sex-Linked Color-Blindness. Red-weak vision is due to mutations in the X-linked opsin gene, *RCP*. A *RCP⁰* null allele is segregating in this family. The arrow points to the propositus. Only males of this family are red-weak, and the children of the red-weak male in generation I have normal color vision (Figure 33.2).

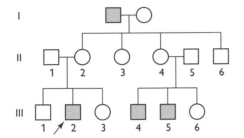

Pedigree Analysis: Cystic Fibrosis. Persons with cystic fibrosis have severe, chronic pulmonary obstruction. The disease is caused by recessive mutations of the autosomal *CFTR* gene, which codes for a chloride channel. $CFTR^0$ (a null allele) is segregating in this family. The arrow points to the propositus. Affected persons do not have affected parents, ruling out a dominant mutation. Males and females with the trait are equally frequent, ruling out sex linkage. III2 and III3 are first cousins; the double line between them denotes marriage between relatives (Figure 33.3).

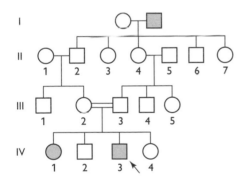

Further Reading

Erlandsen H, Stevens RC. 1999. The structural basis of phenylketonuria. *Mol. Genet. Metab.* 68:103–125.

Scriver CR, Waters PJ. 1999. Monogenic traits are not simple: lessons from phenylketonuria. *Trends Genet.* 7:267–272.

Shoresh M *et al.* 1998. Overexpression *Beadex* mutations and loss-of-function heldup-a mutations in *Drosophila* affect the 3′ regulatory and coding components of the *Dlmo* gene. *Genetics* 150:283–299.

Tobin AJ, Signer ER. 2000. Huntington's disease: the challenge for cell biologists. *Trends Cell Biol.* 12:531–536.

Chapter 34

Locating Genes

Overview

Genetics often entails "detective work." Detectives spend a lot of effort locating suspects, and geneticists spend a lot of effort locating genes. Finding a gene and putting it on a map has several applications:

- To determine that a genetic factor maps to a point on a chromosome (Did one character do the deed, or is somebody in cahoots?)
- To aid in further study of a gene
- To help diagnose a human genetic disorder

This chapter tells about genetic, chromosomal, and physical maps. Genetic maps are based on genetic linkage. Chromosome maps place genes relative to chromosome landmarks. Physical maps are based on DNA sequencing.

Recombination and Chromosome Exchanges

General recombination occurs when paired homologous chromosomes undergo exchanges, or crossovers. Homologous chromosomes recombine in pachynema, the third stage of prophase I of meiosis. General recombination rarely happens in cells undergoing mitotic cycles, but that is not taken up here.

At any chromosome **locus** (location; a gene's address), a crossover is a random event – random in the sense that it is unpredictable and not predetermined. The farther apart two loci, the greater the chance for an exchange between them. Therefore, the probability of an exchange is positively correlated with distance. The probability of recovering a recombinant chromosome, r, measures **genetic distance**; express r as a percentage. One **map unit** separates two linked genes if 1% of the recovered chromosomes are recombinant for these two genes. This definition works no matter what the biological situation. The genetic distance between genes on a T4 phage chromosome is defined the same way as genetic distance between genes of humans, corn, or *Drosophila*.

Genetic distance = % recombinant chromosomes = 100r%.

A map unit is equivalent to 1% recombinant chromosomes.

Now comes a surprise. On average, half the chromosomes that underwent a meiotic exchange between two loci are recombinant for those loci. In meiosis, if no crossover happens between two genes, no recombinants result, but if one or more crossovers happen between two genes, then on average 2/4 chromosomes will be recombinant. Therefore, the percentage of recombinant chromosomes is $\frac{1}{2}$ the percentage of meioses in which crossovers have occurred between two genes. How this works is best explained in diagrams. First, consider no exchange and one exchange (Figure 34.1).

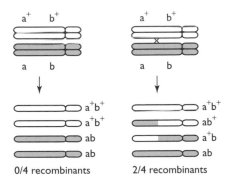

0/4 recombinants   2/4 recombinants

Fig 34.1 No crossover (left). Single crossover (right).

Next, consider the three kinds of double crossover (Figure 34.2). Number the strands 1 through 4, and count the number of ways each kind of double crossover can occur. There are four ways a two-strand double crossover can happen, between 1 and 3, 1 and 4, 2 and 3, or 2 and 4 (crossovers are between homologs, not within a chromosome). There are also four ways four-strand double crossovers can happen, crossovers on the left between 1 and 3, 1 and 4, 2 and 3, or 2 and 4, coupled with crossovers on the right between the remaining two strands. Therefore, the average frequency of recombinant strands observed

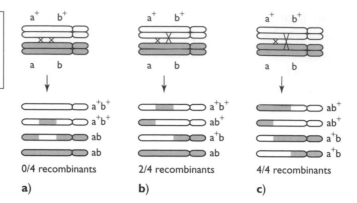

Fig 34.2 Two-strand double crossover (**a**). Three-strand double crossover (**b**). Four-strand double crossover (**c**).

in meioses with double crossovers is 2/4, the same as with a single crossover. No matter how many exchanges occur between two genes, the result is the same: on average, half the chromosomes in a meiotic exchange between two loci are recombinant for those loci.

The biological significance is clear – for genes located on the same chromosome, the maximum frequency of recombinants is 50%, the same frequency of recombinants observed for genes on separate chromosomes. Genes located far apart on a chromosome undergo independent assortment.

Measuring Genetic Distance

In experiments with plants and animals, the testcross (heterozygote × recessive type) is an informative and efficient procedure for measuring the strength of linkage between two or more genes. The object of such a testcross is to count recombinant and nonrecombinant chromosomes. In the following testcross, the a and b loci are $100(n_3 + n_4)/N$ map units apart (Figure 34.3).

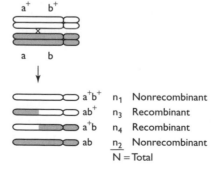

Fig 34.3 The cross $a^+b^+ //ab \times ab//ab$ produces 4 kinds of progeny.

$ec^+ sn^+$		$ec\ sn$	egg haplotype	Phenotype*	Number	Type
$ec\ sn$	×	Y	ec^+sn^+	wild type	1507	Parental
			$ec\ sn$	double mutant	1250	Parental
			$ec\ sn^+$	echinous	231	Recombinant
			$ec^+\ sn$	singed	260	Recombinant
			Total		3248	

*Females received an *ec sn* chromosome from the father and males received a Y; as males and females were phenotypically equivalent they were lumped together.

Fig 34.4 Testcross in *Drosophila* employing *ec* and *sn*.

Here, the genetic distance between a and b is $(n_3 + n_4)/N$. The chromosomes counted in a mapping experiment are a random sample. The aim is to estimate a population mean from a sample. An estimate of high quality requires careful, correct performance of the experiment and a large sample size, N. A deviation from 1:1 segregation, which may happen if a homozygote has low survival, can cause inaccuracy in the estimate of the map distance.

Solved Problem: Linkage between Two Genes. In *Drosophila*, the X-linked recessive mutant *singed* (*sn*) has crooked bristles (they look as though they have been singed by flame) and the recessive mutant *echinous* (*ec*) makes for rough eyes. A testcross was performed (Figure 34.4) to find the map distance.

There were 491 recombinant chromosomes out of 3248 chromosomes inherited from the mother. The fraction of recombinants was 0.151, so the estimate of the genetic distance is 15.1 map units; $s_r^2 = (0.151)(0.849)/3247 = 0.0000395$. The 95% confidence interval (Chapter 32) is:

$$r \pm 1.96s_r + 1/2N = 0.151 \pm (1.96)(0.00628) + 0.00015$$

$$= 0.151 \pm 0.012$$

In other words, we are 95% sure that the true map distance between *ec* and *sn* is between 13.9 and 16.3 map units. Extensive testing has given a distance of 15.5 map units.

Mapping Three Genetic Loci Simultaneously

It is better to map three or four genetic loci at a time than two, because locus order can be established from one experiment, and

Fig 34.5 Testcross in *Drosophila* employing *ec*, *sn*, and *cv*.

$cv^+/cv^+ \ ec^+/ec^+ \ sn^+/sn^+ \times cv \ ec \ sn//Y$ [Gene order unknown!]

↓

$cv^+/cv \ ec^+/ec \ sn^+/sn \times cv \ ec \ sn//Y$

↓

$cv^+ \ ec^+ \ sn^+$	$cv^+ \ ec^+ \ sn$	$cv^+ \ ec \ sn^+$	$cv^+ \ ec \ sn$
$cv \ ec \ sn$	$cv \ ec \ sn^+$	$cv \ ec^+ \ sn$	$cv \ ec^+ \ sn^+$

(all possible haplotypes found in eggs)

one gets information about the rate of double recombination events. The method for mapping three linked loci, given here, can easily be extended to a larger number of genes. On the other hand, scoring a large number of markers simultaneously is tedious and error-prone.

Solved Problem: Mapping Three Genes. This is an extension of the two-gene example, with a recessive X-linked mutation, *crossveinless* (*cv*), added. *cv* makes for a missing wing vein. We know that *sn*, *ec*, and *cv* are on the X chromosome, although at this point we do not know whether *cv* is genetically linked to the other two, nor do we know the gene order. Again, we use a testcross (Figure 34.5).

First organize the haplotypes in pairs. The original parents were homozygous for all three genes, so the nonrecombinant chromosomes are **$cv^+ \ ec^+ \ sn^+$** and *cv ec sn*. For each of the remaining six chromosomes, two genes will be the parental combination of alleles, and the third gene will have recombined with the other two. For example, **cv^+** *ec sn* has the *ec sn* combination of alleles from one chromosome and **cv^+** from the other.

Haplotype	Number	Category
cv^+, ec^+, sn^+	2967	Parental
cv, *ec*, *sn*	2880	Parental
cv^+, *ec*, *sn*	9	"*cv*" recombinant (*cv* recombined)
cv, **ec^+**, **sn^+**	12	"*cv*" recombinant
cv, *ec*, **sn^+**	228	"*sn*" recombinant (*sn* recombined)
cv^+, **ec^+**, *sn*	239	"*sn*" recombinant
cv, **ec^+**, *sn*	279	"*ec*" recombinant (*ec* recombined)
cv^+, *ec*, **sn^+**	270	"*ec*" recombinant
Total	6884	

Step 1. For each of the recombinant haplotypes, determine which two genes have not recombined, and which one has recombined (right, above).

Step 2. Determine the gene order. The double recombinants are the rarest, in this case cv^+ ec sn and cv ec^+ sn^+. Because cv recombined in the double recombinants, it is located between the other two. The gene order is ec-cv-sn (equivalent to sn-cv-ec). Sketch the chromosome arrangement to confirm this:

Step 3. Determine the two pairs of single recombinant types.

$$ec^+ \quad\text{---}\quad cv^+ \quad\text{---}\quad sn^+$$
$$ec \quad\text{---}\quad cv \quad\text{---}\quad sn$$

crossover between ec and cv

279 ec^+—— cv —— sn
270 ec —— cv^+ —— sn

crossover between cv and sn

239 ec^+ —— cv^+ —— sn
228 ec —— cv —— sn^+

Step 4. Sum the number of chromosomes resulting from exchanges between the left and center genes, and compute the map distance. Then do the same for the center and right genes. The number of double recombinants must be added to each total, because a double exchange is represents exchanges in both regions.

For ec —— cv, the map distance is $100[279 + 270 + 9 + 12]/6884 = 8.3$ map units.

For cv —— sn, the map distance is $100[239 + 228 + 9 + 12]/6884 = 7.1$ map units.

Step 5. Optionally, determine if recombination between ec and cv was independent of recombination between cv and sn. Given independence, the expected number of double recombinants is the product of the two recombination rates and the total number of chromosomes: $(0.083)(0.071)(6884) = 40.6$

Map	χ^2 test of independence	Doubles	Other	
8.3 7.1	Expected	40.6	6843.4	
ec----cv---sn	Observed	21	6863	
	d^2/e	9.46	0.06	$X^2 = 9.52$

$\chi^2{}_1 = 3.84$, so reject the hypothesis that recombination in the 2 regions occurred independently.

The observed number of doubles, 21, was smaller than the expected number of doubles, 40.6. This is an example of **interference**. Recombination in the ec-cv region interfered with recombination in the cv-sn region. Perhaps the enzyme complex that causes recombination in one part of a chromosome is "used up" after acting once. In any case, the interference is a deviation from independence; recombination therefore does not occur at completely arbitrary chromosome loci. When there are more recombinants than expected, then the deviation from independence is called by the quaint term negative interference. Negative interference characterizes recombination in phages.

The genetic map we made is a straight line. This fits well with our notion that genes are arranged on a linear DNA molecule. This point may seem ridiculously obvious, but consider that genetic maps were discovered long before it was known that a chromosome is made of DNA and not too long after it was proved that genes reside on chromosomes. The genetic map of E. coli, which is based on the time it takes to transfer genes via conjugation between an Hfr and F⁻ strain (Chapter 18), is a circle, matching the chromosome shape.

Some Practical Considerations

The precision and accuracy of genetic maps depends on careful experimental methods, sample sizes, the penetrance of markers, and the genetic distances between markers. A small pilot study will often yield a rough value for genetic distance. From the rough value, estimate the sample size needed to achieve a desired confidence interval, based on binomial means and variances.

Additivity is a good characteristic for maps. Genetic map distances are additive if distances are small, and otherwise not. From the experiment above, we add the two distances (8.3 + 7.1) to get 15.4 map units between *echinous* and *singed*. This is close to the value of 15.1 map units obtained in the first solved problem.

As a rule of thumb, when making a genetic map, choose markers that are no more than about 10 map units apart. Genetic markers more than 30 map units apart are undesirable, as genetic distance is underestimated. One method of correcting the tendency of large map distances between a pair of markers to underestimate true genetic distance is application of a map function. There are several such functions invented by theoreticians. A map function is no substitute for experiment, though.

Mapping Human Genes

Mapping genes in humans is difficult because controlled crosses cannot be performed, allelic variants may be rare, and families are small, which limits the chromosome yield per heterozygous individual. A major biological problem is ascertaining **haplotype phase** in the parental chromosomes. Haplotype phase is the association of a particular combination of alleles in the heterozygote. The arrangement $a^+ b^+//a\ b$ is called **coupling** or **cis**, and the arrangement $a^+ b//a\ b^+$ is called **repulsion** or **trans**. Phase must be known to measure linkage; it can be inferred from the grandparents' genotypes. Also, if two gene loci are close, then little recombination occurs between them, so that the more frequently observed phase in children is likely the parental phase.

Example. Linkage in Humans. Linkage is measured between two genes on the human X chromosome, XG and OA. Persons with the dominant XG^+ allele (sometimes called Xg^a) test positive for XG blood antigen, while persons having only the XG^- allele test negative. The rare recessive OA^- mutation makes for ocular albinism. In the United States, about 45% of females are heterozygous XG^+/XG^-. Informative families have a heterozygous mother, $XG^+/XG^-\ OA^+/OA^-$ and many sons (Figure 34.6). Alleles of both genes segregate 1:1, so one expects 50% of the sons to have ocular albinism, and 50% of the sons to test positive for XG blood antigen. The gene arrangement in the mother can be either cis or trans.

$$
\begin{array}{cc}
\textbf{CIS} & \textbf{TRANS} \\
\dfrac{XG^+\quad OA^+}{XG^-\quad OA^-} & \dfrac{XG^+\quad OA^-}{XG^-\quad OA^+} \\
\downarrow & \downarrow \\
\approx 86\%\ XG^+\ OA^+ \text{ and } XG^-\ OA^- & \approx 86\%\ XG^+\ OA^- \text{ and } XG^-\ OA^+ \\
\approx 14\%\ XG^+\ OA^- \text{ and } XG^-\ OA^+ & \approx 14\%\ XG^+\ OA^+ \text{ and } XG^-\ OA^-
\end{array}
$$

No matter how closely linked XG and OA are, they will appear to assort independently if the cis and trans mothers are mixed together.

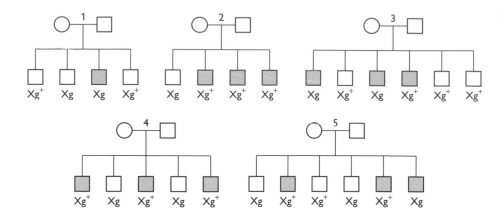

Fig 34.6 Human pedigrees for XG and OA: dark square = OA⁻, only sons are shown.

There appear to be four recombinants, one recombinant from family 1, one from family 3, and two from family 5; the estimate of the genetic distance from this sample (N = 25 chromosomes) is (100)(4)/25 = 16 map units. The sample size is small, but if we assigned haplotype phase correctly, we can infer that OA and XG are linked; the 95% confidence interval is 4.5 to 39.7 map units. Such a large interval is not acceptable. Either more data must be collected, or else nearby, variable DNA sequences (e.g., microsatellite markers) should be used instead.

The method presented here simply requires one to count recombinant chromosomes and to express this as a percentage of the total number of chromosomes observed in the study. A better procedure for detecting linkage in humans is based on the method of maximum likelihood, giving a log-odds ratio, or **LOD score**. This is beyond the scope of this book; consult Ott (1991).

The Method of Affected Sib-Pairs

The method of affected sibs is especially useful for finding molecular markers linked to a human gene, or postulated gene. This method does not require knowledge of haplotype phase. The power of the affected sib-pair method is undiminished by incomplete penetrance but is greatly reduced by the converse, environmental causation of the disorder (phenocopy).

Suppose homozygous a/a individuals tend to develop a rare medical disorder. Consider a chromosomally linked molecular marker m with multiple, codominant alleles (e.g., a microsatellite marker) in a/a children.

$$a\,m^1 /\!/ A\,m^2 \times a\,m^3 /\!/ A\,m^4$$
$$\downarrow$$
$$a\,m^1 /\!/ a\,m^3,\; a\,m^1 /\!/ a\,m^4,\; a\,m^2 /\!/ a\,m^3,\; \text{and}\; a\,m^2 /\!/ a\,m^4$$

If a and m are genetically linked, then more affected sibs will share two m alleles than will share no m alleles. The probabilities that a/a sibs will share 2, 1, or 0 m alleles are given for several degrees of linkage as follows:

Probability that a/a sibs will share m alleles

Shared alleles	$r = 0.5$	$r = 0.2$	$r = 0.1$	$r = 0.01$	$r = 0$
2	0.25	0.46	0.67	0.96	1
1	0.50	0.44	0.30	0.04	0
0	0.25	0.10	0.03	$< \sim 0.01$	0

The stronger the genetic linkage between gene a and marker m, the higher the fraction of sibs sharing two m alleles and the lower the fraction of sibs sharing no m alleles, the higher this ratio. The same basic principle holds for all cases where both parents are heterozygous for both genes, a and m.

Genetic, Chromosomal, and Physical Distances

Any detectable DNA sequences can be mapped genetically. RFLPs and microsatellites are useful for this purpose. The genetic entity being mapped, be it a gene or a nonfunctional DNA sequence, is a **genetic marker** on the genetic map.

The farther apart two loci, the greater the chance for an exchange between them. However, although chromosomal distance is positively correlated with genetic distance, exchanges are less frequent in the heterochromatin-rich regions the centromeres and telomeres. Within and among chromosomes of the same organism, a chromosomal distance (in a mitotic metaphase chromosome examined under a specific set of conditions) in micrometers corresponds to different genetic distances. Furthermore, the overall frequency of recombination varies among species and between the sexes. In female humans, the mean number of chiasmata

Fig 34.7 *Drosophila* chromosome 2, with gene loci (map distances) and corresponding chromosomal loci.

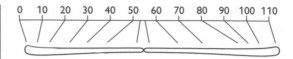

(exchanges) per meiosis is about 80; in male humans, it is about 55. This female-male asymmetry is common in mammals. In many insects, recombination occurs only in the homogametic sex (females in all but the Lepidoptera), and not at all in the heterogametic sex.

Genes are mapped to chromosome loci by using chromosome rearrangements. When a translocation or inversion breakpoint occurs within a gene, the gene is mutated. The chromosomal locus of the gene can be determined by microscopic examination of stained mitotic metaphase chromosomes in individuals heterozygous for the chromosome rearrangement. Genes can be mapped to chromosomal deletions in complementation tests. A deletion heterozygote with one recessive allele for a deleted gene has a mutant phenotype.

Exchange rates vary from region to region of each chromosome, yielding gradual changes in recombination rates through the chromosome as well as recombination "hot spots" and "cold spots" kilobases to megabases in length (Figure 34.7).

The average physical distance (kb of DNA) per map unit varies among life forms. The following gives this relationship for five organisms and a virus.

Life form	DNA, genome	Map units, genome	kb of DNA/map unit
human	3.2×10^6 kb	4155	770*
Zea mays	2.4×10^6 kb	1000	2400
Arabadopsis	1.2×10^5 kb	635	190
Drosophila	1.8×10^5 kb	285	630
yeast	1.2×10^4 kb	4130	3
λ phage	48.5 kb	70	0.7

*Yu *et al.* 2001 measured 600 kb/map unit in females, 1090 kb/map unit in males, and 770 kb/map unit, sex average.

Mapping with DNA Hybridization

A method of chromosome mapping is to hybridize a labeled DNA probe to the DNA of intact chromosomes. This is called chromosomal **in situ hybridization (ISH)**. In one version of this method, the label is a fluorescent tag. In this method, called **fluorescence in situ hybridization** (**FISH**), a fluorescent label is attached to copies of a cloned sequence of DNA, which then becomes a probe. Then the labeled probe is added to the stained chromosomes of cells squashed on a slide, under DNA-denaturing conditions. Subsequently, the DNA reanneals and the labeled DNA hybridizes to homologous sequences in the chromosomes. Excess probe is washed away, and the chromosomes are observed under a fluorescence microscope, revealing the chromosomal position of hybridized fluorescent probe.

The advantages of FISH are several. The chromosomal locus of any cloned DNA sequence can be found, not just polymorphic markers and not just genes. One DNA sequence at a time is located, not two or more markers with respect to each other. No crosses are required, and the method is fast, not depending, as does genetic mapping, on the life cycle of the organism being mapped. The method can be applied to any species whose chromosomes can be stained and examined microscopically. FISH has been used to map human cloned sequences to their chromosomal loci and is being applied to plants such as *Arabidopsis*.

FISH is also being used in medical diagnosis, to locate chromosomes, and to detect aneuploidy. Molecular methods like FISH cannot by themselves identify or confirm the existence of a gene. Rather, they must be coupled with genetic and phenotypic analysis.

Further Reading

Heiskanen M *et al.* 1996. Visual mapping by high resolution FISH. *Trends Genet.* 12:379–382.

Ott J. 1991. *The Analysis of Human Genetic Linkage.* 2nd ed. Johns Hopkins Press, Baltimore.

Rao DC, Province MA (eds.). 2001. *Genetic Dissection of Complex Traits.* Academic Press, San Diego.

Yu A *et al.* 2001. Comparison of human genetic and sequence-based physical maps. *Nature* 409-951–953.

Chapter 35

Finding and Detecting Mutations

Overview

Mutations are essential tools for the science of genetics. Mutations are required to discover a gene and establish its identity, to learn what genes affect a particular phenotype, and to determine the relationship between a gene's structure and its function. Mutations have practical applications in studying genetic disease and in improving organisms used for food, fiber, or valuable molecules.

A mutation is not usable until a strain of organisms carrying that mutation is established. The selection and isolation of mutant strains of any organism requires special procedures; examples are described in this chapter.

Because mutation is a fundamental genetic process, geneticists need sensitive, reliable, efficient methods of measuring mutation rates. One such method, devised to screen chemicals for mutagenicity, is the **Ames test**.

Sources of New Mutations

Natural populations are treasure troves of mutations. Most natural mutations are rare but can be recovered by selection screens, experiments designed to separate organisms that carry mutations in a specific gene or locus.

A second way to get novel mutations is to induce them in organisms, using chemical mutagens, radiation, or genetically engineered transposons. Mutagens increase the rates of random mutation, often by 1000-fold or more. As with natural mutations, a selection screen must be used.

When a gene has been cloned, new mutations can be induced in a controlled way by *in vitro mutagenesis*. The idea here is to target specific DNA sequences within the gene for change, rather than to make random changes.

Mutagens

Mutagens induce mutations at random genomic loci. The experimenter cannot cause mutagens to attack specific genes or to induce exactly predictable changes. Each mutagen produces particular kinds of mutations, so mutagens are chosen on the basis of the type of mutations sought by the investigator.

The **alkylating agents** ethyl methane sulfonate (EMS) and nitrosoguanidine (NG) add single alkyl groups to bases, changing their base-pairing partners and leading to transitions. Alkylating agents also can cross-link the two DNA strands. Misrepair of cross-linked sites can lead to base substitution or small rearrangements.

Base analogs – e.g., 5-bromouracil (5-BU) and 2-aminopurine (2-AP) – are incorporated into newly synthesized DNA and cause transitions via mispairing. 5-BU incorporates in place of thymine and frequently mispairs with guanine, causing TA→CG transitions, while 2-AP incorporates in place of adenine and frequently mispairs with cytosine, causing AT→GC transitions.

Hydroxylating agents, such as hydroxylamine, donate single −OH groups to bases, altering the base-pairing characteristics. Mispairing of bases can lead to transitions or transversions.

Intercalating agents such as proflavin, acridine orange, and ICR compounds (a family of mutagens) insert into the double helix, distorting its shape and leading to single-base insertions and deletions.

Electromagnetic radiation can be a powerful mutagen. UV light induces pyrimidine dimers, which produce base substitutions upon misrepair. X-rays and gamma rays have such high energy that they ionize molecules and tend to break DNA. They can induce chromosomal rearrangements: inversions, translocations, transpositions, and deletions. High-energy ionizing radiation also produces point mutations at high rates, mainly via oxidation of bases. The level of radiation encountered in everyday life ("background radiation") is orders of magnitude lower than levels used in mutagenesis and is not a significant source of mutation.

Transposons can also induce mutations. In *Drosophila*, P elements are used for this purpose. P elements are naturally occurring, 3-kb transposons that contain two genes, one encoding a transposase and another encoding a repressor of the transposase gene. P elements can easily be cloned, modified, and introduced into the nuclear chromosomes of germline cells to make transgenic lines. Once in a genome, a P element can be induced to jump to a new location, where it may affect the functioning of a gene at its new location. The transposon jumping is effected and controlled by the experimenter in a simple genetic cross.

Screening for Mutations

Mutations are rare and mostly cryptic. Inducing mutations at random genomic loci is easy, but selecting and isolating them can be very challenging. For example, to find and grow a clone from one leucine-requiring bacterium in a population of a million nonmutant cells, when each of the nonmutants grows happily on leucine-supplemented medium, is tricky. In diplontic organisms, rare mutations are cryptic because they exist almost entirely in heterozygotes (Chapter 39), and most of them are recessive.

The geneticist who wants to isolate novel mutations from nature or from an experimental population must devise an efficient screen for detecting and collecting them. A desired mutation usually produces a defective phenotype (e.g., a specific nutritional requirement that renders the genotype less able to survive and reproduce in a normal environment), so special methods must be used to favor the genotypes that natural selection eliminates. Mutations are usually screened in a two-step process. The first step enriches for individuals that have the desired defect by discarding most of the nonmutants. In the second step, each potential mutant strain is tested for the specific phenotype of interest.

Enrichment Screens

A strategy that may work for mutations that impair metabolism is to grow organisms in an environment that kills nonmutants but spares mutants. **Prototrophs** are wild type and grow on minimal medium. **Auxotrophs** are enzymatically deficient and have a corresponding nutritional requirement; they grow only in medium containing the required nutrient. Before screening for auxotrophs with a specific phenotype, it is desirable to kill or eliminate as many prototrophs from a population as possible to make the search for a class of auxotrophs easier and more efficient.

Penicillin, an antibiotic against bacteria, kills only growing cells. It binds to an enzyme needed to synthesize peptidoglycan in the growing cell wall and stimulates production of autolysins; cells then lyse. Prototrophs grow and consequently are killed by penicillin, whereas auxotrophs do not grow but do stay alive. After mutagenizing a population of penicillin-sensitive cells, they are grown in minimal medium (MM) + required nutrient + penicillin. After the prototrophs lyse, the remaining auxotrophs are selected, plated on MM + required nutrient, and retested.

Suppose one wishes to find arginine-requiring auxotrophs. After killing most of the prototrophs with penicillin, inoculate a plate of MM + arginine with a low concentration of cells. Discrete clones will grow on the plate. Pick cells from each clone and test it on MM and on MM + arginine. Clones that grow on MM + arginine but fail to grow on MM are arginine-requiring auxotrophs (Figure 35.1).

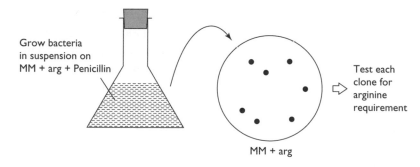

Grow bacteria in suspension on MM + arg + Penicillin

MM + arg

Test each clone for arginine requirement

Fig 35.1
Penicillin-enrichment for auxotrophic bacteria.

The goal may be to find mutants that lack a physiological or behavioral trait, mutants with approximately normal survival. In that case, the initial screen is designed to enrich for individuals that show an aberrant trait, often letting the organisms do the work of sorting, as by growth or activity.

For example, motile organisms can be selected by their response (or lack of response) to external stimuli. Unicellular organisms (*E. coli*, *Paramecium*) react to chemical gradients in the medium by chemokinesis (nonorienting mechanism of responding to chemical concentration). Many mutants defective in chemokinesis have been found in these organisms.

Animal behavior is effected by neurons, which are excitable cells that work by maintaining a potential difference (voltage) across the cell membrane, and changing that voltage in response to outside stimuli. Ion channels are critical components of a neuron's electrical properties. Most of the genes that encode ion channels are essential for life, and severe mutations in such a gene are likely to be lethal. A way around this problem is to select for temperature-sensitive mutations (i.e., the mutant protein functions at a permissive temperature and fails at a restrictive temperature). In *Drosophila*, a cold-blooded animal, genes encoding sodium or potassium channels, or regulating their expression, have been discovered by selecting temperature-sensitive "pass-out" mutants: at 25°C (normal temperature for *Drosophila*) the mutant is normal, and at 40°C it passes out within a few seconds. Cooling the animals to 25°C restores them to normal. Heating wild-type flies to 40°C for a few minutes has no effect. In this way, a large number of individuals can be screened rapidly for a mutation affecting neurons.

Genetic Methods for Selecting Germline Mutations

Selecting mutant strains of bacteria is straightforward, because mutant cells reproduce clonally. Diplontic eukarya (Chapter 22), such as animals, are more difficult to manipulate genetically. One problem is to make a genetic strain that carries a new mutation. New mutations are induced in germline cells, and through a series of crosses the new mutation is detected. If the mutation has a recessive lethal or sterile effect, then it can be maintained in heterozygotes. Maintaining a line heterozygous for a recessive lethal mutation is tedious in most organisms, but in *Drosophila* it is done by using a **balancer chromosome**.

Inbreeding

A simple genetic method is inbreeding, usually done by sib mating: within a strain, the parents of each generation are one pair of sibs; this is continued for several generations. Inbreeding eventually produces completely homozygous strains, and every inbred strain is genetically unique. Strains become homozygous because related individuals tend to share alleles (Chapter 38). A hypothetical example of genetic changes in three lines is given here (two genes, two alleles per gene).

	Generation	Line 1	Line 2	Line 3
Genotypes	1	AaBb × AAbb	aabb × AaBb	AABB × AaBb
of the breeding	2	AaBb × AaBb	aaBb × AaBb	AABb × AABB
pair, each	3	AABb × AAbb	aabb × aaBb	AABB × AABB
generation	4	AAbb × AAbb	aabb × aabb	all AABB
	5	all AAbb	all aabb	

Imagine a genetically heterogeneous starting population in which alleles are segregating for many genes. Many inbred lines are established, each line being started with a randomly chosen breeding pair. When a line becomes homozygous for an allele, that allele is said to be **fixed**. In this hypothetical example, by generation 5, all three lines have become fixed for alleles of genes *A* and *B*.

By a similar process, inbred lines produce individuals that are homozygous for novel mutant alleles. The animals are exposed to a mutagen and then inbred. When an individual heterozygous for a new mutation is mated to its sib, then a quarter of the progeny will be homozygous for the new mutation, and two-thirds of the nonaffected sibs will be heterozygous for the mutation. Hundreds of novel mutations have been recovered by selecting sublines carrying these new mutations.

Special Chromosomes in *Drosophila*

Drosophila melanogaster is wonderful. Special chromosomes, available in this fly and without parallel in other organisms, are powerful tools for manipulating the genotype, finding new mutants, and measuring mutation rates. Two examples are **attached-X** chromosomes and **balancer** chromosomes.

		Eggs: $\widehat{XX}$	Eggs: Y
Sperm	X	$\widehat{XX}$X usually die	XY sons
Sperm	Y	$\widehat{XX}$Y daughters	YY die

Fig 35.2 Segregation of the attached -X chromosome.

Using Attached-X Chromosomes. Attached-X, symbolized $\widehat{XX}$, consists of two X chromosomes joined together (Figure 35.2). Females with such a chromosome are $\widehat{XX}$Y. $\widehat{XX}$Y females donate $\widehat{XX}$ to their daughters and Y to their sons, while their mates donate X to their sons and Y to their daughters; YY progeny die and $\widehat{XXX}$ progeny usually die.

Solved problem: $\widehat{XX}$ as a genetic tool. Suppose you want to identify all the genes on the X chromosome that are essential for normal phototaxis. Your initial mutant screen uses an apparatus in which wild-type flies move toward the light. The letter "x" means "cross (mate)" and "↓" means "produce offspring". The genetic crosses are shown in Figure 35.3.

$\widehat{XX}$Y × XY A big fraction of these males have at least one mutation;

↓ a rare few have an X-linked mutation affecting phototaxis.

$\widehat{XX}$Y × X*Y X* = candidate mutation; at least, this male is non-phototactic.

↓

$\widehat{XX}$Y & X*Y If all males in this family are non-phototactic, the mutation is X-linked. You can maintain the strain without further selection, because X* is the only free X chromosome in the family – all the females are $\widehat{XX}$Y and all the males are X*Y.

Fig 35.3 Example of $\widehat{XX}$ as a genetic tool.

Using Balancer Chromosomes. A balancer is an experimentally derived chromosome with three kinds of mutation: (1) big inversions, (2) a dominant **marker** (a mutation used to detect the presence of a chromosome or chromosome locus), and (3) one or more recessive lethal mutations. The inversions suppress recombination between the balancer and a normal chromosome because recombinant gametes have large parts of the genome deleted or duplicated, which is lethal (Chapter 21). Therefore, in a balancer heterozygote, the haplotype of the entire normal chromosome is preserved intact without danger of recombination with balancer genes (i.e., all the genes on the normal chromosome comprise a supergene).

Solved problem: Using a balancer used as a tool. Balancers from chromosome 2 have the *Cy* mutation, a dominant marker. The asterisk (Figure 35.4) denotes a recessive, potentially sought-after mutation; *Sco* is a dominant mutation on chromosome 2.

Crosses used to isolate mutations on chromosome 2 of *Drosophila*

Cy//*Sco* ♀ × +//+ ♂ (expose males to mutagen, inducing mutations in their
  ↓   germline)
Cy//*Sco* ♀ × *//*Cy* ♂ (1-male family) ... replicate this cross many times
  ↓
*//*Cy* ♀ × *//*Cy* ♂ ... choose only *//*Cy*; all */ chromosomes within one
  ↓   family are identical
*//**, *//*Cy*, (*Cy*//*Cy* dies) ...*//** & *//*Cy* appear in a 1:2 ratio unless *//** does
                 not survive as well as *//*Cy*. Select families having
                 *//** homozygotes with the desired phenotype,
                 which could even be lethality.

Fig 35.4 Example of a balancer used as a tool.

Mutagenesis with Transposons

Transposons make mutations by inserting into genes. As mutagens, transposons have several desirable properties. Transposons:

- Are completely safe to use
- Can be made to carry selectable markers
- Help in molecular analysis of a mutated gene; (sequences adjacent to the P element can be recovered by polymerase chain reaction)
- Can be used to generate small deletions; (when a P element jumps, excision may be imprecise, meaning that adjacent sequences are removed)
- Can help in the study of where and when a gene is expressed; (a gene added to the P element will be expressed in a pattern controlled by nearby enhancers)

P elements transpose in a nonreplicative way. When a P element jumps, transposase causes it to excise from one locus and to insert into another locus.

To induce new mutations with P elements, two strains are used. One strain has several P elements lacking a good transposase gene. The other strain makes transposase, but its P element can no longer transpose. In the heterozygous offspring of the two strains, P elements transpose to new locations within the genomes of germline cells. Each transposition event has the potential of making a new mutation. The P elements carry within them a wild-type allele of a gene for which both strains are homozygous mutant. For example, the P elements may carry w^+ (which makes for normal, red eyes), and both strains are w/w (white-eyed). To isolate mutations on a chromosome of interest, a balancer chromosome is used, and strains are selected in which w^+ allele and the balancer segregate 1:1.

Solved Problem: P element mutagenesis. Suppose one wishes to study genes on chromosome 2 of *Drosophila* that are essential to ovarian development and functioning. A good way to begin the search is to select mutations that cause female sterility. A special XX strain is a source of P elements containing the w^+ gene, denoted *P[W]* here. The source of transposase for the P elements, denoted $\Delta 2$–3 here, is on chromosome 3, which carries a dominant marker *Sb*. Two balancers are used, one for chromosome 2 marked with *Cy* and one for chromosome 3 marked with *Ubx*. In the selection scheme shown in Figure 35.5, the genotypes of three chromosomes (X, 2, and 3) are given. For clarity, semicolons separate the genotypes for each chromosome (X, 2, and 3).

y w P[W]//y w P[W]; +//+; +//+ ♀ × w//Y; +//+; (2–3 Sb//Ubx ♂
↓
w//w; Sco//Cy; +//+ ♀ × y w P[W]//Y; +//+; Δ2–3 Sb//+ ♂
↓
w//w; Sco//Cy; +//+ ♀ × w//Y; P[W]//Cy; +//+ ♂
↓
w//w; P[W]//Cy ♀ × w//Y; P[W]//Cy ♂
↓
w//w; P[W]//Cy & w//w P[W]//P[W] ♀♀ – Test non-Cy ♀♀ for sterility.

Fig 35.5 P element mutagenesis in *Drosophila*, targeting chromosome 2.

Checking Mutagenicity with the Ames Test

Hundreds of new, useful compounds are being discovered every year, but some of them are mutagenic. Cancer is caused by somatic mutations. This is even true of virally induced cancers. It follows that all mutagens are carcinogens , although not all carcinogens are mutagens. To protect human health, new chemicals are screened for mutagenicity. The Ames test is an efficient way to screen for two classes of mutagens, ones that lead to substitutions and ones that lead to small insertions and deletions (Figure 35.6).

The Ames test has four salient features.

• Potential mutagens are pretreated with liver enzymes, because in the body compounds would be exposed to and possibly metabolized by liver enzymes, producing derivative compounds that could be mutagenic.

- Each mutagen is tested for the ability to cause two kinds of mutation, base substitutions and small insertions or deletions (frameshift mutations).
- The test detects wild-type revertants (back mutations) of his^0 auxotrophs of *Salmonella*. This is done because it is much easier to detect a prototrophic clone in a population of auxotrophs than vice versa. The most common revertants of a substitution mutation entail another substitution at that site; e.g., if a substitution was AT→GC, then most of the revertants will be GC→AT. The most common revertants of a 1-nt deletion will be 1-nt insertions and 2-nt deletions close to the deletion – either of these second mutations will restore the reading frame to wild type. Substitution mutations cannot revert frameshift mutations, and frameshift mutations cannot revert substitution mutations.
- Several concentrations of each mutagen are tested, to get a dose-response curve. The rationale is that mutagens have the

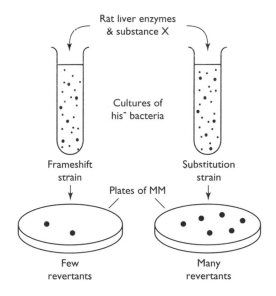

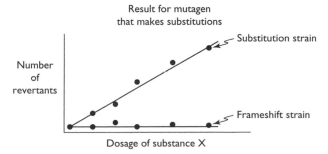

Fig 35.6   Summary of the Ames test. Rat liver enzymes and substance X.

following property: the probability of a mutation increases with increasing exposure to the mutagen.

In Figure 35.6, substance X produces substitution mutations, not deletion mutations. Therefore, substance X causes revertants in the substitution his^0 strain but not in the frameshift his^0 strain. If substance X had been a mutagen that causes small deletions or insertions, revertants would have appeared in the frameshift strain but not the substitution strain. If substance X had not been a mutagen at all, mutation rate would not have been elevated by X in either strain.

Further Reading

Ames BN *et al.* 1973. Carcinogens are mutagens: A simple test system combining liver homogenates for activation and bacteria for detection. *Proc Nat Acad. Sci. USA* 70:2281–2285.

Greenspan R. 1997. *Fly Pushing: The Theory and Practice of Drosophila Genetics.* Cold Spring Harbor Laboratory Press, Cold Spring Harbor, NY.

Salzberg A *et al.* 1997. P-element insertion alleles of essential genes on the third chromosome of *Drosophila melanogaster:* mutations affecting embryonic PNS development. *Genetics* 147:1723–1741.

Williams RW. 1999. A targeted screen to detect recessive mutations that have quantitative effects. *Mammalian Genome* 10:734–738.

Chapter 36

Cytoplasmic Inheritance

Overview

Inherited cytoplasmic factors, unlike nuclear genes, do not segregate or assort regularly in sexual reproduction. This category includes organelle genes, infectious agents, and mRNAs. Extranuclear genes reside in mitochondria, chloroplasts, and cytoplasmic intracellular parasites. Prions infect somatic cells of animals and thereafter are inherited cytoplasmically. This chapter focuses on the modes of cytoplasmic inheritance and methods of analysis.

Mitochondria and Chloroplasts

Characteristics of Mitochondria and Chloroplasts

The mitochondrion is the site of oxidative metabolism in eukarya. There are two membranes separated by a space; the innermost space of the mitochondrion is the matrix. Initial oxidation of fatty acids happens between the two membranes; the oxidation of acetyl-CoA, the citric acid cycle, and β-oxidation of fatty acids take place inside the matrix. The electron transport system and the enzymes of oxidative phosphorylation reside in the inner membrane. The number of mitochondria per cell may be 1 to $\sim 10^5$, depending on cell type and species.

Mitochondria reproduce clonally, by splitting. The mitochondrion has a small chromosome (14 to 2500 kb), present in multiple copies because its replication is not tightly coordinated with organelle division. Chromosomes segregate into daughter mitochondria. Mitochondria have their own machinery for protein synthesis, but nuclear genes encode most mitochondrial proteins.

Mitochondria are distributed to daughter cells during cell division in a haphazard way. Suppose a mutation arises in one of the many mitochondria in an actively cycling cell. The mutant mitochondrion will reproduce side-by-side with nonmutant ones, and both types will disperse randomly to the cytoplasm of daughter cells during cell division. Consequently, some cells derived from the original cell may have only mutant mitochondria, some will have only normal mitochondria, and some will have both mutant and normal mitochondria. Cells containing genetically identical mitochondria are **homoplasmic**, and cells containing diverse mitochondria are **heteroplasmic**. Rarely, mitochondria fuse, whereupon the chromosomes may undergo homologous recombination.

Chloroplasts harvest light energy and produce carbohydrates by photosynthesis. They evolved from photosynthetic bacteria that became symbionts in the cytoplasm of eukarya. Other plastids (leukoplasts and chromoplasts) evolved from chloroplasts.

Inside the two or three closely spaced membranes that enclose the chloroplast sits a system of stacked, membranous sacs called thylakoids. The fluid surrounding thylakoids is stroma. During photosynthesis, carbon is fixed; i.e., CO_2 is reduced to carbohydrate. Light absorbed by chlorophyll in the thylakoids yields (via light reactions) ATP and NADPH, which supply energy for the carbon-fixing dark reactions in the stroma.

In the stroma are multiple copies of one small (120 to 200 kb), circular chromosome, as well as plasmids in some instances. Nuclear genes encode most chloroplast proteins. Chloroplasts reproduce by splitting and cannot arise *de novo*. They sometimes fuse, and chromosomes may recombine. Cells can be homoplasmic or heteroplasmic for chloroplasts.

Deleterious Mutations Affecting Mitochondria and Chloroplasts

Loss-of-function mutations of genes encoding mitochondrial proteins and RNAs are deleterious in a variety of ways, the primary ones being damage to the respiratory chain, impairment of organelle reproduction (also called biogenesis), and loss of the ability to synthesize protein. In vertebrates, mitochondrial mutations may generate novel antigens recognized as "foreign" by the immune system.

In mammals, base substitutions accumulate 10 times faster in mitochondria ($\sim 10^{-8}$ nt^{-1} yr^{-1}) than in the nucleus, in part because of the free radicals produced by oxidative metabolism. Mutations accumulate in the mitochondria of somatic tissues of mammals. The opposite is true of plants, where substitution rates are several times lower in mitochondria than in the nucleus. Plant mitochondria respire more slowly than mitochondria of mammals and therefore produce fewer free radicals.

Mutations in the mitochondrial chromosome can lead to defects in oxidative metabolism. When a cell's mitochondria cannot respire, that cell becomes an obligate anaerobe, relying on the glycolytic pathway for energy. Strains of yeast with nonrespiring mitochondria are **petite** (they make small colonies). Similar strains of *Neurospora* are **poky** (slow-growing).

In humans and other mammals, some mitochondrial mutations cause degenerative diseases, prominently diseases affecting muscles and the nervous system. Human mitochondrial diseases include MELAS (mitochondrial myopathy, encephalopathy, lactic acidosis, and stroke-like episodes), MERRF (myoclonic epilepsy and ragged red fibers), and LHON (Leber's hereditary optic neuropathy). MELAS is due to mutations in the mtDNA tRNA^{leu} gene; MERRF is caused by mutations in the mtDNA tRNA^{lys} gene, and LHON is due to mutations in a respiratory protein (ND4, found in complex I of the respiratory chain).

Mutations in the chloroplast or nuclear genome can impair the organelle's ability to photosynthesize or reproduce. Mutations accumulate faster in chloroplasts than in the nucleus but more slowly than in plant mitochondria (Figure 36.1).

The pattern of inheritance is different for genes in the nucleus and genes in organelles. With maternal inheritance, an affected female transmits the trait to all offspring, and an affected male does not transmit the trait to his offspring. With inheritance of nuclear genes, alleles segregate. A paternally inherited cytoplasmic trait is inherited in the opposite way: by all progeny, both daughters and sons, and invariably from the father, not the mother.

Sometimes the parent is heteroplasmic for a mutation in mitochondria or chloroplasts; i.e., some of these organelles have a mutant genome, and some have a normal genome. Furthermore,

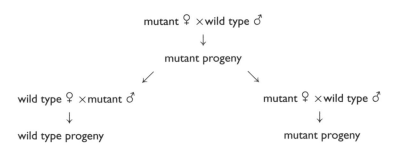

Fig 36.1 Inheritance of mitochondria and chloroplasts in sexual organisms. A maternally inherited (cytoplasmic) mutant phenotype is passed from mother to offspring via egg cytoplasm; it does not segregate, as do nuclear genes.

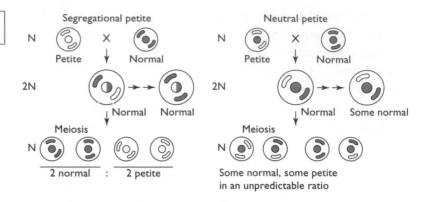

Fig 36.2 Crosses between wild-type and petite strains.

malfunction of mitochondria or chloroplasts can be due to nuclear genes that encode proteins of those organelles. Petite strains of yeast illustrate these two points. There are two kinds of petite yeast strains, segregational petites and neutral petites. Segregational petites are due to mutation of nuclear genes, while neutral petites are due to mutation of mitochondrial genes. Two haploid yeast cells undergoing syngamy both contribute equal numbers of mitochondria to the zygote. Consequently, a cross between a segregational petite strain and wild type yields a normal zygote, which can undergo meiosis to produce two wild-type cells and two petite cells (Figure 36.2). A cross between a neutral petite strain and wild type yields a normal zygote, which can undergo meiosis to produce some wild-type cells and some petite cells, in an unpredictable ratio, by the process described in the foregoing paragraph.

In fungi and in many other simple eukarya, both mating types equally contribute mitochondria to the zygote. In most plants, mitochondria are inherited maternally (only from egg cytoplasm), but in some conifers they are in inherited paternally (only from pollen cytoplasm). In animals, mitochondria are inherited maternally, but a small amount of paternal inheritance has been detected. Rarely, maternal and paternal mitochondria fuse, and their genomes recombine.

Chloroplasts are inherited maternally in most angiosperms, biparentally in some angiosperms, and paternally in conifers. In unicellular photosynthetic organisms, chloroplasts may be inherited either biparentally or through one of the two mating types.

Three main causes of uniparental inheritance of mitochondria and chloroplasts are (1) exclusion of organelles from cytoplasm during meiosis, (2) asymmetry in size of gametes, and (3) destruction of mitochondria or chloroplasts from one gamete after fertilization. During pollen formation in angiosperms (excepting a few genera), all chloroplasts and most mitochondria are excluded from microspore cytoplasm. In some plant species, the remaining mitochondria are destroyed or are expelled from the pollen cytoplasm; in other species, sperm mitochondria are destroyed during fertilization. In animals, one cause of maternal inheritance is the greater volume of cytoplasm in the egg than in the sperm; accordingly, eggs contain orders of magnitude more mitochondria. For example, a human egg carries $\sim10^5$ mitochondria versus $\sim10^2$ mitochondria per sperm. Furthermore, enzymes in the egg destroy the sperm's mitochondria after syngamy.

Green-white **variegation** (mosaicism) of some flowering plants is maternally inherited. Consider the variegated four-o'clock, *Mirabilis jalapa*, which has mixed green and white tissues. A green and white variegated plant has both normal, green chloroplasts and mutant, white chloroplasts lacking chlorophyll. Within a variegated plant there are three kinds of cells: (1) white cells that contain only white chloroplasts, (2) green cells that contain only green chloroplasts, and (3) green cells that contain a mixed population of white and normal chloroplasts.

A zygote containing both green and white chloroplasts develops into a variegated plant (Figure 36.3). Green and white chloroplasts reproduce equally well. When a cell divides, white chloroplasts are apportioned randomly to its daughter cells. If a cell contains green and white chloroplasts, it is green. Because chloroplasts are not apportioned to daughter cells in a manner correlated with genotype, by a stochastic process like genetic drift (Chapter 38), a green, mixed cell will likely give rise to a population of cells, of which some are white and some are green. The green cells have only normal chloroplasts or a mixture of green and white chloroplasts. Once the plant is variegated, white, cells give rise to clones of white cells, cells lacking white chloroplasts

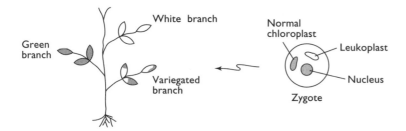

Fig 36.3 A heteroplasmic zygote with both green and white chloroplasts produces a variegated plant.

Fig 36.4 The inheritance of green and white variegation.

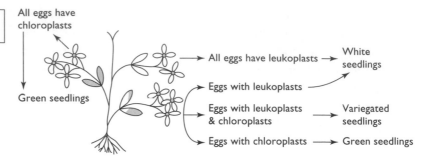

All eggs have chloroplasts

Green seedlings

All eggs have leukoplasts → White seedlings

Eggs with leukoplasts —

Eggs with leukoplasts & chloroplasts → Variegated seedlings

Eggs with chloroplasts → Green seedlings

give rise to clones of green cells, and mixed cells continue to give rise to all three kinds of cells.

How is variegation inherited during sexual reproduction? Eggs from flowers on green branches produce all green progeny, as these eggs contain only normal chloroplasts. Eggs from flowers on white branches produce all white progeny that die soon after the seed sprouts, as these eggs contain only white chloroplasts. Eggs from flowers on variegated branches produce all three kinds of plants (green, variegated, and white), for these eggs may contain only green, only white, or both green and white chloroplasts. Variegated plants produce green, white, and variegated offspring in unpredictable ratios (Figure 36.4).

Maternal-Effect Genes

Maternal-effect genes are nuclear genes transcribed in the oocyte but not in early development of the zygote. The mRNA in the oocyte persists for several cell cycles and is distributed to the cytoplasm in daughter cells during cell division. Maternal mRNAs can therefore influence the early embryo. The genotype of the mother for maternal-effect genes determines the phenotype of the offspring.

A gene may have both maternal and zygotic effects. An example is the *daughterless* (*da*) gene in *Drosophila*; *da* encodes a transcription factor. Female embryos must have da$^+$ protein in order to transcribe the *Sex-lethal* (*Sxl*) gene; Sxl$^+$ protein blocks activation of the male dosage compensation mechanism. To make Sxl$^+$ protein in time to avoid lethal overexpression of X-linked genes, maternally supplied *da*$^+$ mRNA and da$^+$ protein must be present in the egg.

$du^+/da^1 ♀ \times da^1/da^1 ♂$

↓

$da^1/da^1 ♀ \times$ any ♂ $da^1/da+ ♀ \times$ any ♂ $da^+/da^2 ♀ \times da^+/da^2 ♂$

↓   ↓   ↓

daughters die, sons live   progeny live da^2/da^2 die, others live

a)   b)

Fig 36.5 Maternal-effect da^1 mutation (**a**). Zygotic-effect da^2 mutation (**b**).

There are many mutant alleles of da; one of them, da^1, is both a hypomorph (the gene product has reduced activity) and a maternal-effect lethal in females but not males (Figure 36.5). Therefore, all offspring of a da^1/da^1 female and a da^+/da^+ male are da^1/da^+, but only male offspring live, while female offspring from the reciprocal cross, also da^1/da^+, survive. Furthermore, all da^1/da^1 daughters from the cross $da^1/da^+ ♀ \times da^1/da^1 ♂$ survive. In addition to a maternal effect, da exhibits zygotic effects (Figure 36.5). Amorphic (completely nonfunctional) alleles of da (e.g., da^2) are zygotic, non-sex-specific, recessive lethals.

Transposons and Cytoplasmic Inheritance

Interactions between cytoplasm and nucleus can produce strong phenotypic effects. Advanced insects such as *Drosophila* provide examples involving transposons, which have invaded nuclear genomes. Multiple copies of one transposon, the P element, are present in the genome of *Drosophila melanogaster* in wild populations. P strains possess P elements, and M strains do not. The P element encodes both a transposase and a repressor of the transposase gene. The repressor is maternally inherited, present in egg cytoplasm and absent from sperm cytoplasm. If a P ♀ is crossed to an M ♂, the offspring are normal, but the reciprocal cross produces partial sterility in the offspring, a condition known as **hybrid dysgenesis**. The basis of hybrid dysgenesis is the presence of P elements in the zygote in the absence of repressor. Transposase is synthesized and P elements transpose ("jump") to new locations, causing mutation, nondisjunction, and male recombination. The effects are most pronounced in the germ line, so that partial or complete sterility is a common feature of hybrid dysgenesis. Other transposons (e.g., I and H) can cause hybrid dysgenesis in *Drosophila*.

Infectious Inheritance

Bacteria and Viruses

Cytoplasmic inheritance of infectious parasites is prevalent in eukarya. The parasite may modify the host's phenotype or even its genetic machinery. An intracellular bacterium, *Wolbachia*, infects many (by some estimates, $\sim 10^6$) insect species as well as some crustaceans and arachnids. *Wolbachia* is maternally inherited and can be transmitted by mating. It infects reproductive cells and manipulates reproduction of the host by inducing cytoplasmic incompatibility between egg and sperm, parthenogenesis, or feminization of males.

When an uninfected male fertilizes an infected female or when an infected male fertilizes a female infected with the same strain of *Wolbachia*, the cross is compatible, but when an infected male fertilizes an uninfected female, the progeny usually die owing to cytoplasmic incompatibility. Females and males infected by two different strains of *Wolbachia* may exhibit bidirectional incompatibility. These phenomena are strange, because *Wolbachia* inhabit eggs but rarely inhabit sperm. Apparently, bacteria in testes modify sperm and bacteria in eggs rescue the modified sperm. The theory is, either the bacteria add a molecule to sperm that interferes with zygotic mitosis or with syngamy, or the bacteria deplete sperm of a molecule from sperm that is needed in zygotic mitosis or in syngamy. In either case, bacteria in the egg produce molecules that rescue sperm modified by the same strain of bacteria. In incompatible crosses, either sperm fail to fuse with egg pronuclei or paternal chromosomes are damaged. In some cases, paternal chromosomes neither condense nor segregate during mitosis. Loss of paternal chromosomes produces a haploid embryo, which either dies or, in arrhenotokous parthenogenesis, develops into a male (e.g., wasps). In other cases, paternal chromosomes fragment, leading to massive, lethal deletions in the early embryo.

Prions

Prions (proteinaceous and infectious particles) are proteins that cause rare neurodegenerative diseases in mammals:

scrapie, Creuzfeldt-Jakob disease (CJD), kuru, bovine spongiform encephalopathy (BSE = "mad cow disease"), and chronic wasting disease. Some prions can move between species. Prion proteins exist in two isoforms, PrP^C and PrP^{Sc}, where "C" stands for "cellular" and "Sc" stands for "scrapie." PrP^C is rich in α-helices and lacks β-strings, while PrP^{Sc} has a big, four-string β-sheet. A cellular gene encodes PrP^C ; homologs exist throughout eukarya. PrP^{Sc} is heat-stable, insoluble in detergent, and resistant to digestion by proteases. When a mammal eats PrP^{Sc} (present in the tissues of another mammal), some of this protein passes from the gut to the blood, crosses the blood-brain barrier, and infects neurons. PrP^{Sc} converts PrP^C to PrP^{Sc}, by forming a heterodimer, then causing the normal form to change its secondary and tertiary structures. Excess PrP^{Sc} aggregates into amorphous, intracellular deposits. After a long time, infected cells form many vacuoles, experience physiological failure, and may die by apoptosis.

In humans, about 10% of the cases of CJD are familial. The cause is a mutation in the gene that encodes PrP, making for asparagine in place of aspartic acid at amino acid 178. Mutant PrP converts spontaneously from the normal form, which has four α-helices, to the scrapie form, which has a β-sheet (Figure 36.6). The same substitution at amino acid 178 also causes fatal familial insomnia, another prion disease. Mutations at codon 129 determine which disease develops: a methionine at position 129 leads to fatal familial insomnia, while a valine at position 129 leads to familial CJD. In both prion diseases, neurodegeneration begins in middle age, when PrP^{Sc} begins to form. The polymorphism at position 129 does not itself contribute to prion formation: if amino acid at position 178 is aspartic acid, only the PrP^C form is present, regardless of the amino acid at position 129.

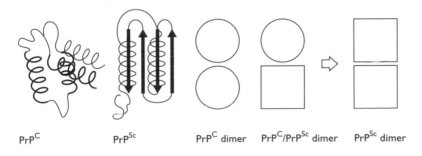

PrP^C PrP^{Sc} PrP^C dimer   PrP^C/PrP^{Sc} dimer   PrP^{Sc} dimer

Fig 36.6 Prion monomer structure and PrP^C→PrP^{Sc} conversion.

Further Reading

Berloco M *et al.* 2001. The maternal effect gene, *abnormal oocyte* (*abo*), of *Drosophila melanogaster* encodes a specific negative regulator of histones. *Proc. Natl. Acad. Sci. USA* 98:12126–12131.

Gillham NW. 1994. *Organelle Genes and Genomes.* Oxford University Press, New York.

Ladoukakis ED, Zouros E. 2001. Direct evidence for homologous recombination in mussel mitochondrial DNA. *Mol. Biol. Evol.* 18:1168–1175.

Ohyama K. 1996. Chloroplast and mitochondrial genomes from a liverwort, *Marchantia polymorpha* – gene organization and molecular evolution. *Biosci. Biotechnol. Biochem.* 60:16–24.

O'Neill SL *et al.* (eds.). 1997. *Influential Passengers.* Oxford University Press, Oxford.

Prusiner SB (ed.). 1996. Prions prions prions. *Curr. Topics Microbiol. Immunol.* 207.

Varmus H. 1983. Retroviruses. In: Shapiro JA (ed.). *Mobile Genetic Elements.* Academic Press, New York, p 411–505.

Wolfe KH *et al.* 1987. Rates of nucleotide substitution vary greatly among plant mitochondrial, chloroplast, and nuclear DNAs. *Proc. Natl. Acad. Sci. USA* 84:9054–9058.

Chapter 37

Genetic Variation in Populations

Overview

Natural populations of most plants and animals are genetically diverse, and a sizeable fraction of their genes have two or more common alleles. Population genetics is about the frequencies of alleles in populations and how these allele frequencies change from generation to generation. This chapter describes the following:

- The relation between allele frequencies and genotype frequencies in populations
- The magnitude of genetic variation in populations, and how it is measured

As with Mendelian theory and experiments, in population genetics one should focus attention first and foremost on gametes and haplotypes.

Populations

A population is a group of organisms of one species, living in one area. An asexual population is a **clone**, while a **population** of sexually reproducing, freely interbreeding eukarya is sometimes called a **deme**. In diplontic species, population size is defined as the number of all diploid individuals in that population, **N**. The copies of a gene in a population comprise a **gene pool**, and the size of the gene pool for diplontic species is **2N**.

Predicting Genotype Frequencies from Allele Frequencies

Populations in Static Equilibrium

Consider gene A, an autosomal gene with two alleles, A^1 and A^2; the gametes in this population carry either A^1 or A^2. The frequency of A^1 gametes is p and the frequency of A^2 gametes is q, where $p + q = 1$.

A population is genetically static under the following assumptions:

- Allele frequencies are the same in female and male gametes.
- The population is large.
- No forces act to change p and q from generation to generation.
- Zygotes form independently of the gametes' haplotypes; this is **panmixia**.

These assumptions define **static equilibrium conditions**. If a two-allele gene, A, experiences these equilibrium conditions, the frequencies of the three A genotypes are predicted by the multiplication rule of probability (Figure 37.1).

The expression, $p^2 + 2pq + q^2$, is the binomial expansion $(pX + qY)^2$. For historical reasons, the prediction of diploid genotype frequencies from allele frequencies is called the **Castle-Hardy-Weinberg (CHW) rule**.

Figure 37.2 illustrates the relationship between allele frequencies, proportional to the width of the row or column of a table, and genotype frequencies, proportional to the area of cells in the table. The cell for A^1A^1 is darkened.

Eggs

	p A^1	q A^2
p A^1	$p^2 A^1/A^1$	pq A^1/A^2
q A^2	pq A^1/A^2	$q^2 A^2/A^2$

Sperm or pollen

Genotypes	A^1/A^1	A^1/A^2	A^2/A^2	
Frequencies	p^2	2pq	q^2	$p^2 + 2pq + q^2 = 1$

Fig 37.1 Gametes uniting at random.

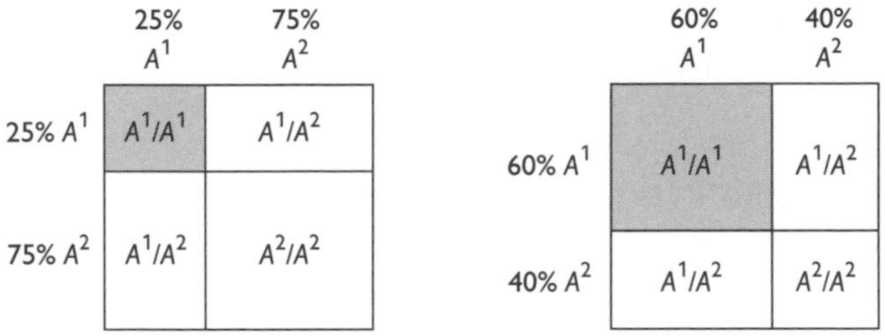

Fig 37.2 Relationship between allele frequencies and genotype frequencies.

Deviations from Castle-Hardy-Weinberg Conditions

What happens if the CHW conditions do not hold?

If allele frequencies for an autosomal gene are different in female and male gametes, then they will become equal in the haploid phase of the next generation. The interested student can easily verify this statement with a fertilization table and a little algebra. However, sex-linked genes behave differently; they are considered later in the chapter.

In a small population, random fluctuations in the sampling of gametes will cause deviations from the predicted frequencies. In a chronically small population, the randomly fluctuating allele frequencies may tend to drift upward or downward. This is **genetic drift** (Chapter 38).

Some genetic forces change allele frequencies in a nonrandom, deterministic fashion. These forces include **natural selection** (Chapter 39), **mutation** (Chapter 38), and **migration** (Chapter 38) from another population.

In wind- and insect-pollinated plants, and in animals with broadcast spawning (e.g., many fishes and aquatic invertebrates release gametes freely into the water), it is easy to see how fertilization can be independent of gametic haplotype – i.e., that panmixia happens. Exceptions include self-incompatibility in plants (the pollen must not share alleles of certain genes with the stigma), and gamete incompatibility in sea urchins. Sometimes panmixia does not hold for animals with internal fertilization, because mating may be nonrandom for certain genotypes. **Positive assortative mating** is the tendency of like genotypes to mate preferentially, while **negative assortative mating** is the tendency of unlike genotypes to mate preferentially. Positive assortative mating occurs frequently in nature, but negative assortative mating appears to be rare. The consequence of positive assortative mating is an inflated number of homozygotes and fewer heterozygotes than would be expected under CHW conditions.

Estimating Allele Frequencies from Genotypic Frequencies

In plants and animals, allele frequencies are usually estimated from genotype frequencies and not the other way round, for the genotypes of multicellular organisms are usually easier to ascertain than the haplotypes of gametes. To estimate the allele frequencies of a population from a sample of diploid organisms, count alleles twice in homozygotes and once in heterozygotes, and divide by twice the number of individuals in the sample. For recessive mutations that cause medical disorders, the heterozygote (say A^+/A^-) is called a **carrier** of the mutant allele.

Solved Problem: estimating allele frequencies for M–N blood types. The L gene controls M and N blood types. In a sample of 1482 persons, the frequencies of the three types was as follows:

Genotype	Phenotype	Number
L^M/L^M	M	406
L^M/L^N	MN	744
L^N/L^N	N	332
		Total = 1482

$f(L^M) = p = [(406)(2) + 744]/(1482)(2) = 0.525$.
$f(L^N) = q = [(332)(2) + 744]/(1482)(2) = 0.475$.

From the statistical toolkit, we estimate the 95% confidence intervals of p and q. Nq > 100, so the binomial is close to normal. $s = \sqrt{pq/N} = \sqrt{(0.525)(0.475)/(1482)}(1.96) = 0.0092$. The 95% confidence intervals are $p = 0.525 \pm 0.018$; $q = 0.475 \pm 0.018$. We are 95% sure that p is between 0.507 and 0.543, and that q is between 0.457 and 0.493.

Phenotype	Expected Number	
M $(0.525)^2(1482) =$	408.5	The expected numbers are close
MN $(2)(0.525)(0.475)(1482) =$	739.1	to the observed numbers,
N $(0.475)^2(1482) =$	334.4	consonant with CHW equilibrium.

Solved Problem: estimating allele frequencies for cystic fibrosis. Cystic fibrosis is a genetic disorder characterized by severe, chronic pulmonary obstruction, due to recessive mutations in the *CFTR* gene, which encodes a chloride channel. More than 600 mutant alleles have been found in the *CFTR* gene, but here all mutant, nonfunctional alleles are lumped together into one category, $CFTR^o$. In a random sample of 343,756 live births in Brittany (Scotet *et al.* 2001), 118 infants had cystic fibrosis. Estimate the frequency of $CFTR^o$ in this population, and the frequency of heterozygous carriers.

$f(CFTR^o) = q = \sqrt{(118)(2)/(343756)(2)} = 0.0185$; again, $Nq > 75$. $s = \sqrt{(.0185)(.9815)/(343756)(2)} = 0.00016$. 95% confidence interval: $0.0185 \pm (1.96)(0.00016) = 0.0185 \pm 0.0003$. We are more than 95% sure that q is between 0.018 and 0.019. The frequency of heterozygotes is $2pq = (2)(0.0185)(0.9815) = 0.036$, or 3.6%. In Brittany, about 3.6% of persons are heterozygous for a mutant allele of *CFTR*, and about 1 in 3000 babies are born with cystic fibrosis.

Sex-Linked Genes

In most vertebrates and insects, and in many plants, females have two X chromosomes, while males have one X chromosome. Consequently, males are hemizygous for X-linked genes, and, at equilibrium, the frequencies of haplotypes are equal to the population allele frequencies, p and q. At equilibrium, female genotype frequencies are predicted by the CHW rule – p^2, 2pq, and q^2. If male and female allele frequencies differ, they will approach equilibrium rapidly, and within 10 generations or less the female and male allele frequencies will be so close it will be difficult to measure the difference. The final equilibrium allele frequency $p_e = (2p_f + p_m)/3$.

Solved Problem: estimating sex-linked allele frequencies. Green-weak vision is due to mutations in the X-linked gene *GCP*, which encodes an opsin with a maximal absorption peak at $\lambda = 530$ nm. In a random sample of Norwegian school children, 551 of 9049 boys were green-weak, and 37 of 9072 girls were green-weak. What is the allele frequency of GCP^- mutations in this Norwegian population?

For males, $q_m = 551/9049 = 0.0609$. For females, $q_f = \sqrt{(37/9072)} = 0.0639$. The overall frequency is $q = [(2)(0.0639) + 0.0609]/3 = 0.0629$. The observed numbers of green-weak boys and girls are both close to the expected numbers, and a χ^2

test fails to reject the hypothesis that the population is in CHW equilibrium.

Overall Genetic Variation in Populations

We have considered single genes, but genomes of organisms have thousands of genes. Population geneticists estimate overall, genome-wide levels of genetic variation in populations by sampling many genes in many individuals. In human genes, there is a nucleotide substitution polymorphism every 200 to 1000 nt, and most of these are in noncoding regions (5′ UTR, introns, and 3′ UTR). Other species have similar levels of DNA variation.

A measure of genetic variation is **P**, the proportion of genes that are polymorphic (have alleles segregating). If the frequency of the most common allele is less than 99%, then a gene is usually said to be polymorphic. A second measure of genetic variation is **H**, the proportion of heterozygotes for a gene. The average heterozygosity $\overline{H}$ is then computed for a large number of genes.

A method for detecting variation in primary protein structure is gel electrophoresis of proteins. Alternative allelic forms of a protein are called **allozymes**; allozymes differ in electrical charge owing to amino acid substitution. For genes that encode proteins, P is between about 15% and 55%, and $\overline{H}$ is between about 4% and 17%, depending on the species and population. In a survey of 71 human enzymes [Harris and Hopkinson 1972], 51 were monomorphic, and 20 had allozyme variants; P = 28% and $\overline{H}$ = 7%.

Allozymes represent only a small fraction of genetic variation for protein-coding genes. A mutation may cause an amino acid substitution that does not change the electrical charge of a protein; it may substitute a synonymous codon (same amino acid); or it may cause a silent change in the 5′ flanking region, the 3′ flanking region, or an intron. In a study of the *Adh* gene of *Drosophila* [Kreitman 1983] 34 nucleotide substitutions were detected in the transcribed sequence, and only one of these caused an amino acid substitution. This huge amount of variation was found in a sample of 11 copies of the *Adh* gene, taken from diverse sources (Figure 37.3).

Fig 37.3 Distribution of polymorphic mutations in the *Adh* gene of *Drosophila*.

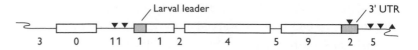

Noncoding DNA sequences are more variable among individuals than DNA sequences that do encode functional gene products. In the genomes of many eukarya, repetitive, noncoding DNA sequences make up a large fraction of the genome. Transposons, transposon-like sequences, and tandem repeats are common kinds of repetitive sequence elements. In humans, microsatellites (tandemly repeated arrays of 1 to 6 nt) make up about 1% of the genome. The number of repeats per array varies substantially in populations. Each microsatellite allele has a different number of repeats per locus; for microsatellites in human populations, H $\approx$ 80%.

Polymerase chain reaction-based methods are widely used to detect genetic variation at the DNA level. These methods, described in Chapter 28, include the following:

- SSCP (single-strand conformation polymorphism)
- RAPD (random amplification of polymorphic DNA)
- AFLP (amplified fragment length polymorphism)
- DNA fingerprinting assays for microsatellite polymorphism

The SSCP assay detects alleles of single genes and can be useful in studies of a single, known gene – for example, *CFTR*, the chloride transporter implicated in cystic fibrosis. RAPD and AFLP are used to measure variation in noncoding sequences and to find genetic correlates of phenotypic variation. DNA fingerprinting, mostly used to identify individuals and to detect and rule out paternity, also provides genetic markers linked to genes affecting variation for any trait of interest.

Further Reading

Cooper DN. 1999. *Human Gene Evolution*. BIOS Scientific Publications, Guildford, UK.

Gauer L, Cavalli-Molina S. 2000. Genetic variation in natural populations of mate using RAPD markers. *Heredity* 84:647–656.

Harris H, Hopkinson DA. 1972. Average heterozygosity in man. *Ann. Hum. Genet.* 36:9–20.

Kreitman M. 1983. Nucleotide polymorphism at the alcohol dehydrogenase locus of *Drosophila melanogaster*. *Nature* 304:403–422.

Roman B *et al.* 2001. Genetic diversity in *Orobanche crenata* populations from southern Spain. *Theor. Appl. Genet.* 103:1108–1114.

Scotet V *et al.* 2001. Neonatal screening for cystic fibrosis in Brittany, France. *Lancet* 356:789–794.

Chapter 38

Mutation, Migration, and Genetic Drift

Overview

Evolution is often subdivided into **microevolution** and **macroevolution**. Microevolution proceeds by baby steps – small, incremental changes in allele frequencies, which over the course of many generations can produce large and dramatic genetic and phenotypic shifts in populations. Macroevolution proceeds by leaps and bounds – sudden, dramatic changes such as polyploidization. New species arise both by microevolution and by macroevolution.

Forces that drive changes in allele frequencies include **mutation**, **migration**, and **inbreeding and genetic drift**. Intragenic mutation, the source of most new genetic variation in populations, is a very weak force that barely contributes to slow directional change. Migration, a strong force, causes populations to become genetically homogeneous. Inbreeding and genetic drift are weak forces that decrease genetic diversity within populations and can cause random changes in allele frequency. **Selection**, another strong force, is considered in Chapter 39.

Mutation

As a rule of thumb for organisms, an average-sized gene mutates at the rate of 10^{-5} per generation, and mutant alleles mutate back to functional alleles at the rate of 10^{-6} per generation. Assume a population with functional allele A^1 at frequency p and mutant allele A^2 at frequency $1 - p = q$. The mutation rate, $A^1 \rightarrow A^2$, is

μ; the rate in the reverse direction, $A^2 \rightarrow A^1$, is v. The change in q is $\Delta q = \mu p - vq$:

$$\Delta q = \mu - q(\mu + v).$$

Imagine the gene pool changing very slowly, with a new allele replacing an old one at one rate and the reverse at another rate. Eventually, the allele frequencies stop changing, and this point is called **equilibrium**, when $\Delta q = 0$, by definition.

To find the equilibrium allele frequency, q_e, set $\Delta q = 0$ and solve for q_e:

$$q_e = \mu/(\mu + v).$$

Assuming that $\mu \approx 10v$ and that $q < 0.10$, then $\Delta q \approx 10^{-5}$ and $q_e \approx 0.90$. This rate of change is so slow that it would take millions of generations to reach equilibrium, and even to change p from 0.01 to 0.10 would take $\sim 10^4$ generations. Mutation could be a sig nificant cause of changing allele frequencies only if stronger evolutionary forces – selection, migration, and genetic drift – were not at work, an unlikely situation.

Mutation is the primary source of novel genetic variation in a population, but otherwise contributes little to genetic changes in populations.

Migration

Migration, also called **gene flow**, has two main effects:

- To inject genetic variation into a recipient population
- To homogenize populations genetically through the exchange of genes

Consider two geographically separated populations with a small amount of migration between them in one direction (Figure 38.1). For some gene the allele frequencies are P and Q in the donor population and p and q in the recipient population. The proportion of migrant genes coming into the recipient population is **m**, and the proportion of resident genes is **1 − m**. In the recipient population, the premigration allele frequency is q_0, the postmigration allele frequency is q_1, and

$$q_1 = (1 - m)q_0 + mQ = q_0 - m(q_0 - Q)$$
$$\Delta q = -m(q_0 - Q).$$

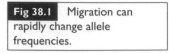

Fig 38.1 Migration can rapidly change allele frequencies.

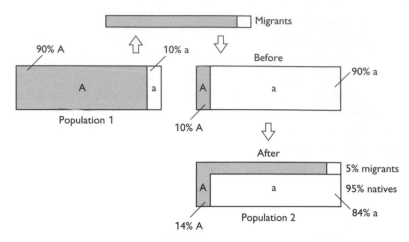

If a species is subdivided into geographically isolated populations with no migration between them, then the populations will become genetically differentiated through the action of selection and genetic drift. Migration between these populations will oppose these evolutionary forces. Even a tiny amount of migration between populations will make their allele frequencies converge to the same values, in the absence of other, opposing forces.

Inbreeding

Inbreeding and outcrossing are deviations from panmixia due to genetic relatedness. In inbreeding, gametes unite at a higher-than-random frequency, or equivalently diploid relatives mate at a higher-than-random frequency. Outcrossing is the opposite. Inbreeding is maximal under pure self-fertilization. A common system of regular inbreeding is one of brother-sister matings. In a large population,

(1) Inbreeding in families increases the fraction of homozygotes over the Castle-Hardy-Weinberg expectations and outcrossing increases the fraction of heterozygotes, but
(2) Neither inbreeding nor outcrossing changes allele frequencies for the population as a whole.

Inbreeding is quantified by **F**, the coefficient of inbreeding. F is the probability that two alleles in two uniting gametes are identical by descent, meaning that they come from the same ancestor. Alleles that are identical by descent are identical in DNA sequence, but the converse is not necessarily true.

This idea deserves further attention. All members of a species are related and share genes they inherited from common ancestors. The number of progenitors of sexually reproducing organisms, N generations in the past, is 2^N. For humans, generation time averages about 30 yr, so 50 generations take about 1500 yr. How many ancestors does one person have, going back 50 generations, or a little over a millenium? $2^{50} \approx 1.1 \times 10^{15}$ great great great . . . (48 greats) grand parents, yet 10^{15} is far bigger than the total number of humans who have ever lived, implying that everyone has ancestors that were related to each other. Therefore, two persons chosen at random from the same population very likely share ancestors more recently than the Dark Ages of Europe. Moreover, analysis of primate DNA sequences points to the descent of all humans from a common ancestral population about 5 million years ago, when chimpanzees and humans diverged. Every person is at least distantly related to every other person. This general principle is just as true of *Penicillium*, penguins, and pine trees as it is of humans. All members of a species share common ancestors and share alleles that are identical by descent.

Now we return to the analysis of inbreeding. When a family is analyzed, all interbreeding individuals of the oldest generation are **defined** as unrelated; hence, within a family, all copies of all genes are considered **not** to be identical by descent except for cases of intermarriage between relatives within the pedigree. F is useful for quantifying inbreeding on a short timescale. To calculate F for an inbred individual, draw a family diagram, and trace the paths from the individual to each common ancestor. For each generation, the probability of transmitting a particular allele is $\frac{1}{2}$, so the probability of transmitting an allele through n generations is $(\frac{1}{2})^n$; raise $\frac{1}{2}$ to the power of the number of segments (generations) in each complete circuit containing the focus individual. F is the sum of these the path probabilities for all ancestral alleles in the diagram.

Solved problem: calculating F for a child born of first cousins. Calculate F for a child born of first cousins. There are two paths from child to ancestor and back, one through the great grandmother and one through the great grandfather (Figure 38.2). Each path contains six parent-to-child segments, so the probability

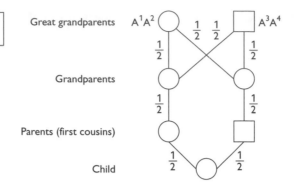

Fig 38.2 Calculation of F for a child born of first cousins.

that the child is homozygous for allele A^1 is $(\frac{1}{2})^6 = 1/64$, and similarly for A^2, A^3, and A^4. Therefore F = (4)(1/64) = 1/16.

Consider gene A with two alleles, A^1 and A^2, whose frequencies are p and q. The probability that two alleles unite at random is the product of their frequencies; for A^1 and A^1, it is p^2. With inbreeding, there are more homozygotes and fewer heterozygotes than with random mating. The genotypic frequencies are as follows:

A^1A^1	A^1A^2	A^2A^2
$p^2 + Fpq$	$2pq - 2Fpq$	$q^2 + Fpq$

Inbred individuals have a higher probability of being homozygous than randomly bred individuals. Therefore, in a small population or an experimental strain, inbreeding can lead to fixation of alleles. Most countries have laws against marriage between close relatives; such laws prevent deleterious recessive alleles from becoming homozygous. The poor survival and weakness of inbred organisms is called **inbreeding depression**. Outcrossing can produce the opposite effect, because the survival and vigor of highly heterozygous individuals often exceeds the average. The high survival and strength of outcrossed organisms is called **heterosis** or hybrid vigor.

Genetic Drift

In a large population, panmixia is theoretically easy to attain, with the proviso that self-fertilization is precluded in most species. In small populations panmixia is not possible; the smaller a population is, the more probable is mating between relatives. To relate the inbreeding coefficient to population size, consider a gene pool of size 2N, or N diploid individuals. If 2N gametes unite at random in the gene pool, then the probability of drawing the same copy of a gene twice is $1/2N$, and the probability of drawing different copies of a gene is $1 - 1/2N$. Let F_t be the inbreeding coefficient in the tth generation. Then

$$F_t = 1/2N + (1 - 1/2N)F_{t-1},$$

or

$$1 - F_t = (1 - 1/2N)(1 - F_{t-1}).$$

The level of heterozygosity can be measured by an index, **H**.

$$H_t = 1 - F_t \quad \text{and} \quad H_t = (1 - 1/2N)H_{t-1}.$$

Imagine a large population in which all the matings are between second cousins, for whom $F = 1/64$ (second cousins share two great grandparents). This large population is therefore equivalent to a random-mating gene pool of $2N = 64$, or $N = 32$ diploid, random-mating individuals, and the two populations have the same heterozygosity. In other words, small population size has the same effect on heterozygosity as inbreeding. A large, inbred population eventually loses its genetic variation because one allele or the other becomes fixed. The same is true of a small, random-mating population, which loses genetic variation each generation by genetic drift. The frequency of heterozygotes, $2pq$, decreases by $1/2N$ each generation, or by $1 - [1 - 1/2N]^t$ in k generations, as one allele becomes more prevalent. At the same time, the frequency of homozygotes increases by $1/2N$.

If a population is founded by a few individuals, then that population will have reduced heterozygosity for the same reason. This phenomenon is the **founder effect**, a form of genetic drift. A population **bottleneck** occurs whenever a relatively large population undergoes a great reduction in N (Figure 38.3).

Genetic drift and Huntington Disease. Huntington disease is a late-onset neurodegenerative disorder due to a dominant allele of the *IT15* gene (locus 4p16.3). In northwestern Venezuela, the frequency of $IT15^{HD}$ is unusually high. This high frequency can be traced to a single woman who moved to northwestern Venezuela in the mid-20th century and left many descendants.

A genetic drift experiment. Kerr and Wright (1954) established 96 lines of *Drosophila*, each carrying the *forked* mutation at a frequency of 50% (Figure 38.4). *Forked* causes a mild deformity of the fly's bristles, and has no measurable effect on survival or

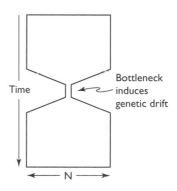

Fig 38.3 Population bottleneck.

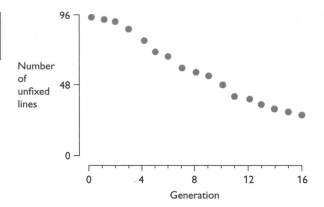

Fig 38.4 Distribution of *forked* gene in Drosophilas, in a genetic drift experiment.

fertility. Each line was maintained by breeding four randomly chosen pairs every generation. Within 16 generations, the f and the f^+ alleles were still segregating in only 30 lines.

Extreme inbreeding can reduce survival and fertility. This can pose a grave threat to a species with very low population numbers and in rare animals kept in zoos. A good example of extinction of local populations is the Glanville fritillary butterfly *Melitaea cinxia*. In Finland, this species is subdivided into very small local populations, as small as a single family per population. Consequently, this butterfly suffers extreme inbreeding, which causes loss of heterozygosity and inbreeding depression. Local populations frequently become extinct, and the probability of extinction is highly correlated with loss of heterozygosity. Conservation biologists monitor genetic variation in endangered species. Inbreeding depression in a captive population or in a small local population can be ameliorated by introducing genetically distinct migrants, thus increasing genetic diversity. While only populations of the Glanville fritillary suffer extinction, the same may apply to whole species whose total numbers are small. On the other hand, many small local populations have little genetic variation owing to genetic drift, yet survival and fertility are not affected adversely.

Further Reading

Edenhamn P *et al.* 2000. Genetic diversity and fitness in peripheral and central populations of the European tree frog *Hyla arborea*. *Hereditas* 133:115–122.

Kerr WE, Wright S. 1954. Experimental studies of the distriution of gene frequencies in very small populations of *Drosophila melanogaster*. I. *Forked. Evolution* 8:172–177.

Lammi A *et al.* 1999. Genetic diveresity, population size, and fitness in central and peripheral populations of a rare plant *Lychnis viscaria. Conserv. Biol.* 13:1069–1078.

O'Brien SJ. 1994. A role for molecular genetics in biological conservation. *Proc. Natl. Acad. Sci. USA* 91:5748–5755.

Saccheri I *et al.* 1998. Inbreeding and extinction in a butterfly metapopulation. *Nature* 392:491–494.

Chapter 39

Natural Selection

Overview

Natural selection is differential reproduction of genotypes in the absence of human intervention. The relative reproductive ability of a genotype is its **fitness**. Selection is a potentially strong force that can change both the population mean and its genetic variation; it is the principal agent of directional genetic change. The speed of evolution under natural selection is a function of fitness and allele frequencies. Rare deleterious alleles are eliminated very slowly from a population. One way genetic variation is maintained in a population is **balancing selection**, which favors heterozygotes.

Kinds of Selection

Selection occurs whenever one genotype reproduces at a higher or lower rate than other genotypes, on average, causing that genotype to contribute more or fewer copies of its genes to the population than other genotypes do. For example, a recessive lethal allele that acts at a prereproductive stage will be selected against because homozygotes will die before they can reproduce. Selection acts on all life forms, regardless of the mode of reproduction or the nature of the life cycle. It is the principal driving force of evolution. The speed of selection and its effect on equilibrium frequencies of alleles and genotypes can be quantified. Selection that acts on survival, fertility, and fecundity is called natural selection, while selection that acts on the mating success of sexual organisms is called **sexual selection**. Fitness (also called **Darwinian fitness**) is the relative ability of a haplotype or genotype to contribute to the next generation.

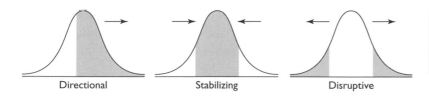

| Directional | Stabilizing | Disruptive |

The three main types of selection, **directional**, **stabilizing**, and **disruptive**, are cartooned in Figure 39.1.

Selection Acting on Haploid Organisms

When natural selection acts during the haploid phase of a sexual organism or on clonally reproducing forms, the fittest haplotype or genotype simply replaces the others during evolution. While selection acts on the entire genotype, only simple cases involving the average effect of single genes can be analyzed rigorously. Here, the additive effects of a single gene are considered. The selection weights w_i are relative fitness values, and the following table shows one generation of selection acting on alleles in a very large population. The average fitness of the population, due to the one gene, is $\overline{W} = \Sigma t_i w_i$.

| | Selection acting in a haploid organism | | | | | |
| | General case | | | Particular case | | |
Allele	A^1	A^2	Total	A^1	A^2	Total
$f(A^1)$ before selection	p	q	1	0.500	0.500	1
Fitness	w_1	w_2	—	1	0.9	—
$f(A^1)$ after selection	$w_1 p/\overline{W}$	$w_2 q/\overline{W}$	1	0.526	0.474	1
		$\overline{W} = \Sigma f_i w_i$			$\overline{W} = 0.95$	

The rate of change in q each generation is $\Delta q = q_1 - q_0 = q[w_2 - \overline{W}]/\overline{W}$. If $w_1 > w_2$, then selection eliminates the less fit allele, A^2, from the population. Elimination of A^2 slows down as this allele gets rarer and rarer, and the average fitness of the population gradually increases. In other words, as $q \to 0$, $\overline{W} \to 1$ and $\Delta q \to 0$. Also, the bigger the difference in fitness between alleles, the faster is evolution. In the example above, the A^2 allele has 10% poorer survival than the A^1 allele in the hypothetical environment; it would take 102 generations until the allele frequency of A^1 changed from 0.5 to 0.9.

If the life form undergoing natural selection is clonal, then the rules above suffice, but if it is the haploid phase of a sexual organism, we must assume a large, panmictic population with no selection in the diploid phase. The allele with lower fitness decreases in frequency every generation.

Selection Acting on Diploid Organisms

Suppose selection acts on the diploid phase of a sexually reproducing organism but not on the haploid phase, in a large, panmictic population. As with selection on the haploid phase, selection acting on a single gene is considered here. Again denote the fitness value of each genotype by w_i and mean fitness by $\overline{W}$. The following table shows the effects of selection, in one generation.

| Genotype | General Model | | | |
	A^1A^1	A^1A^2	A^2A^2	Total
Frequency before selection	p^2	$2pq$	q^2	1
Fitness	w_1	w_2	w_3	$\overline{W}$
Frequency after selection	$w_1p^2/\overline{W}$	$2w_2pq/\overline{W}$	$w_3q^2/\overline{W}$	1

As with selection on haploid organisms, the rate of evolutionary change is Δq.

$$\Delta q = q_0 - q_1 = [pqw_2 + q^2w_3]/\overline{W} - q = [pqw_2 + q^2w_3 - q\overline{W}]/\overline{W},$$

or

$$\Delta q = pq[q(w_3 - w_2) - p(w_1 - w_2)]/\overline{W}.$$

There are three parts to the expression.

(1) Δq varies directly with pq, which is maximal when $p = q$, and minimal if either allele is fixed.
(2) Δq varies inversely with $\overline{W}$, the population mean, which is low when the population is far from equilibrium, and which reaches a maximum at equilibrium.
(3) Finally, Δq varies directly with the average effect of substituting one allele for another, which is the long expression in brackets. This expression is large if the fitness weights are very different from each other – that is, if selection is strong.

Selection operating on diploid organisms can have distinct outcomes for allele frequencies, depending on the type of selection:

(1) Directional selection: Allele frequency changes until one allele becomes fixed.

(2) Stabilizing selection: Alleles are maintained at the same frequencies (genetic polymorphism).

(3) Disruptive selection: Genetic polymorphism is maintained when extreme types are at an advantage. This can occur when genotypes encounter different environments within the species range, and the population is subdivided in some way or other. In the case of balancing selection, the heterozygote is favored, and allele frequencies are maintained at an intermediate frequency.

New Mutations

When a new allele arises in a population by mutation, its fate is strongly influenced by selection. If it is selectively disadvantageous, it is highly likely to be lost. If the allele is **neutral** (no selective value), then the probability it will survive one generation is $1 - e^{-2} = 0.63$; the probability it will survive two generations is $1 - e^{-(1-.63)} = 0.47$. After 100 generations, there is only a 2% chance the mutant allele will still exist. If the new allele is selectively advantageous with fitness w_3, then the probability that it will ultimately survive in the population is $2(w_3 - w_2)$, if the other two genotypes have fitness $w_1 = w_2 < w_3$. To have at least a 50% chance of surviving in the population, the fitness of the homozygote for a new recessive mutation must be at least 25% higher than the fitness of the other genotypes, an unlikely situation.

Directional Selection

Directional selection shifts the population mean and reduces variance. Selection acting on one gene is directional if one of the homozygotes has the greatest fitness. If selection favors a recessive allele over other alleles, the favored allele tends to increase in frequency. Alternatively, a detrimental recessive allele tends to decrease in frequency. At equilibrium, one allele is fixed, $q_e = 0$ or $q_e = 1$.

$$\text{If } w_1 \leq w_2 < w_3 \quad q_e = 1$$
$$\text{If } w_1 > w_2 \geq w_3 \quad q_e = 0.$$

It is instructive to work through examples of directional selection acting on one gene.

Model example. Selection favors a recessive allele, $p = 0.9$, $w_1 = w_2 = 0.7$, $w_3 = 1.0$, preselection; 1, one generation postselection; 2, two generations postselection.

			Generation	
Genotype	w_i	0	1	2
A^1A^1	0.7	0.81	$(0.7)(0.81)/0.703 = 0.807$	$(0.7)(0.807)/0.704 = 0.802$
A^1A^2	0.7	0.18	$(0.7)(0.18)/0.703 = 0.179$	$(0.7)(0.179)/0.704 = 0.178$
A^2A^2	1	0.01	$(1)(0.01)/0.703 = 0.014$	$(1)(0.014)/0.704 = 0.020$
$\bar{W}$		1	0.703	0.704
q		0.1000	0.1035	0.1090

Example. Industrial melanism. Industrial melanism is the ascendance of dark morphs in populations of insects during the industrial revolution of the 19th century, when so much coal was burned that the resting places of insects became soot-covered. On dark backgrounds, dark-colored insects are less conspicuous to insectivorous birds than light-colored insects, and because of this, genes that make for melanization were selected during the industrial revolution. In most cases, a single gene has been implicated, with either partial quantitative dominance of dark over light, or else no quantitative dominance (qualitatively, M^D/M^D is very dark, M^D/M^L is dark; M^L/M^L is light). The following table gives estimates of fitness weights for an English population of moths *Biston betularia*.

In the 20th century, the use of coal diminished, as did deposition of soot on trees and other resting places for insects. As a consequence of the reverse in directional selection, dark morphs of *B. betularia* and other insects disappeared.

A change in environment can cause a change in allele frequency. How fast do allele frequencies change? Eliminating a dominant allele by selection in an organism's diploid phase is as rapid as elimination by selection in a haploid organism. Eliminating a recessive allele takes much longer, and Δq slows as the allele frequency q becomes small (Figure 39.2). The smaller q becomes, the greater the fraction of recessive alleles are found in heterozygotes. The ratio of recessive alleles in heterozygotes to those found in homozygotes is $pq/q^2 = (1 - q)/q$; when q is small, the ratio is very close to $1/q$. The number of generations, t, to change the frequency of a deleterious recessive allele is $t = 1/(1-w_3) [(q_0 - q_t)/q_0q_t + \ln(q_0p_t/q_tp_0)]$.

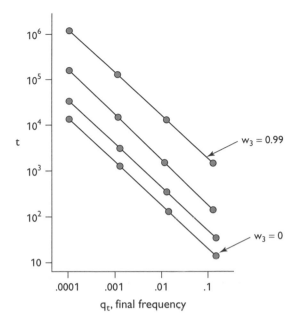

$w_3 = 0.99$

$w_3 = 0$

q_t, final frequency

Fig 39.2 Speed of eliminating a deleterious recessive allele by natural selection. The number of generations, t, required to reduce q, the frequency of a deleterious recessive allele, from 0.5 to q_t, a final frequency, is shown for several fitness values of the recessive homozygote. Both the X and Y axes are logarithmic scales. When the recessive allele is lethal, it takes 10,000 generations to reduce q from 0.5 to 0.0001. When the recessive type has a 1% disadvantage, it takes a million generations to reduce q from 0.5 to 0.0001.

Why is cystic fibrosis so common in Western Europe? Cystic fibrosis is a recessive disorder due to mutations in a gene encoding a chloride transporter, *CFTR*. In Western Europe and the United States, mutant $CFTR^0$ alleles collectively have a frequency of about 0.02. If the mean fitness of persons with CF were 10% of the relative mean fitness of persons without CF and heterozygotes were equal in fitness to normal homozygotes, how long would it take to change the mutant allele frequency by a factor of 10, from 0.02 to 0.002?

$$t = [(0.02 - 0.002)/(0.02)(0.002) + \ln(0.02)(0.998)/(0.002)(0.98)]/0.1$$

$$= 4523 \text{ generations, or about } 136{,}000 \text{ years.}$$

Why are $CFTR^0$ mutations so common? Likely it is because there is, or was in recent human history, a heterozygote advantage. Assuming this to be true, and letting the constant fitness weights be 1, w_2, and 0.1,

$$0.02 = (w_2 - 1)/(2w_2 - 1 - 0.1), \quad w_2 \approx 1.02.$$

Heterozygous carriers of $CFTR^0$ alleles may be resistant to cholera and other infections of the gastrointestinal tract. The advantage to heterozygotes does not need to be large to explain the frequency of $CFTR^0$ alleles in European populations.

Stabilizing Selection

Stabilizing selection resists change in the population mean and reduces the variance. It can act either directly by eliminating homozygotes that make for extreme phenotypes or indirectly by producing and maintaining genotypes that act epistatically to suppress phenotypic variation (**canalization**).

Stabilizing selection in fig wasps. Fig wasps are the sole pollinators of figs, and each species of fig wasp specializes on one species of fig tree. The number of females that founds a brood within each fig varies among species. Female fig wasps reproduce by arrhenotoky (unfertilized eggs develop into males) and control the sex ratio in their offspring. The more males in a brood, the fewer dispersing daughters, each of whom potentially produces a brood in the next generation. Too few males in a single-foundress brood may doom the females to death inside the fig, as males chew an exit hole for their sisters to leave the fig, and a minimum number of males is required for this task. There is good evidence that stabilizing selection optimizes the sex ratio in fig wasps. The higher the proportion of single-foundress broods in a species, the lower the variance in sex ratio among broods.

Canalization in Drosophila. In wild *Drosophila melanogaster*, the number of large scutellar bristles is exactly four; $\sigma^2 \approx 0$. If flies are heat shocked during the pupal stage, then a few adults will develop three or five bristles. The heat shock reveals hidden genetic variation for bristle number, in part by reducing the effectiveness of chaperonins, which help proteins assume the correct tertiary structure. By heat shocking, then selecting extreme types, more and more flies that deviate from the norm appear. After a few generations, the heat shock is no longer needed to reveal the formerly hidden genetic variation. In normal animals, development is carefully canalized to reduce the expressed genetic and environmental variation for traits. Many phenotypes of animals are canalized. In the instances in which canalization has been studied genetically, many genes, each with small effect, contribute to canalization. Furthermore, there is genetic variation for the strength of canalization itself. In theory, populations that encounter anticanalizing stresses should reveal hidden genetic variation, which is then subject to natural selection. In a couple of cases, canalization of a single gene has been studied (*Bar* and *crossveinless* in *Drosophila*).

Disruptive Selection

Genetic polymorphism can be established and maintained by disruptive selection, which is usually locally acting selection on different populations within a species.

Disruptive selection in butterflies. In North America, the tiger swallowtail butterfly, *Papilio glaucus*, has a female-limited genetic polymorphism for two morphs (forms), yellow and dark. The dark morph mimics the pipevine swallowtail, *Battus philenor*, while the yellow morph is nonmimetic. The polymorphism is controlled by a sex-linked gene on the W chromosome. (Recall that in butterflies, males are ZZ and females are ZW.) The dark morph enjoys protection against predation because the model species is distasteful to some predators. The frequency of the dark morph varies geographically, correlating with the abundance of the model species. On the other hand, males prefer yellow females, suggesting that sexual selection for the yellow morph opposes selection for survival of the dark females.

Balancing Selection

Balancing selection, which favors the heterozygote over the homozygotes, maintains single-gene polymorphisms. If $w_1 < w_2 > w_3$, then

$$0 \leq q_e \leq 1 \text{ and } q_e = (w_2 - w_1)/(2w_2 - w_1 - w_3).$$

Balancing Selection in Humans. In every human population where malaria has been a long-term health problem, there exist recessive alleles of the β-hemoglobin gene (*HBB*) that protect against malarial parasites. Most of the protective alleles are recessive lethals that raise the fitness of heterozygotes. An example is the allele for sickle-cell anemia (*HBB^S*), a single nucleotide change that causes a substitution of glutamate for serine at amino acid 6; the normal allele is *HBB^A*. This allele is present in many populations in Africa, Europe, and Asia, as it has arisen many times by mutation. In a West African area with high rates of *Plasmodium falciparum* malaria, the following situation obtains:

	HBB^A/HBB^A	HBB^A/HBB^S	HBB^S/HBB^S	
f_i	0.81	0.18	0.01	
w_i	0.89	1.00	0	$q_e = 0.1$

The heterozygote survives better than the HBB^A/HBB^A homozygote. Even though persons with sickle cell anemia do not survive to maturity without medical intervention, the HBB^S allele has a rather high equilibrium frequency, 10%. The price of maintaining this polymorphism by balancing selection is the death of 1% of the population every generation, not by malaria but by lethal genotype.

Further Reading

Endler JA. 1986. *Natural Selection in the Wild.* Princeton University Press, Princeton, NJ.

Holzapfel CM, Bradshaw WE. 2002. Protandry: The relationship between emergence time and male fitness in the pitcher-plant mosquito. *Ecology* 83:607–611.

Katz LA, Harrison RG. 1997. Balancing selection of electrophoretic variation of PGI in two species of field cricket. *Genetics* 147:609–621.

Mueller U, Mazur A. 2001. Evidence of unconstrained directional selection for male tallness.*Behav. Ecol. Sociobiol.* 50:302–311.

Scriber JM *et al.* 1996. Genetics of mimicry in the tiger swallowtail butterflies, *Papilio glaucus* and *P. canadensis. Evolution* 50:222–236.

West SA, Herre EA. 1998. Stabilizing selection and variance in fig wasp sex ratios. *Evolution* 52:475–485.

Chapter 40

Quantitative Genetics

Overview

In natural populations of sexually reproducing organisms, individuals differ phenotypically in a quantifiable way. Most variable phenotypes of interest (e.g., size, fertility, and longevity) can be measured on a continuous scale and are influenced by many genes and a host of environmental factors. Usually, no single gene is a major source of natural variation for a phenotype (major gene), but it is possible to assess the magnitude and nature of the multigenic causes of natural variation for that phenotype, where many genes, each with a tiny effect, act together.

The goal of quantitative genetics is to measure the relative contributions of genotype and environment to phenotypic variation in a population of organisms, breaking down phenotypic variance into genetic and environmental components. The **heritability** of a trait is the fraction of phenotypic variance due to genes for a particular population of organisms in a particular environment. This genetic variation can be subdivided into components, notably **additive effects**, **dominance**, and **epistasis**. A further component of the environment is **genotype-environment interaction**. Natural selection can bring about evolutionary change in a trait if and only if a population has additive genetic variation for that trait. Genes that underlie natural quantitative variation for a phenotype can be mapped genetically to **quantitative trait loci** (QTL).

Quantitative Models

For any quantitative phenotype, there is usually natural genetic variation. Most of the genes contributing to this variation remain anonymous, as the effects of any one of them on quantitative variation are too small to observe directly, even though the concerted effects of many genes are big enough to measure. In this context, if a gene's effects on a quantitative phenotype can be observed directly, it may be termed a major gene. The relationship between gene and phenotype is assumed to be the same for genes with minor influence as for genes with major influence, and all genes are inherited in the same fashion. An excellent way to gain insight into how multigenic variation arises is to make models.

Additive Genetic Effects

In an additive model, each copy of each gene in the system adds up to a total influence on the phenotype. Consider a diploid population in which a single gene with two alleles influences a quantitative trait; A^1 adds 1 unit to the trait, and A^2 adds 2 units to the trait; $A^1A^1 = 2$, $A^1A^2 = 3$, and $A^2A^2 = 4$. The mean and variance of the population depend on the allele frequencies, p and q. Calculate the mean μ and the variance σ^2 with the following formulas, where f_i is the ith fraction of phenotypes:

$$\text{Mean:} \qquad \mu = \Sigma f_i x_i$$
$$\text{Variance:} \qquad \sigma^2 = \Sigma f_i x_i^2 - \mu^2, \text{ all } f_i < 1.$$

Under Castle-Hardy-Weinbery conditions, the genotype frequencies, f_i, are p^2, 2pq, and q^2.

When $p = q = 0.5$, then
$$\mu = (0.5)^2(2) + (2)(0.5)(0.5)(3) + (0.5)^2(4) = 3$$
$$\sigma^2 = (0.5)^2(2)^2 + (2)(0.5)(0.5)(3)^2 + (0.5)^2(4)^2 - (3)^2 = 0.5.$$

A population enriched for allele A^1 will have a lower mean, and a population enriched for allele A^2 will have a higher mean; the variance reaches a maximum when the allele frequencies are equal. For example, if $p = 0.9$ and $q = 0.1$, then the genotype frequencies are 0.81, 0.18, and 0.01, and $\mu = 2.2$ and $\sigma^2 = 0.18$. It is easy to expand the model to include several genes, if there is no epistasis: simply add the means and variances due to each gene to get the population mean and variance.

Compare this 1-gene model to 2-, 3-, and 100-gene models; the allele frequency is 0.5 (Figure 40.1).

As the number of genes increases, the number of phenotypes increases and the histogram of phenotypes approaches a normal

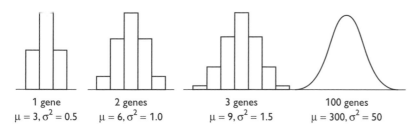

Fig 40.1 For each gene, the value of one allele is 1.0, and the other is 2.0.

curve. With n genes and 2 equal, additive alleles per gene, there are 2n + 1 phenotypes.

Additive Environmental Effects

Environmental effects e_i can be put in the model, so that each individual's phenotype is the sum of e_i and an allelic value a_i, making for even more phenotypes and a better approximation to the normal distribution. Take the simple case of three environmental factors, $e_1 = 4$, $e_2 = 6$, and $e_3 = 8$, where half the individuals in a population encounter the medium-sized factor e_2, and the other half are divided equally (0.25 and 0.25) between e_1 and e_3. The environmental component has a mean of 6 and variance of 2. Putting this together with the 2-gene example, we get 9 phenotypes instead of 5. Genotype and environment add up to the phenotype, giving a phenotypic mean of 12 and variance of 3 (Figure 40.2).

Dominance and Epistasis

In the previous sections, the effects of each allele and each environment simply add up to give a phenotype. With dominance or epistasis, however, a phenotype cannot be predicted from the average effects of alleles and environment; the deviation from additivity is **interaction**. The two main kinds of genetic interaction are dominance, an interaction between alleles, and epistasis, an interaction between genes. Both dominance and epistasis increase individual differences and therefore contribute to genetic variance.

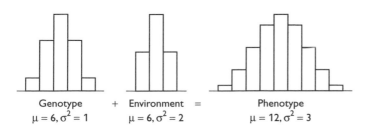

Genotype $\mu = 6, \sigma^2 = 1$ + Environment $\mu = 6, \sigma^2 = 2$ = Phenotype $\mu = 12, \sigma^2 = 3$

Fig 40.2 Genotype and environment add up to phenotype. Additive environmental effects make for a more realistic model, and make the histogram approach the normal distribution.

Multigenic models can be expanded to include interactions. In the 2-gene model, let A^1 be partially dominant to A^2 and let B^2 be partially dominant to B^1:

$$A^1A^1 = 2 \qquad B^1B^1 = 2$$
$$A^1A^2 = 2.5 \qquad B^1B^2 = 3.5$$
$$A^2A^2 = 4 \qquad B^2B^2 = 4$$

Adding epistasis, let B^2 be an enhancer of A, so that if the B^2 allele is present, then 0.25 is added for each A^1 allele, and 0.50 is added for each A^2 allele. Putting dominance and epistasis together, the genotype values are now:

$$A^1A^1\,B^1B^1 = 4 \qquad A^1A^2\,B^1B^1 = 4.5 \qquad A^2A^2\,B^1B^1 = 6$$
$$A^1A^1\,B^1B^2 = 6 \qquad A^1A^2\,B^1B^2 = 6.75 \qquad A^2A^2\,B^1B^2 = 8.5$$
$$A^1A^1\,B^2B^2 = 6.5 \qquad A^1A^2\,B^2B^2 = 7.25 \qquad A^2A^2\,B^2B^2 = 9$$

Again, assume intermediate allele frequencies ($p_i = q_i = 0.5$). The distributions of phenotypes follow; additive gene effects (A), dominance (D), and epistasis (I) (Figure 40.3).

Note that when dominance was added to the model, there was no shift in population mean, even though both genes exhibited dominance. This often happens in real populations, but it need not be the case. In a cross between two types that differ for many genes, the F_1 may resemble one parent more than the other, in which case we say there is **directional dominance**, meaning that most of the dominance effects act in the same direction.

In this example, adding epistasis shifted the population mean. This would not have happened had there been two epistatic genes with opposite effects and with the same allele frequencies. Again, if we cross two types that differ for many genes, the F_1 mean may be intermediate between the two parents, although epistasis variance may abound.

A small number of interaction effects can cause deviations from normality, but when a large number of interactions between alleles and genes are present, they tend to make for a normal distribution of phenotypes.

> **Fig 40.3** Additive gene effects (A), dominance (D), and epistasis (I).

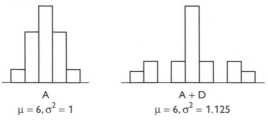

A
$\mu = 6, \sigma^2 = 1$

A + D
$\mu = 6, \sigma^2 = 1.125$

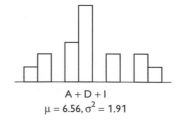

A + D + I
$\mu = 6.56, \sigma^2 = 1.91$

Model Summary

There are several sources of variation for any phenotype in any population. In this model, the total variation for the trait is the sum of the additive genetic variance, dominance variance, epistasis variance, and environmental variance:

$$\sigma^2{}_P = \sigma^2{}_A + \sigma^2{}_D + \sigma^2{}_I + \sigma^2{}_E.$$

Often, the total genetic variation, which includes $\sigma^2{}_A$, $\sigma^2{}_D$, and $\sigma^2{}_I$, is symbolized $\sigma^2{}_G$. Here, $\sigma^2{}_G = \sigma^2{}_A + \sigma^2{}_D + \sigma^2{}_I$, although in reality and in complex models $\sigma^2{}_G$ has more components.

	$\sigma^2{}_G$			$\sigma^2{}_E$
$\sigma^2{}_P$	$\sigma^2{}_A$	$\sigma^2{}_D$	$\sigma^2{}_I$	
3.91	1.00	0.13	0.78	2.00
100%	26%	3%	20%	51%

Remember, these values come from one particular hypothetical model, devised merely to illustrate principles. In a real example, the proportions of the components might be quite different. Typically, the environmental variance is rather large, and the additive genetic variance is often below 20%.

Heritability

Heritability is the ratio of additive genetic variance to phenotypic variance in a population; in other words, it is the proportion of phenotypic variance caused by natural genetic variation upon which selection can act. Heritability is symbolized h^2 because of the mathematical theory underlying its derivation, a procedure called path analysis. The reader may encounter another concept, "broad-sense heritability," which is total genetic variance divided by total phenotypic variance; this concept is virtually useless and should be avoided.

For the preceding example (model, above),

$$h^2 = \sigma^2{}_A/\sigma^2{}_P = 1.00/3.91 = 0.26, \text{ or } 26\%.$$

Suppose a population having additive genetic variation were subjected to natural selection, favoring high phenotypes. Each generation, individuals with a high phenotype and therefore enriched for high alleles would leave more offspring than individuals with a low phenotype. The result would be an increased frequency of high alleles and therefore of high phenotypes. The population mean consequently would increase in response to

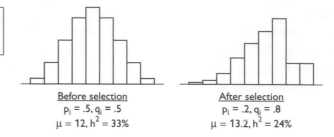

Fig 40.4 Histograms showing frequencies of phenotypes in a model population.

Before selection
$p_i = .5, q_i = .5$
$\mu = 12, h^2 = 33\%$

After selection
$p_i = .2, q_i = .8$
$\mu = 13.2, h^2 = 24\%$

selection. Figure 40.4 shows the frequencies of phenotypes in a model population (two genes, additive genetic effects, environmental variation). Before selection, the allele frequencies are all 50%; after generations of selection, the selected alleles are 80%.

If selection had favored low phenotypes, then the mean would have shifted downward and again h^2 would have been reduced.

It is wise to pause and consider what heritability does and does not mean. First, heritability measures the relative amount of genetic variation in one population for one trait in one environment. It is specific to that trait, that population, and that environment. Second, heritability does not reflect gene action. Heritability may be very low for a trait whose proximal causes are mainly genetic and very high for a trait whose proximal causes are mainly environmental. An example of a genetically programmed trait with low heritability is the number of limbs in frogs. In a normal environment, virtually all frogs have four legs; the population variance is zero and therefore heritability is zero. In an environment containing teratogens, frogs with five or more legs may be common, but heritability may still be zero. These multilimbed frogs are not genetically aberrant, but rather are victims of teratogens that severely perturb highly canalized developmental programs. An example of a trait whose proximal causes are mainly environmental but that may have a high heritability is the response of animals to a toxin. There is often genetic variation for susceptibility to a toxin, although of course without the toxin there could be no response at all.

Genotype–Environment Interaction

Genotype–environment interaction happens if genotypes react differently to environmental variation. Imagine two cows, Flossie and Bossie, of different breeds. Flossie is very sensitive to feed and

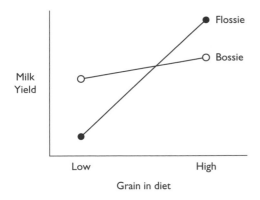

increases milk yield substantially when her diet is supplemented
with grain. Bossie is less sensitive to feed and increases milk yield
little in response to grain supplements. Such interaction adds to
the total phenotypic variation and normally is considered part
of environmental variation (Figure 40.5). It is often difficult to
measure, and usually is not explicitly considered in heritability
studies.

Measuring Genetic Variation

Measuring quantitative, multigenic variation, relative to environmental variation, has a single purpose – to assess a population's
potential for evolutionary change. In the case of natural populations, the interest is mainly academic, while in the case of animals
and plants that are grown to supply food and fiber, the interest
is primarily practical.

Three approaches are used to measure multigenic variation
for quantitative traits: (1) analysis of resemblance between relatives, (2) controlled crosses between inbred lines, and (3) artificial
selection.

Resemblance between Relatives

Parent–Offspring Regression
A straightforward method for estimating h^2 is parent–offspring
regression. Regression is a statistical procedure in which a line is
fitted to points on a graph. If there is additive genetic variation
for a trait, then a scatter plot of the offspring's values against

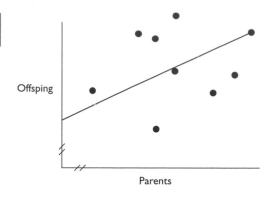

Fig 40.6 Flower length in offspring and parent.

midparent values (the parents' mean) approximates a straight line with a positive slope. Linear regression finds the line that "best" fits the points $Y = a + bX$; Y is the offspring's value, and X is the midparent value. The slope b measures h^2. In a carefully controlled experiment where the parents and offspring have equivalent, uncorrelated environments, $b = h^2$.

Example. The example in Figure 40.6 is about the inheritance of flower length (Chapter 32). Flower length (mm) is measured for parents (X, the average of the two parents) and for offspring (Y). For one sample, the values were as follows:

$$Y\ 12\ 15\ 14\ 15\ 13\ 13\ 16\ 15\ 19\ 15$$

$$X\ 10\ 12\ 13\ 13\ 14\ 14\ 14\ 16\ 17\ 18$$

$$Y = 14.7,\ s^2_Y = 3.79,\ X = 14.1,\ s^2_X = 5.66,$$

$$\mathrm{cov}(X, Y) = 2.81$$

$$b = \mathrm{cov}(X, Y)/s^2_X = 2.81/5.66 = 0.50.$$

The regression equation is true for the X and Y means, so

$$14.7 = a + (0.50)(14.1),\ a = 7.7$$

$$Y = 7.7 + 0.50X.$$

The heritability is 0.50 (50%). In a heritability study, the required sample size is usually several hundred X,Y pairs.

Siblings

Heritability can be determined from the covariance among siblings or half siblings. For humans, a popular approach is to use twins; this has severe drawbacks, and reports of heritabilities from human twin studies should be viewed with skepticism.

Half-sib families are designed in experiments to include mat-ings between one male and two or more females, each female having two or more offspring. Analysis of variance estimates the variance among sires (fathers), variance between dams (mothers) within sires, and within progenies. The component of variance due to sires, σ^2_S, estimates $\sigma^2_A/4$.

$$h^2 = 4\sigma^2_S/\sigma^2_T.$$

Crosses between Inbred Lines

Inbreeding fixes alleles, so that after many generations of strong inbreeding (e.g., sib mating), inbred lines will be homozygous for most genes. Only deleterious alleles will be eliminated selectively from inbred lines, or else they will be maintained in a heterozy-gous condition nonrandomly. The experimenter can sample from a genetically heterogeneous, random-fertilizing population to es-tablish many inbred lines. These inbred lines will be genetically distinct and may represent a large fraction of the genetic vari-ation in the original population from which the lines were es tablished. Under this scenario, the average amount of variation within inbred lines estimates σ^2_E. Estimates of additive genetic variance and dominance variance come from the F_2 and backcross (B_1 and B_2) variances. B_1 is $F_1 \times P_1$ and B_2 is $F_1 \times P_2$.

$$\sigma^2_E = s^2_P, \sigma^2_A = 2[2s^2_{F2} - s^2_{B1} - s^2_{B2}], \text{ and } \sigma^2_D$$
$$= 4[s^2_{F2} - \sigma^2_E - \sigma^2_A/2].$$

Solved Problem. Males of two strains of mouse *Mus musculus* differed in open field activity (DeFries and Hegmann 1970). In Mendelian crosses, the means and variances were

Generation	N	Mean	Variance	Generation	N	Mean	Variance
P_1	53	15.7	4.9	F_2	327	12.0	16.1
P_2	30	4.7	6.3	B_1	159	15.2	8.7
F_1	198	14.5	9.7	B_2	126	9.1	14.0

Check for directional dominance with a t test comparing the F_1 to the midparent value, where the variance of the difference is the pooled parental variance.

$$t = (14.5 - 10.2)/\sqrt{(5.40/83 + 9.7/198)} = 12.73, P < 0.0005.$$

Also, the F_1 mean is smaller than the P_1 mean ($t = 3.19$, $P < 0.005$). Clearly, then, the active strain showed partial directional dominanance. Calculate heritability from the formulas above.

$$\sigma^2_E = 5.4, \sigma^2_A = 2[(2)(16.1) - 8.7 - 14.0] = 19.0,$$
$$\sigma^2_D = 4[16.1 - 5.4 - 9.5] = 4.8$$
$$h^2 = 19.0/(5.4 + 19.0 + 4.8) = .65.$$

Artificial Selection

A third approach to the measurement of additive genetic variation is artificial selection, which yields an estimate called **realized heritability**. In preparation for the selection experiment, a genetically heterogeneous base population is established, usually by interbreeding many captured wild individuals from a single natural population. It is reasonble to assume that a large base population made in this way contains roughly the same amount of genetic variation as the natural population. The goals of the selection experiment are to test for additive genetic variation in the base population and to produce extreme genotypes for further study or for economic gain, as with cultivated plants and domesticated animals.

Several lines are established by sampling randomly from the base population. Normally, each line is a closed population, with no interbreeding allowed between lines. Every generation, parents of the next generation are chosen based on phenotype within selection lines, and they are chosen at random in one or more unselected control lines. High lines are selected for high mean phenotype, and low lines are selected for low mean phenotype. A sample is measured from each generation, and a fixed proportion of that sample is selected, say the extreme 50%. Within each selection line, the difference between the generation mean and the mean of the selected group, in standard deviation units, is called the **selection differential**, or S. The **response** to selection **R** is the mean difference, in standard deviation units, between generation means. Figure 40.7 illustrates R and S.

Extra scutellar bristles (large bristles on the thorax) were selected in two lines of *Drosophila simulans*. During generations one to five (dark bar), larvae were given a heat shock to reveal hidden genetic variation. The mean realized $h^2 = 0.25$ (Figure 40.8).

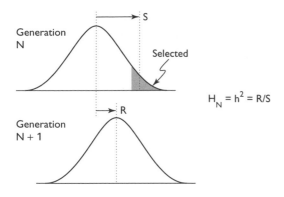

Fig 40.7 Selection differential, S, and response to selection, R.

$$H_N = h^2 = R/S$$

Mapping Genes That Affect Quantitative Traits

It is also possible to map the chromosomal loci where genes that influence a quantitative trait reside; these are quantitative trait loci (QTL). QTL contain genes or clusters of genes with measureable effects on the phenotype. The idea is to search for associations between the trait of interest and genetic markers – usually arbitrary molecular markers – whose alleles are segregating in the target population. The markers can be RFLPs or PCR-based markers such as microsatellites. Preliminarily, markers spanning the genome are identified and mapped. Following that, associations between markers and the quantitative trait are sought.

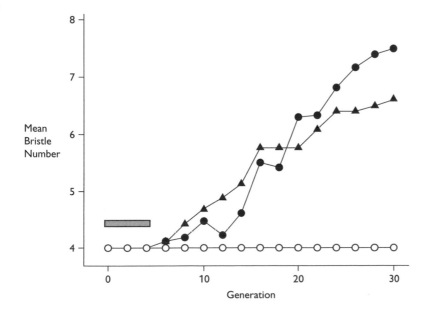

Fig 40.8 Artificial selection for extra scutellar bristles in *Drosophila simulans*.

QTL analysis has two major uses, basic and applied. In basic research, the idea is to find out the number and distribution of genetic factors with alleles segregating in a population, which affect some phenotype of interest. Applications include medical diagnosis and breeding of plants and animals for agriculture.

Further Reading

Belknap JK *et al.* 2001. QTL analysis and genomewide mutagenesis in mice. *Behav. Genet.* 31:5–16.

DeFries JC, Hegmann JP. 1970. Genetic analysis of open-field behavior. In: Lindzey G, Thiessen DD (eds.). *Contributions to Behavior-Genetic Analysis: The Mouse as a Prototype*, p. 23–56. Appleton-Century-Crofts, New York.

Falconer, DS. 1989. *Introduction to Quantitative Genetics,* 3rd ed. Longman, New York.

Kamin LJ, Goldberger AS. 2002. Twin studies in behavioral research: A skeptical view. *Theoret. Pop. Biol.* 61:83–95.

Kearsey MJ, Farquhar AGL. 1998. QTL analysis in plants; where are we now? *Heredity* 80:137–142.

Le Roy I *et al.* 2001. Genetic architecture of testis and seminal vesicle weights in mice. *Genetics* 158:333–340.

Chapter 41

Speciation

Overview

For all the sexually reproducing eukarya, the fundamental taxonomic unit is the **species** (s. and pl.). **Biological species** are populations of organisms between which reproduction is prevented biologically, by **reproductive isolating barriers**.

The birth of new species is **speciation**, during which one species splits into two reproductively distinct new species. The death of species is **extinction**. Speciation and extinction have produced, over the past two billion years, earth's diverse assemblage of extant species of eukarya, numbering $\sim 10^7$. The species of today are perhaps at most 1% of all species that have ever lived. Speciation is an inevitable by-product of genetic evolution. The exact genetic events that cause speciation are under intensive investigation.

Species Concepts

Phenotypic or Typological Species Concept

In the phenotypic concept, members of one species may vary quantitatively in form but are clearly distinct from members of another species. Individuals within species are considered variants of the same type. The phenotype is usually morphological, but it can be molecular. The Linnaean system of classification and binomial nomenclature for species (e.g., *Homo sapiens*) is purely phenotypic.

The main difficulty of this concept is that there is no evidence for the reality of a "true type," from which individuals deviate, even though it is commonplace for species to appear distinct.

Biological Species Concept

According to this concept, biological species are **reproductively isolated** from each other. This concept applies only to sexually reproducing eukarya. One or more kinds of intrinsic reproductive isolation – between species that are not separated by physical barriers – may operate to stop reproduction between species. Common isolating mechanisms are classified as follows:

Reproductive Isolating Barriers (RIBs)

A. Prezygotic RIBs
   1. Habitat isolation (organisms choose different habitats)
   2. Temporal isolation (different breeding times or seasons)
   3. Behavioral isolation (in animals, behavioral block to interspecific mating)
   4. Mechanical isolation (gametes physically prevented from contact)
   5. Gametic isolation (fertilization is blocked)
B. Postzygotic RIBs
   1. Hybrid inviability (hybrid zygotes die)
   2. Hybrid sterility (hybrid offspring do not produce functional gametes)
   3. Hybrid breakdown (F_2 and backcross generations are inviable or sterile)

Tempo of Speciation

How long does speciation take? In theory, speciation can be nearly instantaneous ($\sim 10^0$ to 10^1 generations) but only by mechanisms thought to operate in a minority of speciation events. The most prevalent mode of rapid speciation is polyploidization. In most cases, speciation requires the slow, gradual change in allele frequencies of many genes – microevolution that commonly takes $\sim 10^5$ to 10^7 generations to complete. However, there are examples of fast, microevolutionary speciation; for example, many new species of cichlid fishes in African lakes appear to have arisen via multigenic changes within $\sim 10^3$ generations.

Evolution of Reproductive Isolation

Speciation by Gradual Multigenic Changes

When a species splits into two new species, the new species are genetically distinct, and the genetic differences make for one or more of the RIBs listed above. To become genetically distinct, the species-to-be have to be separated reproductively long enough for RIBs to evolve; this usually does not happen if there is **gene flow** – migration – between incipient species. Gene flow between populations can be stopped by geographic or other physical barriers. When populations or species are physically separated, they are called **allopatric**, while populations or species that live together without physical barriers are called **sympatric**. Species that are mostly separated physically but inhabit a small zone of overlap are said to be **parapatric**. There is considerable indirect evidence that gradual speciation usually occurs in allopatric populations and occasionally in sympatric or parapatric populations.

In allopatric speciation, two geographically isolated populations accumulate genetic differences owing to mutation, natural selection, and genetic drift. As an incidental by-product of genetic differentiation, RIBs develop. Post-zygotic RIBs result from unfavorable epistatic interactions. At a minimum, two interacting **speciation genes** are required to produce postzygotic isolation. If the ancestral species is $A^1A^1B^1B^1$, and one population evolves to $A^2A^2B^1B^1$ and another population evolves to $A^1A^1B^2B^2$, and furthermore if the A^2 and B^2 alleles interact to make for inviability or sterility, then speciation will have occurred. That is, the $A^1A^2B^1B^2$ hybrid is inviable or sterile.

A good example of hybrid breakdown (inviability of some F_2 and backcross progeny), caused by a two-gene difference, occurs in two closely related fish, platyfish and swordtails. The platyfish has a two genes absent in the swordtail: one encoding a tyrosine kinase (TK), and a dominant suppressor of the *Tk* gene, *Su(Tk)*. Some of the F_2 and backcross individuals inherit the *Tk* gene but not *Su(Tk)*, and hybrids with these genotypes subsequently develop lethal melanomas.

Some of the genes responsible for reproductive isolation between species of *Drosophila* have been discovered. *Drosophila melanogaster* and *Drosophila simulans* are very closely related species

Fig 41.1 Reproductive isolation due to two genes in platyfish and swordtails. X^- = X chromosome laching the Tk gene; X^{Tk} = X chromosome carrying the Tk gene.

$$X^-X^- \ +/+ \ \times \ X^{Tk}Y \ Su[Tk]/Su[Tk] \quad P$$

$$\text{Swordtail} \quad \downarrow \quad \text{Platyfish}$$

$$X^{Tk}X^- \ Su[TK]/+ \ \times \ X^-Y \ Su[Tk]/+ \quad F_1$$

$$\downarrow$$

$$X^{Tk}X^- \ +/+ \ \text{and} \ X^{Tk}Y \ +/+ \ \text{die} \quad F_2$$

$$\dots \text{other genotypes live}$$

that are reproductively isolated by three mechanisms: behavioral barriers, hybrid inviability, and hybrid sterility. Both the males and females of *D. simulans* are extremely reluctant to mate with *D. melanogaster*. When mating takes place (and it does occur rarely in nature), all the progeny of the sex of the *D. simulans* parent die – hybrid females as embryos and hybrid males just before adulthood. The few surviving hybrid adults are sterile.

Five mutations have been found that partially rescue interspecific hybrids between the two species, three in *D. melanogaster* and two in *D. simulans*. It appears that a few genes with large effects, as well as many unidentified genes with small effects, contribute to the three isolating mechanisms between these species. Reproductive isolation between other pairs of *Drosophila* species is also multigenic, with speciation genes scattered throughout the genome.

Once two incipient species form, selection for the reinforcement of reproductive isolation may occur. Suppose a population becomes geographically isolated, then undergoes genetic changes that produce, as a by-product, postzygotic reproductive incompatibilities with its parent species. If the daughter population, now an incipient species, comes into contact with the parent species, selection will act strongly against hybridization, because reproduction between the two populations will produce fewer offspring than reproduction within populations. This can lead to the evolution of prezygotic isolating barriers. There is empirical evidence that this happens in nature.

Speciation by Polyploidy

More than half of all flowering plant species and a few animal species have arisen through polyploidization. In these cases, extant species have chromosome numbers that are multiples of the ancestral chromosome number. Polyploid species are of two

wild *Triticum searsii* **BB** X wild *T. monococcum* **AA** → domesticated *monococcum*

2X = 14 ↓   2X = 14   (einkorn)

AB

↓   ...genome doubled*

wild *T. turgidum* **AABB**

↓   4X = 28

wild *T. tauschii* **DD** X domesticated *turgidum* **AABB** (emmer)

↓   2N = 4X = 28

ABD

↓ ...genome doubled*

domesticated *T. aestivum* **AABBDD**

2N = 6X = 42.

*The chromosome number doubled, either by endomitosis (mitosis without nucleardivision) early in development, or by formation of rare balanced gametes.

kinds, **autopolyploids** (multiple copies of one ancestral genome) and **allopolyploids** (two or more ancestral genomes come together), mostly the latter. When the diploid genome of one species doubles, an autotetraploid is produced, and when two species' diploid genomes come together, an allotetraploid is produced (Figure 41.2).

Interspecific hybridization between ordinary diploids, A × B → AB, almost never works because the AB hybrid is virtually sterile. In meiosis I of diploid organisms, homologous chromosomes pair, but because the A set and B set are not homologous, they do not pair in AB hybrids. Therefore, each A and each B chromosome moves to one pole or the other randomly, which nearly always results in unbalanced gametes. The probability that all 14 chromosomes move toward the same pole in anaphase I is $(1/2)^{14} \approx 6 \times 10^{-5}$, and the probability that two balanced gametes will fertilize is the square of this probability, or 4×10^{-9}. If the AB zygote or one of the cells of the early embryo undergoes endomitosis, then it produces an AABB cell, and the entire germ line could be AABB. In that case, meiosis would be normal, because the A chromosomes would pair with each other, as would the B chromosomes, so that all gametes would receive a full set of all chromosomes and would be AB. For the same reason that AB hybrids are nearly always sterile, the AABB plant could reproduce with other AABB plants but not with AA or BB plants. The AABB plant would represent a new species, produced in one generation.

Fig 41.2 Hypothesized origin of modern hexaploid wheat, *Triticum aestivum*.

Speciation and Selfish Genetic Elements

Selfish genetic elements include transposons, "segregation distorters," and infectious agents, mainly the cytoplasmically inherited bacterium *Wolbachia*. Their hallmark is nonmendelian transmission from one generation to the next, so that they spread through populations. Selfish genetic elements can reduce fitness in crosses between populations and hence are candidates for agents of speciation.

Wolbachia, an endosymbiotic bacterium that infects millions of species of insects, mites, spiders, crustaceans, and nematodes, can cause cytoplasmic incompatibility between infected males and uninfected females (or females infected with another strain of *Wolbachia*), blocking fertilization. The bacteria cause the loss of paternal chromosomes after the sperm enters the egg. Furthermore, the sterility caused by cytoplasmic incompatibility is specific to the strain of the endosymbiont. This phenomenon has been proposed to cause infectious speciation. If two populations are infected with different strains of *Wolbachia*, then reproduction will be normal within the populations, but crosses between the populations are sterile. Infectious speciation could be virtually instantaneous. There are good biological species of wasp, both carrying specific strains of *Wolbachia*; when the wasps are treated with antibiotics that kill the bacteria, then the two species are completely interfertile, with no prezygotic or postzygotic RIBs between the species.

Selfish genetic elements appear not to be a major cause of speciation. Most speciation events occur either by gradual multigenic changes or by macroevolutionary changes such as polyploidization.

Patterns of Extinction and Speciation

Both speciation and extinction occur continuously but at widely varying rates over geological time. Several large mass extinctions, each wiping out 40 to 95% of the species on the planet, have been followed by periods of explosive speciation. Even so, 90 to 95% of extinction has occurred **between** mass extinctions. Massive environmental changes are thought to have caused each mass extinction. Under one theory, there are periodic spikes of extinction that happen, on average, every 26 million years.

The Ratchet of Speciation

In the long run, the rate of speciation has exceeded the rate of extinction. Geographically isolated populations adapt to local conditions and easily become genetically differentiated. Why, though, does genetic differentiation so often result in reproductive difficulties between populations? Why do postzygotic RIBs so rapidly accumulate as a by-product of evolutionary change? Apparently, there is a strong tendency for sexual reproduction to be disturbed by epistatic interactions. The haploid genomes of eukarya are sufficiently large – 10^7 to 10^{11} nt – to make the number of possible genotypes virtually infinite and the potential for genetic differentiation huge. At the same time, sexual reproduction is possible only if many interacting molecular systems work together perfectly. Relatively small genetic change can block sexual reproduction with other genotypes and thus lead to speciation. Furthermore, once a daughter population becomes reproductively isolated from its parent population, selection favors increased reproductive isolation and hinders the evolution of decreased reproductive isolation. This is an evolutionary ratchet, biased toward speciation and against its reversal.

Further Reading

Benton MJ. 1986. The evolutionary significance of mass extinctions. *Trends Ecol. Evol.* 1:127–130.

Bordenstein SR *et al.* 2001. *Wolbachia* induced incompatibility precedes other hybrid incompatibilities in *Nasonia*. *Nature* 409:707–710.

Coyne JA. 1992. Genetics and speciation. *Nature* 355:511–515.

Hurst GDD, Schilthuizen M. 1997. Selfish genetic elements and speciation. *Heredity* 80:2–8.

Hutter P. 1997. Genetics of hybrid inviability in *Drosophila*. *Adv. Genet.* 36:157–185.

Kornfield I, Smith PF. 2000. African cichlid fishes: Model systems for evolutionary biology. *Annu. Rev. Ecol. Syst.* 31:163–196.

Nei M, Zhang J. 1998. Molecular origin of species. *Science* 282:1428–1429.

Orr HA. 2001. The genetics of species differences. *Trends Ecol. Evol.* 16:343–350.

Soltis DE, Soltis PS. 1999. Polyploidy: recurrent formation and genome evolution. *Trends Ecol. Evol.* 14:348–352.

Chapter 42

Molecular Evolution and Phylogeny

Overview

Genomes in organisms evolve primarily by microevolutionary changes in DNA sequences – substitutions, deletions, and insertions. Genomes can expand by gene duplication.

Genes evolve at different rates, depending on how the functions of RNAs and proteins are constrained by structure. The more a protein's structure is constrained, the more slowly does its gene evolve. Any given protein tends to evolve at a more or less characteristic rate, in extreme cases for hundreds of millions of years, making each protein useful in estimating when evolutionary lineages diverged.

All organisms arose from ancestral organisms, implying a single ancestor at the beginning. The organisms living today can be arranged as the twigs at the ends of a tree-like diagram whose trunk represents the original ancestor. The pattern of branching in the evolutionary tree is **phylogeny**. Much can be learned about phylogeny by comparing the nucleotide sequence of **homologous genes** – genes that descended from a common ancestral gene.

Evolution of Homologous Genes

Genes in different species that descended from a common ancestral gene are said to be **homologous**. There are two kinds of homology, **orthology** and **paralogy**. Orthologs are copies of a gene, present in two species and inferred to have existed in their most recent common ancestor – e.g., the β-hemoglobin genes of mice and humans. Paralogs are copies of a gene that arose by gene duplication – e.g., α-hemoglobin and β-hemoglobin. Homologous genes differ in nucleotide sequence because substitutions, deletions, and insertions have occurred within lineages, between speciation events.

Rates of Protein Evolution

Change in homologous genes is slow and gradual, and the rate of evolution for one particular gene tends to be approximately the same in many related lineages of organisms. For example, the gene encoding cytochrome c has evolved at about the same rate in animals, plants, and fungi, a 10% change in amino acid sequence about every 200 million years. The hemoglobins of diverse vertebrates have undergone a 10% change in amino acid sequence about every 55 million years. Both cytochrome c and the globins have been said to act like stochastic molecular clocks, each ticking at a characteristic rate. Approximate dates of divergence of some lineages of organisms are well known from examination of the fossil record (fossils are mineralized remains of organisms, whose age can be determined by analyzing isotopes of certain atoms of the matrix in which fossils are embedded). Then, the number of amino acid differences between pairs of species representing long-diverged lineages is counted; the average difference is then divided by twice the divergence date to get number of amino acid changes per million years.

There are plenty of exceptions to this "molecular clock" model. A protein may evolve at different rates in distant lineages; e.g., *Gpdh* (encoding glycerol 3 phosphate dehydrogenase) has evolved ten times as fast in mammals as in *Drosophila*. Even within a genus, evolutionary rates can vary a lot; e.g., the rate of *Sod* (encoding superoxide dismutase) evolution varies by an order of magnitude among lineages within the genus *Drosophila*. The exceptions do not invalidate the use of protein evolution to estimate times of phylogenetic divergence but force the geneticist to use many proteins simultaneously to evaluate the uncertainty of the estimates.

Speed of Evolution Relates to Function

The speed of evolutionary change in a DNA sequence depends heavily on what that sequence does. For a protein-coding gene, the coding region evolves most slowly. Introns and flanking regions, both 5' and 3', undergo rapid evolution. Within coding regions, nucleotides at the third position of codons may evolve as rapidly as noncoding sequences, because many synonymous codons (coding for the same amino acid) are the same at positions 1 and 2 but differ at position 3. Pseudogenes also evolve very fast. A pseudogene is homologous to a gene but during evolution has lost the ability to be expressed – for example because of a mutated promoter or the presence of chain-terminating mutations in an exon.

Gene Duplication

When a gene is duplicated, the two new copies of that gene tend to evolve different functions. Gene duplication increases the complexity and size of the genome. Related genes that evolved by gene duplication exist in all the domains of life and range from the elongation factors of translation to vertebrate olfactory receptors. The cytochrome P450 oxidases arose by gene duplication, as did the opsins (light-sensitive proteins in the retinal cone cells). A good example of evolution by gene duplication is the globin gene family in vertebrates. Myoglobin is the ancestral globin. The hemoglobins (α, ζ, β, δ, γ, and ε) evolved by successive gene duplication, starting with myoglobin (Figure 42.1).

Unequal Crossing-Over

The most common way genes are duplicated is by unequal crossing-over in meiosis. Misalignment of homologous chromosomes followed by an exchange at the point of misalignment can yield one recombinant with a duplication and another recombinant with a deletion (Chapter 14). The genes encoding

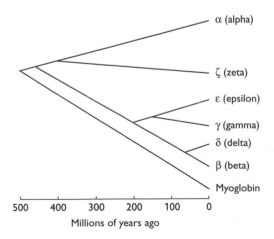

Fig 42.1 Evolution of the globins.

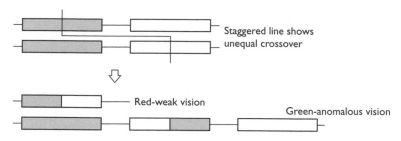

Staggered line shows
unequal crossover

Red-weak vision

Green-anomalous vision

Fig 42.2 Unequal crossing-over between X-linked opsin genes. *OPSR* (hatched) is on the left; *OPSG* (clear) is on the right.

red-absorbing opsin gene (*OPSR*) and green-absorbing opsin gene (*OPSG*) are paralogs that arose by gene duplication. They are adjacent genes in Xq28, whose proteins are 96% identical in amino acid sequence. In humans, all normal X chromosomes have one copy of *OPSR* but may have one, two, or three tandem copies of *OPSG*. Red-weak and green-weak vision can be caused by mutation or unequal crossing-over. An example of the latter is depicted in Figure 42.2.

Taxa and Trees

A **phylogenetic tree** depicts the relationships of species, or of larger taxonomic groups, or **taxa** (sing., **taxon**). The tree attempts to represent evolutionary history and not a taxonomic classification based on characteristics of the organisms in the tree. The branching pattern of a phylogenetic tree is dichotomous (forks have pairs of branches, not triplets or more), reflecting the idea that speciation is bifurcative (each speciation event yields exactly two daughter species). The assumptions of making a phylogenetic tree are

- One ancestral organism, long extinct, gave rise to all extant species.
- A single common ancestor gave rise to each pair of species.
- Organisms become more and more different genetically, through the evolutionary time after divergence.

A phylogenetic tree is based on inferences about the genotypes of taxa in the tree. DNA sequences of genes, amino acid sequences of proteins, and clear-cut morphological traits are three widely used types of information to make such inferences and guesses. We cannot know the DNA sequences or amino acid sequences of extinct ancestral species, with a few exceptions. Fossils preserve some of the morphology of extinct species, but one cannot know whether a fossilized species was ancestral to present-day forms or whether it represents an extinct branch of the tree of life. By and large, we must use the DNA sequences, amino acid sequences, and "higher" phenotypes of extant species.

The tree is made of nodes connected by branches. The nodes at the ends of branches are extant species, while internal nodes represent ancestral organisms. The tree is rooted if an inferred ancestral node is included; all paths to extant species branch from that root. The tree is unrooted if species are merely connected, but no ancestral node is included.

Principles of Building Trees

Tree building is complex and employs methods that are well beyond the scope of this book. However, the basic principles are summarized here, using sequences of nucleic acids.

Species are chosen for phylogenetic analysis, including an outgroup – a species outside the group being studied. Each species must possess the set of genes used in the analysis. For example, humans and all apes, plus rhesus macaque (the outgroup) may comprise the species under study. The DNA sequences to be compared might include the gene encoding β-hemoglobin.

There are two approaches to studying the evolutionary relationships of extant organisms using DNA sequences, **phenetic** and **cladistic**. Phenetics is based on the degree of similarity among DNA sequences. Cladistics is based on shared nucleotides that differ from the ancestral nucleotides in a sequence.

In the phenetic approach, calculate evolutionary distances between species by counting the proportion of nucleotides in a sequence that differ between pairs of species; a distance statistic is then derived from that proportion. Make a matrix with the species as both rows and columns, with the distances entered in the cells of the matrix. Build a tree connecting species so that the distances in the branches of the tree correspond to the distances in the matrix.

The cladistic approach uses one of two methods, parsimony or maximum likelihood. The parsimony method finds a tree requiring the minimum number of evolutionary changes required to produce the minimum total number of substitutions. The maximum-likelihood method makes a statistical model based on trees, and the tree that makes for the model of greatest likelihood is chosen.

Some problems that must be addressed by any method are multiple hits at a nucleotide site, the unequal probabilities of transitions and transversions, and synonymous versus nonsynonymous substitutions. If multiple substitutions ("hits") occur in two divergent branches for some nucleotide (character), then the number of substitutions may exceed the estimated number of substitutions. Transitions are much more common than transversions, so that a transversion should be given more weight. Synonymous substitutions do not cause an amino acid substitution and therefore are likelier than nonsynonymous substitutions; hence, the latter should be given more weight. Statistical procedures can be used to correct for multiple hits and to weight some substitutions more heavily than others.

Two biological problems are the existence of multiple substitutions that cannot be detected and horizontal genetic transfer of genes between species. Suppose an AT pair undergoes two substitutions, AT → GC → AT; then it will appear that no change has taken place, when in fact two changes have. Suppose a bacterial gene is transferred to an archaeon by transformation. Then it will appear that the archaeon's descendents are more closely related to bacteria than they really are.

Solved Problem: Making a Cladogram. Make a cladogram for four species, based on a 6-nt sequence (Figure 42.3).

In this example there are four species, X, Y, W, and O (known outgroup). Assume the smallest number of single transitions needed to produce the extant sequences.

A real analysis would use a large sequence, consisting of hundreds or thousands of nucleotides. Often, sequences differ by additions and deletions as well as substitutions. To align homologous

Fig 42.3 Cladogram for four species.

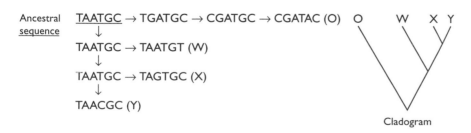

Cladogram

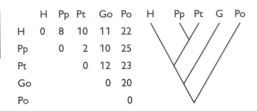

Fig 42.4 Phenogram for β-hemoglobin gene in six species.

	H	Pp	Pt	Go	Po
H	0	8	10	11	22
Pp		0	2	10	25
Pt			0	12	23
Go				0	20
Po					0

sequences of different length, gaps must be introduced into some of the sequences.

Solved Problem: Making a Phenogram. Make a phenogram using differences in nucleotide sequences of six species, for the β-hemoglobin gene (Figure 42.4). The species are human (H), pigmy chimp (Pp), common chimp (Pt), gorilla (Go), and orangutan (Po).

Evolutionary trees should be based on as much genetic information as is available and useful. For closely related species, slowly evolving genes are not useful, because these species will be identical or nearly so for such genes. For distantly related species, rapidly evolving genes are not useful, because multiple substitutions will have occurred at many points along a DNA sequence, obscuring the true number of changes in each evolutionary lineage. Thus, mitochondrial DNA, which evolves rapidly, is useful for studying the evolutionary relationships among populations within species. At the other end of the spectrum, very slowly evolving genes, such as those encoding rRNA or elongation factors, are useful for studying the phylogeny of the three domains or the major lineages within domains.

Further Reading

Ayala F. 1995. The myth of Eve: molecular biology and human origins. *Science* 270:1930–1936.

Fitch WM. 2000. Homology. *Trends Genet.* 16:227–231.

Lynch M, Conery JS. 2000. The evolutionary fate and consequences of duplicate genes. *Science* 290:1151–1155.

Nathans J, *et al.* 1986. Molecular genetics of inherited variation in human color vision. *Science* 232:203–210.

Scherer S. 1990. The protein molecular clock. *Evol. Biol.* 24:83–106.

Yoder AD, Yang Z. 2000. Estimation of primate speciation dates using local molecular clocks. *Mol. Biol. Evol.* 17:1081–1090.

Glossary

abortive initiation Termination of transcription during the initiation step, after synthesis of a very short RNA chain.

acceptor splice site The nucleotide at the 5′ end of an exon, to which an upstream exon is joined during splicing of pre-mRNA.

acentric chromosome A nuclear eukaryal chromosome lacking a centromere and thus incapable of segregation during mitosis or meiosis.

achiasmate Referring to meiosis lacking chiasmata.

acrocentric chromosome A nuclear eukaryal chromosome with one relatively long arm and one extremely short arm.

activator A protein that stimulates the binding step of transcription, thereby turning on (activating) gene expression.

adapter Double-stranded DNA oligonucleotide that is ligated to the ends of a DNA fragment to facilitate its amplification or cloning.

addition rule If X and Y are independent events, the probability that X or Y will happen is the sum of their probabilities minus the probability of their joint occurrence: $P[X \text{ or } Y] = P[X] + P[Y] - P[X \text{ and } Y]$.

additive Referring to a factor whose effect is free of interaction.

additive genetic variance The portion of genetic variance in a population due to additive effects of alleles (all genetic effects not entailing interactions, such as dominance).

adenine (A) A purine base commonly found in DNA and RNA.

AFLP (amplified fragment length polymorphism) A polymorphism detected by a PCR-based method of the same name, based on variation for multiple microsatellite loci; many polymorphisms are displayed on one gel.

A-form A right-handed configuration of a nucleic acid double helix, common in RNA and in RNA-DNA hybrids.

alkylating agent A mutagenic chemical that adds alkyl groups to DNA bases.

allele An alternative version of a gene.

allele frequency In a population, the number of copies of one allele divided by the total number of copies of a gene; by convention, copy number is determined by the population size (number of individuals at a particular phase of the life cycle).

allele switching The replacement of one allele by another via site-specific recombination.

allelic exclusion The expression of only one copy of a gene in a diploid cell, as in immunoglobin (Ig) expression in B cells.

allopolyploid Referring to a species derived from hybridization of two ancestral species and having two or more copies of each ancestor's haploid genome.

allosteric Referring to an interaction between a ligand and a protein, which changes the shape and activity of the protein.

allozyme An allelic form of an enzyme, identifiable by electrophoresis.

alpha (α) helix In a polypeptide, a right-handed helix with an average of 3.6 amino acids per twist, held in place by hydrogen bonds.

alternative splicing Variation in the pattern of splicing pre-mRNA, yielding more than one species of mRNA per gene.

ambiguous Referring to a code in which one symbol stands for two different things; the genetic code is nearly free of ambiguity.

ambisense virus A single-stranded RNA virus in which both the parental and complementary strands contain mRNA sequences.

Ames test An assay of a compound's mutagenicity, using his^- strains of *Salmonella* to detect reversion from his^- to his^+ in response to the compound.

amino acid A carboxylic acid (compound with $-COOH$) containing an amino group ($-NH_3$). Polymers of amino acids comprise polypeptides (proteins).

aminoacylation The covalent addition of an amino acid to a tRNA.

aminoacyl site (A site) The place in the ribosome where a charged tRNA enters at the beginning of each elongation cycle.

aminoacyl tRNA A tRNA with an amino acid attached to its 3' end.

amphidiploid Allotetraploid species (AABB genome).

aminoacyl-tRNA synthetase A family of enzymes that catalyze the addition of amino acids to tRNA molecules.

amino terminus The end of a peptide or polypeptide having a free amino group, also called the N terminus. Polypeptides grow N→C.

amorph An allele that codes for a nonfunctional product.

amplicon A DNA fragment amplified by polymerase chain reaction.

anaphase The phase of mitosis or of meiosis II in which chromatids disjoin, or the phase of meiosis I in which homologous chromosomes disjoin. In mitosis, anaphase has two parts: A and B.

aneuploid Having a chromosome number that is not an integer multiple of the haploid chromosome number.

anticodon Three adjacent nucleotides in tRNA, which base-pair with mRNA.

AP endonuclease An enzyme that cuts one strand of double-stranded DNA 5' to a site where a base is missing.

AP lyase An enzyme that cuts one strand of DNA 3' to an abasic site.

AP repair Repair of an AP site, in which an AP-containing sequence is excised.

AP site A place in a DNA molecule where a base is missing.

apoptosis (ăpōtō′sĭs) Programmed cell death.

apoptosome (ăpō′tō sōm) A large protein multimer that activates caspases.

archaea (sing., archaeon) One of the three domains of organisms.

Archaea are one-celled organisms lacking nuclei, superficially resembling bacteria but sharing characteristics with eukarya.

ascospore A haploid spore produced by one mitotic division following meiosis in the **ascus** (narrow sac containing eight ascospores) of sac fungi (Ascomycota).

asexual reproduction The formation of progeny without a sexual process; the term applies to organisms.

assisted base-pairing An action of DNA polymerase, which facilitates the correct pairing of nucleotides before the polymerization reaction.

assortment The movement of different genes, or of nonhomologous chromosomes, to daughter cells during meiosis. In an organism heterozygous for two genes, the four possible allele combinations are produced in a predictable ratio. If the two genes assort independently, that ratio is 1:1:1:1.

attached-X chromosome A mutant chromosome in *Drosophila* that contains two copies of the X chromosome's euchromatic region.

attenuation A kind of genetic regulation in bacteria, in which transcription can be halted early in the elongation phase by interaction between mRNA and a ribosome.

AURE (AU-rich element) A sequence found in some mRNAs, promoting mRNA degradation.

autopolyploid Referring to a species derived from duplication of an ancestral genome and having three or more copies of the ancestral haploid genome.

autoradiograph A photograph in which the sole source of light is a radioactive label.

autosome Any chromosome other than a sex chromosome. In dioecious species with chromosomal sex determination, both sexes have identical sets of autosomes.

auxotroph A cell or strain of cells whose growth depends on a dietary supplement.

Aviemore model A hypothesis of how general recombination works at the molecular level. At an intermediate step in this model, a Holliday structure forms.

BAC (bacterial artificial chromosome) A circular cloning vector that can carry 100 to 500 kb of inserted DNA.

backcross A cross between one parental type and the offspring of that type and another parental type.

back mutation Mutation of an abnormal allele back to the wild type.

bacteria (sing., bacterium) One of the three domains of organisms. Bacteria lack nuclei, and most are one-celled.

bacteriophage Any virus that infects bacteria.

bait In a yeast two-hybrid system, a protein used to detect (bind to) searched-for proteins (=prey).

balancer A mutant chromosome containing multiple inversions and a dominant marker, used to suppress recombination in breeding experiments.

balancing selection Natural selection favoring the heterozygote.

Barr body A densely staining, heterochromatized, genetically inactive X chromosome of mammals, normally observed in nuclei of somatic cells of females.

base A purine or pyrimidine derivative, making part of a nucleotide.

base analog An analog of a DNA base; it may be used to label a chromosome or in mutagenesis.

base excision repair Removal of a damaged or modified base from a nucleotide in otherwise intact DNA, followed by AP repair.

base modification Posttranscriptional processing of rRNA or tRNA, in which single bases are structurally altered.

base pair (bp) Two bases joined by hydrogen bonds and normally in two separate strands of nucleic acid.

base substitution Replacement of one base pair by another in DNA.

β-galactosidase An enzyme that splits lactose into glucose and galactose.

beta (β)-ribbon A flat segment of polypeptide; when β-ribbons closely align, they form a β-sheet .

B-form A common, right-handed configuration of a DNA double helix.

binding step The first step of transcription, in which RNA polymerase binds to the promoter; usually this step is rate-limiting.

bioinformatics Methodology for acquiring and analyzing large databases of DNA sequences.

binomial distribution A probability distribution based on independent trials of an experiment with a pair of outcomes whose probabilities are constant.

biological species One or more populations of sexual organisms; within a species reproduction is free, but between species reproduction is prevented biologically.

bivalent A pair of synapsed chromosomes in prophase I of meiosis.

bottleneck A time of small population size in the history of a population.

bouquet An arrangement of chromosomes in zygonema; telomeres attach to one region of the nuclear envelope.

breed 1. (v.) To conduct a controlled genetic cross. 2. (n.) An artificially produced, closed genetic population of animals.

canalization Developmental buffering that hides the effect of genetic variation.

cancer A class of diseases characterized by uncontrolled cell proliferation and caused by multiple somatic mutations.

cap A 7-methyl guanosine added to the 5′ end of eukaryal mRNA, linked to the next nucleotide by a 5′-5′ triphosphate.

CAP (catabolite gene activator protein) A bacterial protein that helps to activate transcription of operons encoding catabolic enzymes.

CAP site A site in the promoter of certain bacterial genes, to which CAP binds.

capsid The protein coat of a virus.

catabolite repression Reduction of a gene's transcription by the action of a catabolite produced by the enzyme that gene encodes.

C terminus (carboxyl terminus) The end of a peptide or polypeptide having a free carboxyl group.

carcinogen An agent that causes cancer; it is either a mutagen or a mitogen.

carrier A heterozygote, usually applied to a recessive genetic disease of humans.

caspase A protease essential to apoptosis.

Castle-Hardy-Weinberg (CHW) rule In a population at genetic equilibrium, if there are two alleles at frequencies p and q, the genotype frequencies are p^2, 2pq, and q^2.

cDNA (complementary DNA) A DNA molecule synthesized from an mRNA template with reverse transcriptase.

cDNA library A large collection of cloned cDNA made from mRNA derived from a specific source (e.g., one tissue).

cell autonomous Referring to the expression of a gene only within a cell containing it, without transfer of the phenotype to other cells via a diffusible factor.

cell cycle The reproductive cycle of a cell. In archaea and bacteria it has C and D phases; in eukarya it has G_1, S, G_2, and M phases.

cell cycle checkpoint Any critical point in a cell cycle at which passage through the cycle is stopped until essential events or processes have been completed.

cell envelope The outside covering of a cell; minimally, an enclosing membrane.

cell lineage In the development of an animal or plant, an ancestral stem cell and all the cells that descend from it.

centromere A DNA sequence in a eukaryal chromosome, essential for attachment of spindle fibers to the chromosome.

centrosome An organelle in eukarya necessary for cell division. The spindle originates from a pair of centrosomes.

chain elongation The successive addition of amino acids to a growing polypeptide.

chain initiation The beginning phase of polypeptide synthesis.

chain termination The ending phase of polypeptide synthesis.

chaperonin A protein that assists nascent polypeptides to assume their secondary and tertiary structures.

charged tRNA A tRNA molecule with an amino acid covalently attached to it.

checkpoint A critical phase of the cell cycle at which the cycle is stopped if there are problems such as DNA damage or incomplete spindle development.

chiasma (pl., chiasmata) An X-shaped configuration of paired chromosomes in diplonema of meiotic prophase I, resulting from recombination.

chiasmate Referring to meiosis in which chiasmata form.

chimera An organism derived from embryonic stem cells of two or more zygotes.

chimeric gene A gene containing parts of two or more genes and produced by recombination, usually *in vitro* recombination.

chi-square (χ^2) A family of statistical distributions, each having v degrees of freedom. χ^2 is used to test genetic hypotheses about ratios.

chloroplast A photosynthesizing organelle in the cytoplasm of plant cells. Chloroplasts have their own genomes and reproduce by splitting.

chromatid (also called sister chromatid) One of the twin replicates of a eukaryal chromosome in G_2 or M phase. Chromosomes in G_1 and G_0 do not possess chromatids.

chromatin The $\approx$30-nm thread comprising a eukaryal chromosome, made of DNA and proteins (largely histones).

chromatin activation Local decondensation of chromatin preparatory to transcription.

chromatin silencing Local maintenance of chromatin structure to switch off transcription.

chromophore A light absorbing pigment. Photolyases and some transcription factors are associated with chromophores.

chromosome A long, thin thread made of DNA, or in some cases RNA; proteins may also be present. Chromosomes contain genes, and are replicated.

chromosome arm In a linear eukaryal chromosome having a localized centromere, the part on one side of the centromere. In some species, the two arms are designated p (petite) and q (long).

chromosome band A darkly staining area seen in a mitotic chromosome.

chromosome diminution The elimination of chromosomes or parts of chromosomes, or the fragmentation of chromosomes, during development.

chromosome elimination The loss of whole chromosomes from a genome. This is normal in the development of certain insects.

chromosome puff An expanded region of a polytene chromosome, promoting transcription at that locus.

cis-acting Of regulatory DNA sequences that control transcription of nearby genes in the same chromosome.

cis (coupling) configuration The arrangement of genotypes on two copies of a chromosome, in which one chromosome has both mutant alleles: $a\ b//a^+\ b^+$.

cladistic Referring to a method that attempts to reconstruct an evolutionary tree by inferring ancestral genotypes or traits.

class switching DNA rearrangement in B cells of the vertebrate immune system. In class switching, the constant region of the heavy chain gene is replaced.

cleaving and trimming Posttranscriptional RNA processing in which the transcript is cut into pieces and the pieces are subsequently shortened by an exonuclease.

clone 1. A population of genetically identical life forms. 2. A subpopulation of cells within a multicellular organism. 3. To produce a clone, by a natural or an artificial process. 4. To produce copies of a piece of DNA inserted artificially into a vector (carrier DNA) by inducing some life form to make the copies.

cloned DNA A piece of DNA inserted into a vector, introduced into some life form (e.g., bacterial cell), which reproduces to make a genetically identical population.

cloning vector A segment of DNA into which foreign DNA is inserted for cloning. A cloning vector has an origin of replication and conveniently placed restriction sites; it may also have a selectable marker.

closed population A population that is genetically isolated from migration.

coding region The part of mRNA that encodes protein.

codominant For dichotomous traits, referring to two alleles that are not dominant (neither allele hides the other allele).

codon A triplet of nucleotides in mRNA that codes for an amino acid.

codon usage The average frequencies of codons used in protein synthesis, for one species.

complementary Referring to the base pairing ability of two strands of nucleic acid. A:T and G:C are complementary base pairs.

complementary strand The nontemplate strand of DNA – i.e., the strand that is not a template for RNA synthesis.

complementation The complementary action of two nonallelic, recessive mutations (say x and y), to produce a wild-type phenotype in the heterozygote x +//+ y. If x and y fail to complement (x +//+ y has the mutant phenotype), then x and y must be alleles, and the converse is usually true.

concatamer A DNA molecule made of multiple copies of a sequence in tandem, destined to be cut into pieces – e.g., rolling-circle-replicated λ phage chromosome.

conditional mutation A mutation that produces a mutant phenotype under restrictive conditions and a wild-type phenotype under permissive conditions.

conditional probability The probability of one event given that another event has occurred.

confidence interval A range of values, having a known probability of containing some parameter such as a mean.

conjugation The transfer of DNA between connected cells. In bacterial conjugation, a plasmid is transferred nonreciprocally; in ciliate conjugation, micronuclei are exchanged reciprocally.

conjugative plasmid A small, nonessential chromosome transferred between bacteria.

consensus sequence A sequence of nucleotides that most closely resembles all members of a set of homologous sequences.

constant region A domain of an immunoglobin polypeptide that shows little sequence variation.

constitutive Referring to a permanent condition: 1. constitutive gene expression is unregulated, and the gene is transcribed constantly; 2. constitutive heterochromatin is present at all stages of development.

contig A set of overlapping, cloned DNA segments containing no gap.

copy choice A kind of nucleic acid synthesis in which the polymerase enzyme switches from one template molecule to another, mid-synthesis.

copy error A mistake of replication.

core nucleosome A particle in chromatin, made of DNA wrapped ≈1.7 times around an octamer of histones.

co-repressor A small molecule binds to a repressor protein, rendering it functional.

correlation coefficient (r) A statistical measure of association between two random variables; r ranges from −1 to +1.

cosmid A cloning vector prepared from the ends of the λ phage genome.

CpG island A region of DNA containing several 2-nt sequences, 5′CG3′.

C phase The part of a bacterial cell cycle when the chromosome is replicated.

cross To breed one genotype with another in a controlled genetic experiment.

crossing-over An exchange between homologous chromosomes paired in meiosis.

cyclin A kind of protein essential to the regulation of the cell cycle.

CDK (cyclin-dependent kinase) A protein that binds to cyclin and acts with it to regulate the eukaryal cell cycle.

cyclobutane pyrimidine dimer Covalently connected pyrimidine bases of two adjacent nucleotides of one DNA strand, making a rectangular ring.

cytokinesis Cell division that accompanies mitosis or meiosis.

cytological map A map that places genes at locations that are recognizable in stained metaphase chromosomes.

cytosine (C) A pyrimidine base.

dark repair Excision repair of DNA; a short, lesion-containing segment of one strand is excised and replaced.

daughter cell One of two cells produced by cell division.

daughter chromosomes The chromosomes produced by disjunction of chromatids in anaphase of mitosis or in anaphase of meiosis II.

deficiency A mutation in which a segment of a chromosome is lost.

degenerate code Redundant code, in which more than one symbol (e.g., codon) may encode one represented element (e.g., amino acid).

degree of dominance (d) In quantitative dominance, the difference between the heterozygote's mean and the midparent value.

degree of freedom (df or ν) A parameter associated with a statistical distribution such as χ^2, t, or F, and needed to test statistical hypotheses using that distribution.

deletion A loss of nucleic acid from a chromosome. Big deletions are sometimes called deficiencies.

deme A population within a sexually reproducing species. Individuals within the deme have the potential to interbreed.

DNA (deoxyribonucleic acid) A polymer whose subunits are nucleotides; DNA of an organism's chromosome is double-stranded (dsDNA) and helical. DNA is sometimes single-stranded (ssDNA).

deoxyribose A 5-carbon sugar of DNA, whose 2' carbon lacks an OH group.

diakinesis The fifth stage of prophase I in meiosis, when chiasmata terminalize and chromosomes condense maximally.

dicentric chromosome An aberrant chromosome having two centromeres.

dideoxyribose A modified form of ribose, lacking OH groups at the 2' and 3' carbons.

differential gene expression Tissue-specific or developmental stage-specific synthesis of gene products.

differentiation Cell specialization during development.

dihybrid Heterozygous for two pairs of alleles.

diminution The loss of chromosomes or chromosome parts, or fragmentation of chromosomes as a regular feature of development.

dioecious Referring to a species with female and male individuals; more often used to describe plants than animals.

diploid Having two homologous copies of the genome; the terms haploid and diploid apply only to sexually reproducing organisms.

diplonema (adj., diplotene) The fourth stage of meiotic prophase I, when chiasmata appear.

diplontic Referring to a sexual life cycle with mitosis restricted to the diploid phase.

direct repair Any DNA repair in which the sugar-phosphate backbone is not cut.

direct repeat A copy of a short nucleic acid sequence, appearing twice or more in tandem in the same orientation.

directional dominance The tendency of an F_1 hybrid to resemble one parental type more than the other.

directional selection Natural or artificial selection favoring one extreme of a distribution (high or low, not both).

discriminator region The DNA sequence just upstream of the transcription start site of a bacterial gene, between -10 and $+1$.

disjunction The separation of chromosomes in anaphase of meiosis I, and the separation of chromatids in anaphase of meiosis II or of mitosis.

disruptive selection Natural or artificial selection favoring both extremes of a distribution and disfavoring phenotypes near the mean.

dizygotic twins Siblings arising from two zygotes and growing together in the uterus.

D-loop (displacement loop) A region of melted dsDNA in the chromosome of mammalian mitochondria, where the primer for heavy-strand synthesis sits.

D-loop replication A pattern of replicating a circular chromosome in which each strand is replicated separately, starting from different origins.

DNA chip (DNA microarray) An array of different DNA molecules fixed to a tiny rectangular surface, used to screen sequences by hybridization.

DNA damage checkpoint A critical point at the end of G_1 or G_2 when the cell cycle is arrested if DNA damage is detected.

DNA fingerprinting A method of identifying individuals by determining their genotypes for highly variable microsatellite loci.

DNA methylase An enzyme that adds methyl ($-CH_3$) groups to nucleotide bases, notably in bacterial genomes at restriction sites.

DNA methyltransferase An enzyme that methylates cytosine at CpG sites.

DNA polymerase An enzyme that catalyzes the polymerization of a DNA polymer, using a DNA template strand; DNA polymerases add nucleotides to a primer.

DNA replication The synthesis of two double-stranded daughter DNA molecules from one original molecule.

domain 1. A portion of a polypeptide that is distinct in function and secondary structure from other parts of the polypeptide. 2. A large division of organisms; the three domains of life are bacteria, eukarya, and archaea.

dominance An interaction between two alleles in which the dominant allele hides phenotypic effects of the recessive allele.

donor splice site A nucleotide at the 3′ end of an exon, which is joined to a downstream exon during splicing of pre-mRNA.

dosage compensation A mechanism ensuring that the ratio of X-linked to autosomal gene product is equal in both sexes.

double-strand break repair The special kind of DNA repair that joins broken ends.

downstream In a 3′ direction in the nontemplate strand of DNA.

Down syndrome A cluster of traits in humans having three copies of chromosome 21, or at least three copies of the long (q) arm of chromosome 21.

D phase The part of a bacterial cell cycle when daughter chromosomes segregate and the cell divides.

duplex DNA A double helix of DNA formed by base pairing; one strand of DNA can fold over and base pair with itself to make a duplex region.

duplication A mutation in which a DNA sequence occurs twice, usually in tandem.

editing Posttranscriptional RNA processing of pre-mRNA, in which single nucleotides are inserted, deleted, or oxidized.

EF (elongation factor) A protein required in the chain elongation step of translation.

effector caspase A protease in the caspase family that is activated during apoptosis and that cleaves proteins throughout the cell.

electrophoresis A method for separating macromolecules (e.g., DNA) by inducing them to migrate in a gel placed in an electric field.

electroporation Transfer of DNA into a cell using an electric field.

elongation step The phase of RNA or protein synthesis when the polymer grows.

embryo An early developmental stage of a multicellular organism.

end-joining repair A special kind of DNA repair that joins two ends at a double-strand break.

endomitosis A cell cycle lacking nuclear division and cell division; it results in polyploidy.

endonuclease An enzyme that cleaves an interior phosphodiester linkage in a polynucleotide chain.

endosperm A triploid tissue of plant seeds, which nourishes the embryo.

enhancer 1. A short segment of DNA, separate from the promoter, that helps regulate transcription of a gene. 2. A gene that epistatically increases the phenotypic effects of another gene.

enhancer trap A mutant search in which a reporter transgene is introduced into random loci of the genome; the reporter gene identifies nearby tissue-specific enhancers.

enrichment screen A preliminary step in a mutagenesis experiment, when wild-type individuals are eliminated.

environmental variance The variation in a population for a trait, caused by environmental factors.

equational division The meiotic division in which sister chromatids segregate; meiosis II is usually equational.

equilibrium The state of a system not undergoing change.

epigenetic Referring to modifications of chromosomes that do not entail mutation – e.g., methylation of DNA and chromatin remodeling.

epistasis An interaction between genes, whereby one gene modifies the action of another.

ES cell (embryonic stem cell) A totipotent cell of the early embryo.

E site (exit site) The place in the ribosome where an uncharged tRNA binds just before leaving the ribosome.

essential chromosome A chromosome that is absolutely required for the survival or reproduction of a life form.

EST (expressed sequence tag) A short sequence within a cloned cDNA fragment, used to identify genes and to determine their position.

estrogen A class of female sex hormones; estrogen induces genes that promote female characteristics.

ethidium bromide A fluorescent, DNA-binding chemical used to tag DNA in a gel.

euchromatin The part of eukaryal chromosomes that stains lightly and contains most functional genes (in contrast to heterochromatin).

eukarya (sing., eukaryon) One of the three domains of organisms. Eukarya have nuclei and their genetic machinery differs markedly from that of bacteria.

euploid Having a chromosome number that is an integer multiple of the haploid number. The term applies only to eukarya.

excision repair A kind of repair of damaged DNA using exinuclease.

exinuclease An enzyme that nicks one strand of DNA on both sides of a lesion; also known as excision endonuclease.

exon A sequence in pre-mRNA that is retained in mRNA after splicing.

exonuclease An enzyme that digests a strand of nucleic acid from one end.

expression vector A cloning vector that promotes expression of a polypeptide encoded by the DNA insert.

expressivity The degree to which a genotype produces a phenotype's intensity; a genotype with variable expressivity yields phenotypes ranging in strength.

extein Part of a polypeptide that remains after posttranslational splicing.

extranuclear inheritance Transmission of traits from parent to progeny, via cytoplasmic genes (e.g., genes in mitochondria or chloroplasts).

F Coefficient of inbreeding; the probability that two alleles in one individual are identical by descent.

F_1 generation First filial generation, the offspring of a cross between two parental genotypes.

F_2 generation Second filial generation, the offspring of $F_1 \times F_1$.

facultative heterochromatin A condensed form of chromatin, restricted to some cells and genotypes; e.g., Barr bodies in female mammals.

female (♀) Mating type that produces eggs.

female pathway gene A type of gene required for the determination of female sexual traits of mammals.

fertilization The union of two gametes to form a zygote.

fitness The relative ability of a genotype to contribute copies of itself in the next generation – i.e., to survive and reproduce.

5′ end The end of a strand of DNA or RNA whose 5′ carbon is unlinked to another nucleotide.

5′ untranslated region The part of mRNA, near its 5′ end, that is not translated.

fixed allele An allele whose population frequency is very close to 1.

FISH (fluorescence *in situ* hybridization) A method of mapping a unique DNA sequence to a chromosome locus.

founder effect Genetic drift that results from the founding of a new population by a small number of individuals.

F plasmid A low-copy-number, conjugative plasmid of *E. coli*, ≈100 kb.

fragile site A chromosome locus where breakage occurs under certain *in vitro* conditions.

frameshift A shift from the normal translational reading frame, leading to an altered polypeptide or even premature chain termination.

frameshift mutation A small insertion or deletion within the protein-coding region of a gene.

free radical A highly reactive molecule with a single electron in one shell; e.g., OH·.

GAL4 An activator that binds to an upstream activating sequence (UAS) of genes of the *gal* operon of yeast.

gamete A haploid reproductive cell – i.e., ovum or sperm.

gametophyte In plants, a haploid individual or structure that produces gametes by mitosis.

G banding A pattern of chromosome staining produced by Giemsa dye.

gene A segment of nucleic acid whose immediate function is to encode a piece of RNA; it includes both coding and regulatory sequences.

gene amplification The cellular production of extra copies of a gene by replication, independent of normal chromosome replication.

gene conversion A change in allelic state, not caused by mutation, and occurring within regions of heteroduplex DNA resulting from general recombination.

gene dosage The number of copies of a gene at one locus in a genome.

gene expression The synthesis of gene products.

gene family A set of homologous genes within a genome (paralogous genes).

gene knockout The silencing or mutation of a gene via transgenic DNA.

gene flow The migration of genes from one population to another.

gene pool All copies of a genome, present in a population's reproductive cells. By convention, a population of N diploid individuals has a gene pool of size 2N.

gene product For every gene, the RNA transcript is the initial gene product; for most genes, the final gene products are peptides or polypeptides.

gene regulation Control of the rate at which a gene's products are synthesized.

gene therapy To treat a human genetic disease by modifying the genotype of human cells.

generalized transduction Any transduction in which the viral capsid contains only DNA from the (killed) host.

general recombination Natural recombination between paired homologous DNA molecules, not restricted to particular sites within these molecules.

generation A full turn of a life cycle. A sexual generation includes all development from meiosis in the parent to meiosis in the offspring.

genetic background Genetic makeup, exclusive of a gene under consideration, that characterizes a strain, line, or population.

genetic code The correspondence between mRNA codons and the amino acids they encode during translation.

genetic distance A measure of separation between two loci on a genetic map, based on the frequency of recombination between those loci.

genetic drift Random change in allele frequency over generations in a small population.

genetic instability An increased tendency to accumulate mutations, owing to compromised repair systems. Advanced tumor cells are genetically unstable.

genetic linkage The degree to which recombination between genes is restricted.

genetic map A line marked with points corresponding to genes; distances between close points are proportional to the frequency of recombinants for adjacent genes.

genetic marker A gene used to mark a chromosome or locus.

genetic polymorphism Allelic variation for a gene in a population, so that the frequency of the most common allele is less than 99%.

genetic variance Phenotypic variation in a population, caused by genes.

genome One full set of genes in a life form; a haploid set in sexual organisms.

genomic library A large, representative collection of cloned DNA fragments derived from an organism's DNA.

genomics The study of the structure and functioning of whole genomes.

genotype An individual's genetic makeup.

genotype-environment interaction Joint effects of genotype and environment on phenotype, not predictable from the average effects of the two.

germ line A lineage of sex cells, distinct from somatic cell lineages.

glycosylase An enzyme that removes a base from DNA, producing an AP site.

G_1 phase The part of the eukaryal cell cycle following M and preceding S.

G_2 **phase** The part of the eukaryal cell cycle following S and preceding M.

Gram negative and Gram positive Types of bacteria, based on staining properties. The envelope of Gram-negative bacteria is two membranes separated by a thin wall, and that of Gram-positive bacteria is a thick wall outside a single membrane.

group I and group II introns Types of intervening sequence in the pre-mRNA of mitochondria, chloroplasts, and other simple organisms. Many group I and group II introns are self-splicing.

growth rate control Regulation of the synthesis of stable RNAs in bacteria, in response to carbon starvation.

guanine (G) A purine derivative commonly found in nucleic acid.

guide RNA An RNA template used in RNA editing.

gynandromorph A mosaic that is part female and part male.

gyrase A kind of topoisomerase II that reduces supercoiling of duplex DNA.

hairpin loop A structure formed by base pairing within one strand of DNA or RNA.

haplodiplontic Referring to a sexual organism in whose life cycle mitosis occurs in both the haploid and diploid phases.

haploid Having the gametic number of chromosomes (one copy of the genome); the terms haploid and diploid apply only to sexually reproducing organisms.

haplontic Referring to a sexual life cycle with mitosis restricted to the haploid phase.

haplotype An allele combination present in one chromosome of a sexually reproducing organism.

heat shock protein A protein induced by exposing the organism to high temperature. Heat shock proteins assist protein folding.

helicase An enzyme that unwinds double-stranded DNA in replication.

hemizygote (adj., **hemizygous**) A zygotic genotype that is haploid for the genes being considered; e.g., male mammals are hemizygous for X-linked genes.

heritability The ratio of additive genetic variance to phenotypic variance, for one trait in a specific population.

hermaphrodite An animal having both ovaries and testes (see monoecious).

heterochromatin Densely staining part of a eukaryal chromosome; heterochromatin lacks active genes.

heteroduplex A region of DNA derived from separate DNA molecules, whose two strands are not perfectly complementary. It is produced by general recombination.

heterogametic In species with sex chromosomes, the sex that makes two chromosomal types of gamete; e.g., male mammals.

heteromultimer A protein made of two or more different polypeptides.

heteroplasmy (adj., **heteroplasmic**) The presence of genetically different mitochondria or chloroplasts in an organism.

heterosis (adj., **heterotic**) Increased fitness of heterozygotes: hybrid vigor.

heterothallic Referring to an ascomycete fungus with two mating types.

heterozygote (adj., **heterozygous**) A genotype with two different alleles.

Hfr (High frequency of recombinants) A strain of *E. coli* having an F plasmid integrated into the main chromosome.

his operon A negative, repressible operon of bacteria, encoding enzymes controlling the biosynthesis of histidine.

histone Small, basic, highly conserved, DNA-binding protein found in chromatin.

Holliday structure A cross-shaped structure resulting from the joining of two double-stranded DNA molecules in general recombination.

holocentric Of a chromosome, having many points of attachment for spindle fibers.

homeobox A 180-bp DNA sequence coding for a DNA-binding domain of a regulatory protein.

homeotic gene A gene, mutation of which causes one structure to replace another during development.

homogametic In organisms with sex chromosomes, referring to the sex that makes only one chromosomal type of gamete; e.g., female mammals.

homolog (homologous chromosome) A nonidentical copy of a chromosome; homologs pair in meiosis.

homologous Referring to two things or attributes derived from a common ancestor by evolution.

homomultimer A protein made of two or more identical polypeptides.

homoplasmy (adj., **homoplasmic**) Genetic uniformity of mitochondria or chloroplasts within an individual organism.

homozygote A genotype with two identical alleles.

horizontal gene transfer The transfer of DNA from one organism to another but not from parent to offspring. Usually it refers to interspecific transfer.

H (heavy) strand The purine-rich strand of animal mitochondrial DNA.

hybrid The offspring of two different strains or species.

hybrid dysgenesis Genetic abnormalities (mutation or aberrant recombination) triggered when transposons are introduced into a genome.

hybrid vigor High fitness of hybrids due to genetic interactions.

hydroxylation Covalent addition of a hydroxyl group to a molecule, notably to a DNA base.

hypermorph An allele whose phenotype is stronger than normal.

hypersensitive site A locus within native chromatin, whose DNA is susceptible to attack by endonucleases.

hypomorph A recessive allele whose phenotypic effect is weak.

illegitimate recombination Rare recombination that is neither homologous, nor site-specific, nor transpositional. It is enhanced by DNA damage and causes chromosome rearrangements.

immunoglobin (Ig) A tetrameric antibody produced in the B cells of vertebrates and consisting of two light chains (about 220 amino acids) and two heavy chains (about 440 amino acids). Each chain has a variable region and a constant region.

imprecise excision Transposition of a mobile element, during which flanking DNA sequences are mistakenly cut out and move with the transposon.

imprinting Epigenetic inactivation of a gametic gene; only the copy of the gene contributed by the other parent is expressed in the early stages of development.

inbreeding Union of genetically related gametes. Degree of inbreeding is measured by F, the probability that two alleles of a zygote are identical by descent.

inbreeding depression Low fitness of inbred populations, mainly due to homozygosity of deleterious alleles.

incomplete dominance For a quantitative trait, the partial masking of one allele by another, causing the heterozygote to resemble one parent more than the other.

incomplete penetrance A kind of phenotypic variation within a genotype, so that only a fraction of individuals with that genotype have an expected phenotype.

independent assortment The behavior of two genes in meiosis, resulting in no association between haplotypes in the gametes.

inducer A small molecule that triggers transcription by binding to an activator.

inducible regulation A type of control of gene expression in which a small molecule, the inducer, is necessary for switching on transcription.

IF (initiation factor) A protein that assists in the formation of a functional ribosome, bound to mRNA and an initiator tRNA.

initiation step The step of transcription, replication, or translation when polymer formation begins.

initiator tRNA A tRNA necessary for the initiation of translation; its anticodon binds either to AUG or to GUG, and it carries either methionine or a modified methionine.

inosine A purine derivative found in RNA.

insertion mutation A mutation in which DNA is inserted into a chromosome.

IS (insertion sequence) A small (1 to 3 kb) transposable element lacking a selectable marker; an IS has short inverted repeats at its ends.

in situ hybridization Formation of nucleic acid duplexes between a labeled, ssDNA probe and a ssDNA target (e.g., chromosomal DNA) that is fixed in place.

integrase A kind of recombination enzyme that can insert a temperate phage's chromosome into the host chromosome.

integration Insertion of foreign DNA into a chromosomal locus by site-specific recombination; e.g., prophage formation in bacteria.

intein A sequence of amino acids within a polypeptide, removed in posttranslational splicing.

interaction The joint effect of two or more factors, as in epistasis, which is not predictable from the effect of each single factor.

intercalation Binding of molecules (**intercalating agents**) to double-stranded DNA in the grooves.

interference The effect of a recombination event on nearby recombination events; in meiosis, a cross-over at one locus decreases crossing-over at nearby loci.

interphase The part of the cell cycle of eukarya excluding mitosis: G_1, S, and G_2.

intrinsic termination In bacteria, termination of translation caused by interaction between mRNA and the ribosome.

intron (intervening sequence) A segment of RNA removed during splicing.

inversion A chromosome mutation in which a segment's orientation is reversed.

inversion loop A structure resulting from the pairing of an inverted chromosome segment and a normal segment in an inversion heterozygote, during chromosome synapsis, as in zygonema of meiosis I.

invertase-resolvase A bacterial enzyme that catalyzes inversion of DNA segments.

inverted repeat A repeated DNA sequence; two copies have opposite orientation. A transposon has IRs at its ends.

in vitro In glass, referring to experiments with cell extracts.

in vivo In life, referring to processes that take place with cells.

in vitro mutagenesis A procedure that produces mutations in a cloned piece of DNA; subsequently, the mutated DNA is introduced into an organism's genome, making a mutant strain.

isoaccepting tRNAs Families of tRNA molecules that can be charged with the same amino acid.

isochore In a vertebrate genome, a GC-rich or GC-poor region of about 300 kb.

J region A short region of immunoglobin that joins the constant and variable domains.

karyokinesis Nuclear division that accompanies mitosis and meiosis.

karyotype The set of chromosomes in an organism, observed in a photographic display of stained mitotic metaphase chromosomes arranged by size.

kilobase (kb) 1000 nucleotides or nucleotide pairs in DNA or RNA.

kinesin A contractile protein that helps spindle fibers to move in mitosis and meiosis.

kinetochore A protein structure attached to a chromosome at the centromere; spindle fibers attach to the kinetochore.

kinetochore spindle fiber A type of spindle fiber that connects to the kinetochore in metaphase of mitosis or meiosis; cf. polar spindle fiber.

kinetoplast A disk-shaped body in the mitochondria of certain ciliates and containing thousands of tiny, circular, interconnected chromosomes.

Klinefelter syndrome A medical condition in human males, caused by having an extra X chromosome (XXY).

knockout mouse A transgenic mouse in which the foreign DNA is inserted into a gene, thereby rendering the gene nonfunctional ("knocked out").

lac operon A set of genes and regulatory sequences in *E. coli*, needed for lactose metabolism. Two important parts of the operon that encode RNA are:

lac I – A gene encoding the *lac* operon repressor.

lac Z – Part of the *lac* operon that encodes β-galactosidase.

lac Y Part of the *lac* operon that encodes permease.

lac A Part of the *lac* operon that encodes transacetylase.

lagging strand The discontinuously copied strand of DNA at a replication fork.

lariat A loop formed during the splicing of exons in RNA.

leader The untranslated segment of mRNA at the 5′ end.

leading strand The continuously copied strand of DNA at a replication fork.

leaky allele A mutant allele encoding a protein with a small amount of wild-type activity.

leptonema (adj., leptotene) The initial stage of meiotic prophase I, when chromosomes condense sufficiently to appear as thin threads under the light microscope.

lethal allele An allele, recessive for lethality (homozygotes for the allele die).

library A large, representative collection of cloned DNA fragments.

life form An assemblage of large molecules capable of reproduction and including at least one chromosome; e.g., cell, virus, mitochondrion.

ligand A molecule that binds specifically to a receptor.

ligase An enzyme that covalently joins adjacent DNA nucleotides at a break.

light repair The repair of cyclobutane pyrimidine dimers by photoreactivation.

LINE (long interspersed nuclear element) A kind of repeated DNA sequence: either a retrotransposon lacking long terminal repeats or else a sequence that evolved from such a retrotransposon; LINEs are found in mammals; cf. SINE.

line Genetic variety, breed, or strain; this term applies to plants or animals.

linkage (adj., **linked**) The tendency of nearby genes on one chromosome to assort together.

linkage group A set of genetically linked genes – the genes on one chromosome.

linker 1. A segment of DNA included in a cloning vector, which contains restriction sites. 2. The region between core nucleosomes.

liposome A tiny water-filled lipid sphere that is useful for delivering DNA to an animal cell.

locus (pl., loci) A gene's location on a chromosome or genetic map.

LCR (locus control region) A regulatory region upstream of a eukaryal gene cluster, which facilitates chromatin activation.

LOD score A statistic measuring the likelihood of some event, such as linkage between two genes, based on the logarithm of odds ratio.

L (light) strand The pyrimidine-rich strand of animal mitochondrial DNA.

lysogen A clone of bacteria that contain a bacteriophage chromosome (prophage) integrated into the cell's main chromosome.

lysogenic Referring to the noninfectious phase of a temperate phage's life cycle.

lytic Referring to the infectious phase of a temperate phage's life cycle.

macroevolution Evolution that proceeds by sudden large changes.

macronucleus In ciliates, a large nucleus that divides nonmitotically.

major groove In a nucleic acid double helix, one of two channels running the length of the molecule; in B-form DNA, the wider one.

male ($\male$) Mating type that produces small gametes (sperm or pollen).

male pathway gene A type of gene required for the determination of male sexual traits of mammals; the male pathway is triggered by testis-determining factor.

malignant Referring to the property of tumors that invade adjacent tissues.

map function A mathematical function that describes the relationship between recombination frequency and genetic distance.

MAP kinase Mitogen-activated protein kinase, which activates a transcription factor; MAP kinases regulate development.

map unit A measure of genetic distance, corresponding to 1% recombinant chromosomes in a genetic mapping experiment.

marker (= genetic marker) A gene used to mark a chromosome or locus.

matrix attachment region (MAR) The part of a eukaryal chromosome that attaches to the nuclear envelope.

maternal-effect gene A nuclear gene expressed in the mother before fertilization, which affects embryonic development and the zygote's phenotype.

maternal inheritance Inheritance of factors in egg cytoplasm.

maxicircle A circular chromosome in the mitochondria of certain protozoans. Such a mitochondrion has 40 to 50 maxicircles, 25 to 35 kb long.

megabase (Mb) One million nucleotides or nucleotide pairs in DNA or RNA.

meiosis The two successive nuclear divisions (meiosis I and II) whereby haploid sex cells are produced from diploid cells.

mendelize In sexual reproduction, to conform to the rules of segregation and assortment, rules originally conceived by Gregor Mendel.

mRNA (messenger RNA) A kind of RNA that is translated into a polypeptide.

metacentric Referring to a eukaryal chromosome with a central centromere; the two arms are about equal in length.

metaphase The stage of mitosis or meiosis between prophase and anaphase; chromosomes, attached to the spindle, are midway between the two centrosomes.

metaphase plate An imaginary disk containing metaphase chromosomes; it is sometimes called the equator.

methyltransferase An enzyme that methylates DNA.

microevolution Evolution that proceeds by small, incremental steps.

microRNA (miRNA) A class of small, double-stranded RNA (20-24 nt) serving regulatory functions in eukarya.

micronucleus In ciliates, a small nucleus that undergoes mitosis.

microsatellite A moderately repeated eukaryal DNA sequence, 1 to 9 bp long and located in variable-length tandem repeats of 15 to 100 copies.

microtubule A cytoplasmic tube (d = 25 nm), made of α- and β-tubulin; microtubules make the mitotic spindle, maintain the cell's shape, and help organelles to move.

migration The movement of genes between populations.

minicircle A circular chromosome in the mitochondria of certain protozoans. Such a mitochondrion has $\sim 10^4$ minicircles, 0.7 to 2.5 kb long.

minisatellite A moderately repeated eukaryal DNA sequence, 10 to 70 bp long and located in variable-length tandem repeats of $\sim 10^2$ copies.

minor groove In a nucleic acid double helix, one of two channels running the length of the molecule; in B-form DNA, the narrower one.

(−) end of spindle fiber The centrosomic end of a spindle fiber.

(−) strand In a virus, a strand of RNA or DNA that is complementary to mRNA.

misexpression The synthesis of mRNA at the wrong time or in the wrong tissue.

mismatch repair A kind of DNA repair in which a mismatched base is removed from a newly synthesized strand of the double helix.

mispaired bases Noncomplementary bases opposite each other in duplex DNA.

missense mutation A mutation in a protein-coding gene, causing an amino acid substitution.

mitochondria (sing., mitochondrion) Semiautonomous organelles found in most eukarya; they are the site of oxidative metabolism.

mitogen A molecule that signals a eukaryon to make a transition from G_1 to S phase.

mitosis Nuclear division in the eukaryal cell cycle, in which chromatids segregate.

mobile element A transposon.

moderately repeated sequence A DNA sequence whose copy number in a genome ranges from $\sim 10^1$ to $\sim 10^5$ copies.

modification enzyme A bacterial enzyme that methylates nucleotides at restriction sites, rendering these sites invisible to the cell's own restriction enzymes.

modifier gene A gene with a small phenotypic effect.

modulon A large system of coordinately regulated operons in bacteria.

monocentric Of a chromosome having one localized centromere.

monoecious Referring to a species whose female and male sexual organs are housed in one individual; most commonly a plant term.

monogenic Referring to one gene; e.g., a monogenic cross involves a single gene.

monomer Consisting of a single unit; e.g., insulin is a monomer of a single polypeptide; cf. multimer.

monopartite Referring to a viral genome packed into a single chromosome: unsegmented.

monoploid In a polyploid species, one full set of chromosomes; also 1X.

monosomy Aneuploidy in which the normally diploid organism has but one copy of one of its chromosomes ($2N + 1$).

monozygotic twins Siblings arising from one fertilized egg (zygote).

mosaic A multicellular organism arising from one zygote but whose cells differ in genotype.

M phase Mitosis, usually coupled with karyokinesis and cytokinesis.

mtDNA Mitochondrial DNA.

microtubule organizing center (MTOC) The centrosome.

multigenic Referring to the joint influence of many genes on a trait; typically the effect of each gene is too small to detect.

multimer A molecule consisting of several subunits; e.g., RNA polymerase is a multimer made of several polypeptides.

multipartite Referring to a viral genome packed into two or more chromosomes: segmented.

multiple alleles Referring to a gene with three or more relatively common alleles.

multiple cloning site (MCS) A DNA sequence in a cloning vector, consisting of several restriction sites arranged in tandem.

multiplication rule The probability that two independent events X and Y will both happen, P[X and Y], is the product of their separate probabilities, P[X]•P[Y].

mutagen A substance or agent that produces mutations.

mutagenesis An experiment or procedure in which mutations are induced and mutant strains are recovered.

mutant An organism resulting from a mutation.

mutation A permanent, stably inherited alteration of a genome.

mutation rate The frequency at which mutations arise in a specific gene, in a population of individuals or cells within an individual.

natural selection Differential reproductive success among individuals.

negative assortative mating The tendency of unlike genotypes to interbreed preferentially.

negative regulation Control of gene expression by a protein that prevents the binding step of transcription.

neomorph A mutation causing a novel phenotype.

neutral mutation A mutation whose fitness is the same as that of the wild-type allele.

nick An enzyme-induced break in one strand of double stranded DNA.

noncoding sequence A sequence of DNA or RNA that does not encode a final gene product.

nondisjunction The failure of chromosomes or chromatids to separate in anaphase of meiosis or mitosis.

nonhistone protein A protein component of chromatin other than a histone.

nonreciprocal translocation An exchange between nonhomologous chromosomes, in which one chromosome donates and the other receives DNA.

nonreplicative transposition The movement of a transposon, in which the transposon is excised from one locus and inserted in another locus.

nonsense mutation A mutation that substitutes a stop codon for an amino acid-coding codon.

nontemplate strand The strand of DNA that does not act as a template for transcription. The base sequence of this strand is therefore the same as that of the transcript, except that the DNA version has thymine instead of uracil.

normal distribution A bell-shaped curve; most phenotypes are approximately normally distributed in populations.

Northern blot A procedure for identifying RNA transferred from an electrophoresis gel, by hybridizing a DNA probe to the RNA.

N terminus The amino end of a peptide or polypeptide.

nuclear envelope A double membrane enclosing a cell's nucleus.

nuclear pores Elaborately structured holes in the nuclear envelope through which molecules move between the nucleus and the cytoplasm in eukarya.

nuclease An enzyme that digests nucleic acid.

nucleic acid A polymer of nucleotides – DNA or RNA.

nucleic acid hybridization The formation of double-stranded nucleic acid from complementary single-stranded molecules of nucleic acid.

nucleoid The region of a bacterium or archaeon containing its genome.

nucleolar organizer region (NOR) A region of a eukaryal genome containing tandemly repeated copies of the large rRNA gene.

nucleolus (pl. nucleoli) An organelle inside the nucleus, where ribosomes are assembled.

nucleoside A molecule consisting of a purine or pyrimidine base linked covalently to the 1′ carbon of ribose or deoxyribose.

nucleosome The fundamental unit of chromatin, its **core** consists of a segment of DNA ($\approx$145 nt) wrapped around an octamer of histones (H2A, H2B, H3, and H4); histone H1 binds to the **linker** region (10 to 60 nt) between core nucleosomes.

nucleotide A nucleoside whose 5′ carbon is phosphorylated. There may be up to three phosphates per nucleotide, designated α, β, and γ, from proximal to distal.

nucleus The compartment of a eukaryon that contains the cell's genome.

null allele An allele from which no functional gene product is synthesized.

nullosomy A kind of aneuploidy in which one of the chromosomes is not present in the diploid phase (2N − 2).

Okazaki fragment A piece of DNA produced during replication, in the lagging strand.

oligonucleotide A strand of nucleic acid containing a few nucleotides.

oligo(T) primer An artificial primer of DNA synthesis, used in reverse transcriptase polymerase chain reaction to make DNA copies of mRNA.

oncogene A gene, gain-of-function mutations of which, or overexpression of which, contribute to tumor development.

oncogenesis The process of tumor development.

oöcyte A precursor of the egg. A primary oöcyte is the diploid cell that enters meiosis I; its haploid daughter cells are a secondary oöcyte and a polar body.

oögenesis The development of a mature egg from a primary oöcyte.

oögonium (pl., oögonia) A germ cell that gives rise to primary oöcytes by mitosis.

oötid The haploid precursor of an egg, produced by meiosis II from a secondary oöcyte.

open complex RNA polymerase bound to melted DNA upstream of a transcription start site.

operator In bacteria, a regulatory sequence of DNA near the start site of transcription, to which a repressor protein binds.

operon A set of two or more genes and regulatory sequences; the proteins encoded by an operon are coordinately expressed and have closely related functions.

open reading frame (ORF) A sequence of nucleotides in the nontemplate strand of a gene, corresponding to codons for amino acids and lacking a stop codon.

organism A life form consisting of one or more cells.

origin of replication (ori) A chromosomal locus where DNA synthesis begins.

ortholog A homologous gene or protein found in different species; e.g., the β-hemoglobin genes of human and chimpanzee are orthologous.

outcross To cross unrelated individuals, often from different populations.

overdominance A kind of quantitative dominance in which the heterozygote has a more extreme phenotype than either parent.

overwound Referring to DNA that is supercoiled in the same direction as its twist.

ovum (pl., ova) A mature egg.

pachynema (adj., pachytene) The third stage of meiotic prophase I, when homologs undergo recombination.

palindrome In double-stranded DNA, a short segment whose complementary base pairs have the same sequence if both are read $5' \rightarrow 3'$.

panmixia In a population, zygote formation in a pattern that is random with respect to gamete haplotypes; random mating is necessary but not sufficient for panmixia.

paracentric inversion A chromosome rearrangement in which a segment has reversed orientation; the breakpoints are in one chromosome arm.

paralog A homologous gene or protein found in one genome; e.g., the human genes encoding β- and δ-hemoglobin are paralogous.

parental strand The strand of chromosomal DNA that served as a template during replication.

parental type The genotype or phenotype of a parent used in a genetic cross.

p arm The smaller portion of a monocentric chromosome, beyond the centromere.

parthenogenesis An asexual mode of reproduction in which gametes are produced but fertilization does not take place.

partial dominance For a quantitative trait, dominance in which the heterozygote is intermediate between one parent and the midparent value.

patch A heteroduplex DNA segment produced by Holliday-type recombination, so that outside markers do not recombine.

pedigree A family tree using symbols to denote relationship and lineage.

P element A transposon of *Drosophila* ; it is a useful transformation vector.

penetrance The proportion of individuals with a genotype that have the phenotype expected for that genotype.

peptide A short linear polymer of amino acids.

peptide bond A covalent bond between two amino acids.

peptidyl site (P site) The part of a ribosome where a peptide bond is formed during protein synthesis.

peptidyltransferase The ribosomal enzyme that catalyzes formation of a peptide bond during protein synthesis.

peptidyl tRNA During translation, the peptidyl tRNA is attached to the carboxyl end of the growing polypeptide chain.

pericentric inversion A chromosome rearrangement in which a segment has reversed orientation; the breakpoints occur in both chromosome arms.

permissive condition An environment in which a conditional mutant has a normal phenotype.

P generation The parental generation of a controlled cross.

phage Bacteriophage: a virus that infects bacteria.

phage arms The two distal fragments of the linearized λ bacteriophage genome.

phagemid A cloning vector constructed from parts of a plasmid and a phage chromosome.

phenetic Referring to a method used to construct an evolutionary tree based on differences between the genotypes or traits of extant taxa.

phenocopy A phenotype, produced by subjecting a genotype to an abnormal environment, that mimics the phenotype normally produced in another genotype.

phenotype Trait (attribute) or set of traits of an individual or population.

phenotypic variance The population variance for a phenotype.

phenylketonuria (pku) A human genetic disease caused by recessive null mutations in the *PAH* gene, which encodes phenylalanine hydroxylase.

phosphodiester bond The covalent link between two nucleotides via a phosphate molecule in nucleic acid.

phosphorylation The covalent addition of a phosphate group to a molecule.

photolyase An enzyme that repairs cyclobutane pyrimidine dimers, using visible light as an energy source.

photoreactivation The repair of cyclobutane pyrimidine dimers by photolyase.

phylogenetic tree A bifurcated diagram connecting taxa and reflecting evolutionary history and relationships.

physical map A map of DNA sequences.

pilus (pl., pili) A thin tube projecting from the envelope of a bacterium.

plaque A clear circular area in a lawn of bacteria, where phages have lysed cells.

plasmid A relatively small, inessential chromosome; plasmids are usually circular.

plastid A chloroplast or one of its relatives.

pleiotropy The multiple phenotypic influences of a gene or genotype.

pluripotency The ability of a cell to develop into many specialized cell types.

(+) end of spindle fiber The end that attaches to a chromosome's kinetochore.

(+) strand In an RNA virus, the protein-coding strand of RNA (it acts as mRNA).

point centromere A localized centromere consisting of a few hundred base pairs of DNA.

point mutation A mutation entailing change in a single nucleotide.

polar bodies In oögenesis, the small haploid cells that do not develop into ova.

polarity Directionality; in genetics, it usually refers to the tendency of a mutation to influence expression of downstream sequences (in the 3' direction).

polar spindle fiber A spindle fiber that runs pole to pole and does not connect to a chromosome; cf. kinetochore spindle fiber.

pole One of two cellular regions during mitotic telophase, near the centrosomes.

pollen The male gamete of a flowering plant.

polyadenylation Adding a segment of repeating adenosines to the 3' end of RNA.

poly(A) polymerase (PAP) The enzyme that polyadenylates RNA.

poly(A) tail A segment of tandemly repeated adenosines added to pre-mRNA after transcription.

polycistronic mRNA An mRNA molecule that encodes two or more polypeptides.

polygenic Referring to a phenotype that is influenced by many genes, each having a small effect.

polylinker A multiple cloning site or sequence within a cloning vector that contains several restriction sites.

polymer A chain of identical or similar molecules.

polymerase chain reaction (PCR) A rapid *in vitro* method for making many copies of a segment of double-stranded DNA.

polymorphism The presence of two or more relatively common (having a frequency greater than 1%) alleles in a population.

polypeptide A large, linear polymer of amino acids.

polyploid Referring to a eukaryon having three or more copies of a genome; triploids have three copies, tetraploids have four copies, etc.

polysome An mRNA molecule being translated simultaneously by several ribosomes.

polytene Referring to a giant chromosome consisting of many identical copies synapsed point for point. Polyteny is a special case of polyploidy.

population A group of conspecific individuals. In sexual species, a group of potentially interbreeding individuals.

positive assortative mating The tendency of similar genotypes to interbreed preferentially.

positive regulation The control of gene expression by an activator protein.

posttranscriptional splicing The joining of exons after intron removal.

posttranslational splicing The joining of exteins after intein removal.

precise excision The exactly correct removal of a transposon from some locus, yielding a wild-type reversion.

pre-mRNA A transcript whose final, processed product will be mRNA.

prey The protein, in a yeast two-hybrid screen, chosen to bind to unknown "bait" proteins.

primary gene product Freshly transcribed, unprocessed RNA.

primary sex determination In vertebrates, the development of ovary or testis from the undifferentiated gonad.

primary structure In a protein, the sequence of amino acids in a polypeptide.

primase An enzyme that synthesizes the primer used in DNA replication.

primer An oligonucleotide, complementary to a template strand of DNA, to which DNA polymerase adds nucleotides.

primosome A protein complex, including primase, that synthesizes a primer for DNA replication.

prion A proteinaceous and infectious particle that causes rare neurodegenerative diseases in mammals. Prion protein is **PrP** , existing in a normal cellular form, PrP^C, and in an abnormal, scrapie form, PrP^{Sc}.

probability A number between 0 and 1 that predicts the frequency of an event.

proband A subject of a human genetic study.

probe A labeled fragment of DNA used to locate nucleic acid of interest by hybridization.

programmed cell death A cell's self-destruction via a genetically encoded plan.

programmed frameshifting Rare single-nucleotide shifts in reading frame during translation, programmed by interactions between mRNA and the ribosome.

prometaphase The part of mitosis immediately following prophase; the spindle forms during prometaphase.

promoter A short segment of DNA to which RNA polymerase binds; it is located near the start site of transcription.

promoter strength The binding affinity of a promoter for RNA polymerase.

proofreading The error-reducing activity of DNA polymerase, which corrects mispaired nucleotides during replication.

prophage The chromosome of a temperate phage, integrated into the host bacterium's main chromosome. Prophage **induction** is the deintegration of a prophage, leading to lytic reproduction.

prophase The first stage of mitosis or meiosis, when chromosomes become visible under the light microscope.

propositus (f., proposita) Proband.

proteome All the proteins encoded by a genome.

prototroph A wild-type cell not requiring nutritional supplements.

pseudoautosomal gene A gene located on both sex chromosomes (e.g., X and Y).

pseudogene A segment of DNA that is homologous to a gene but incapable of expression.

purine base A double-ring basic compound present in nucleic acid.

pyrimidine base A single-ring basic compound present in nucleic acid.

pyrimidine dimer A compound produced by the formation of covalent bonds between pyrimidine bases in adjacent nucleotides of a strand of DNA.

q arm The larger portion of a monocentric chromosome, centromere to telomere.

quantitative dominance Genetic dominance that applies to continuously varying traits.

quantitative trait A continuously varying trait, measured on a scale; e.g., length.

quantitative trait locus (QTL) A genetic locus to which a small effect, for a polygenic trait, has mapped.

quaternary structure In a protein, the aggregation of two or more polypeptides.

R banding A pattern of chromosome staining produced with Giemsa; R bands are the reverse of G bands.

reading frame The phase of mRNA during translation; mRNA has three possible reading frames, only one of which is used in a ribosome.

receptor A protein that binds specifically to a molecule; as a result of binding, it initiates a particular cellular response.

receptor tyrosine kinase (RTK) A eukaryal protein that spans the cell membrane; a mitogen binds to the exterior part and the interior part phosphorylates tyrosine residues of itself, thereby activating a signal transduction pathway.

recessive Referring to the propensity of an allele's phenotypic effects to be hidden by another (dominant) allele, in a heterozygote.

reciprocal cross A breeding experiment between two strains (say A and B), with a balanced design: A♀ × B ♂ and B♀ × A ♂ [♀ = female, ♂ = male].

reciprocal translocation An exchange of material between two non-homologous chromosomes, each chromosome being donor and recipient.

recombinase An enzyme that directs recombination between two DNA molecules.

recombination The process of shuffling DNA sequences from two (usually homologous) DNA molecules. The resulting DNA molecules are **recombinant** .

recombination fraction The fraction of chromosomes that are recombinant, in an experiment to measure recombination rate.

recombination repair The repair of damaged DNA after replication, using recombination between newly synthesized gapped strands and parental strands.

reductional division The meiotic division in which homologs segregate; meiosis I is normally reductional.

redundancy A property of the genetic code, wherein an amino acid may have more than one codon. Some biologists call this "degeneracy."

regional centromere A localized centromere consisting of 1 to 10^3 kb of DNA.

regression A statistical procedure useful in estimating heritability, that estimates a functional relationship between variables.

regulatory vector A cloning vector that contains a regulatory gene such as *lacI*.

regulatory gene A gene that encodes an activator or repressor.

regulon A small system of coordinately regulated bacterial operons.

relaxed DNA Duplex DNA that is neither overwound nor underwound.

RF 1. Release factor, a protein that causes the ribosome to dissociate from mRNA at the end of translation. 2. Replicating form, the double-stranded DNA made as an intermediate molecule in the replication of single-strand DNA phages.

remodeling complex A large protein multimer, essential to gene expression in eukarya, that alters nucleosome structure.

replica plating A method of inoculating several microbial plates in an identical spatial pattern, reproducing the pattern of clones growing on a master plate.

replicase A viral RNA-dependent RNA polymerase, which synthesizes an RNA chromosome.

replicating form (RF) A double-stranded nucleic acid molecule synthesized during the replication of some viruses with single-stranded chromosomes.

replication The synthesis of two daughter chromosomes from one chromosome.

replication fork The Y-shaped arrangement of partially unwound DNA during replication.

replicative transposition The movement of a transposon, in which the old, resident copy remains at its locus, and a newly synthesized copy jumps to a new locus.

replicon A chromosomal region that is synthesized simultaneously with other such regions during chromosome replication.

replisome A large protein multimer of enzymes used to synthesize cellular chromosomal DNA.

reporter gene A gene that is included in a transfecting vector, used to tag transfected cells; these cells are identified by the reporter gene's product.

repressible regulation A type of control of gene expression in which a small molecule, the co-repressor, is necessary for switching off transcription.

repressor A protein that switches off transcription by binding to DNA or to positive transcription factors.

reproductive isolation The inability of two species to reproduce with each other.

response element A regulatory DNA sequence (enhancer) to which hormone-bearing proteins bind.

response to selection The difference between parental mean and progeny mean in a selection experiment.

restriction endonuclease An enzyme that binds to a specific sequence of duplex DNA and cuts both strands within that sequence. The resulting pieces are **restriction fragments**.

restriction fragment length polymorphism (RFLP) A genetic polymorphism for the binding site of a restriction endonuclease, so that the enzyme cuts one allelic form but not the other and thereby produces two short fragments versus one long one.

restriction site A binding and cutting site for a restriction endonuclease.

restrictive condition An environment in which a conditional mutant expresses the mutant (abnormal) phenotype.

retinoblastoma A rare genetic disease whose first sign in early childhood is growth of tumors in the retina.

retrogene A pseudogene that lacks introns and can move from locus to locus by retrotransposition using an external source of transposase.

retrotransposon A transposon that moves through the synthesis of an RNA intermediate.

retrovirus An RNA virus of eukarya, which synthesizes a double-stranded DNA copy of its chromosome, using a reverse transcriptase.

reverse transcriptase An enzyme that synthesizes DNA, using an RNA template.

reverse transcriptase PCR (RT-PCR) An *in vitro* method of synthesizing DNA from mRNA and then amplifying some of the DNA by PCR.

reversion Mutation of an abnormal allele back to the wild-type allele, which is termed **revertant**.

rho-dependent termination In bacteria, termination of translation caused by interaction between a small protein, rho (ρ), and the ribosome.

ribonucleic acid (RNA) Nucleic acid made from nucleotides containing ribose. It may be single-stranded or double-stranded.

ribose A five-carbon sugar that makes part of an RNA nucleotide.

ribosomal RNA (rRNA) A kind of RNA, a major component of the ribosome.

ribosome A structure made of rRNA and proteins, where translation takes place.

ribozyme An enzyme made of RNA.

RNA helicase An enzyme that unwinds RNA duplexes.

RNA polymerase An enzyme that synthesizes RNA from a DNA template. In the nuclei of eukarya, there are three RNA polymerases:

RNA polymerase I–transcribes the large pre-rRNAs.

RNA polymerase II–transcribes pre-mRNA and some pre-snRNAs.

RNA polymerase III–transcribes pre-tRNAs and other small RNAs.

RNA polymerase holoenzyme The version of RNA polymerase capable of full transcription in cells.

RNA processing Posttranscriptional modification of RNA; e.g., base modification, splicing, and polyadenylation.

RNA surveillance Degradation of aberrant mRNAs with internal nonsense codons.

rolling-circle replication A mechanism for replicating a circular chromosome in which the leading strand uses one unbroken strand of the entire chromosome as a template.

R point (restriction point) The critical point in G_1 phase; the R point must be passed for the cell to enter S phase.

sample mean and sample variance Statistics calculated from samples; they estimate the population mean and variance.

satellite DNA Highly repeated sequences of DNA in a eukaryal genome, differing in buoyant density from the genomic average.

scaffold In interphase eukarya, a network of proteins in the nucleus, to which chromatin fibers attach.

screen A procedure for selecting organisms with certain phenotypes or genotypes.

secondary sex determination In vertebrates, the development of sexual traits other than gonad type.

secondary structure Shapes assumed by strands of a polypeptide, mainly α-helices and β-ribbons.

segmented viral genome A virus genome with two or more chromosomes.

segregation 1. The separation of alleles or homologs in meiosis and distribution to daughter cells. 2. The separation of chromosomes during cell division.

selectable marker A gene used to separate one type of cell from others by virtue of differential survival or observable phenotype.

selection An evolutionary force that changes allele frequencies as a result of differential survival or reproduction of certain genotypes.

selection coefficient A quantity that measures how much the fitness of one genotype is reduced or increased relative to the fitness of another genotype.

selection differential The difference between the parental generation's mean and the mean of the parents selected in that generation in a selection experiment.

selective medium A growth medium that allows the growth of some cells but not others.

selfish genetic element A gene, chromosome, or intracellular parasite that interferes with normal genetic processes, to the detriment of the element's host and to the reproductive benefit of the element.

semiconservative Referring to DNA replication in which each of the two daughter molecules contains one old and one new strand.

semidiscontinuous replication DNA replication in which one strand is synthesized continuously from a single primer, and the other strand is synthesized discontinuously by using many primers.

sense codon A triplet of nucleotides in mRNA complementary to the anticodon of a tRNA; there are 61 sense codons and three nonsense (stop) codons.

separin A protease that unglues chromatids during anaphase A of mitosis.

sex chromosome One of two chromosomes in a dioecious species, for which females and males differ; in humans, the X and Y chromosomes.

sex pilus (pl., pili) A slender, protein tube made by F^+ E. coli cells; it attaches F^+ and F^- cells as a first step in conjugation.

sexual generation One full turn of a sexual life cycle, e.g., from egg to egg.

sexual reproduction The production of offspring from the union of haploid gametes of two mating types; the life cycle alternates between haploid and diploid.

sexual selection Differential reproduction of genotypes within one sex through the agency of the other sex.

shotgun sequencing A method of determining the DNA sequence of an entire genome by cutting cloned fragments into a random collection of smaller, overlapping fragments for sequencing.

sigma (σ) A promoter-binding subunit of bacterial RNA polymerase.

signal sequence A short segment at the N terminus of a polypeptide, which tags the polypeptide for secretion.

signal transduction pathway A sequence of cellular events, featuring the phosphorylation of proteins. Binding of an extracellular ligand to a receptor triggers the pathway and results in the activation of one or more transcription factors.

silencer A short DNA sequence in a eukaryal genome, to which negative regulatory proteins bind, thereby switching off gene expression.

silent mutation A mutation that does not result in an amino acid substitution.

simple dominance The dominance of one allele over another for dichotomous traits.

SINE (short interspersed nuclear element) A kind of repeated DNA sequence that can be a retrotransposon; SINEs lack long terminal repeats; cf. LINE.

single-nucleotide polymorphism (SNP) A genetic polymorphism resulting from a single transition or transversion.

sister chromatid One of two identical, connected copies of a eukaryal chromosome. The term does not apply to separated copies.

site-specific recombination Exchange between two DNA molecules at a specific site (sequence); the two molecules need have no other homology.

6,4 pyrimidine dimer A pyrimidine dimer with a covalent bond between the 6 carbon of one base and the 4 carbon of the other.

small interfering RNA (siRNA) A class of small, double-stranded RNA (20–22 nt) that regulate mRNAs.

snRNA Small nuclear RNA, including uRNAs, which never leave the nucleus.

snRNP A ribonuclear particle, consisting of small nuclear RNA and protein.

soma Collectively, animal cells that make body parts other than gametes.

SOS response A response of bacteria to massive DNA damage: repair enzymes are induced and DNA replication becomes error-prone.

Southern blot A method of transferring denatured DNA from a gel to a membrane and then hybridizing a labeled DNA probe to the blot-transferred DNA.

spacer region The sequence within a bacterial promoter between −35 and −10.

specialized transduction The transfer of one bacterium's DNA to another bacterium by a lysogenic phage; the capsid contains both phage and host DNA.

speciation genes Genes that underlie speciation; e.g., genes that affect reproductive isolation between species.

species Two sexually reproducing organisms are species if they do not reproduce with each other in nature. Other criteria are sometimes used to define species.

speciation The process by which one species splits into two.

sperm The male gamete, usually a motile, flagellated cell.

spermatid A haploid precursor of sperm; in meiosis II, a secondary spermatocyte gives rise to two spermatids.

spermatocyte A precursor of the sperm. A primary spermatocyte is the diploid cell that enters meiosis I; its haploid daughter cells are secondary spermatocytes.

spermatogenesis The development of mature sperm from primary spermatocytes.

spermatogonium (pl., spermatogonia) A germ cell that gives rise to primary spermatocytes by mitosis.

sperm nucleus In a pollen tube, one of two gametic nuclei that fuse with egg nuclei to make the diploid embryo nucleus and the triploid endosperm nucleus.

S phase The part of the eukaryal cell cycle during which chromosomes are replicated.

spindle In M phase or in meiosis, the apparatus that moves chromosomes; the spindle is made of microtubules (spindle fibers).

spindle checkpoint An arrest of M phase until the spindle is properly attached to all chromosomes.

spindle fiber A microtubule consisting primarily of α- and β-tubulin; part of the spindle.

splice 1. A heteroduplex DNA segment produced by Holliday-type recombination, so that outside markers recombine. 2. To join two exons of a pre-mRNA molecule after intron removal.

splice acceptor The sequence at the 5′ end of an exon.

splice donor The sequence at the 3′ end of an exon.

spliceosome A large particle, made of proteins and RNA, that splices pre-mRNA.

spore A specialized resting cell. In many organisms, spores are remarkable for their resistance to environmental extremes, including high-energy radiation.

spore 1. In bacteria, a specialize resting cell. 2. In plants, a reproductive cell.

sporophyte The diploid, spore-forming structure in a plant; also, the diploid phase of a plant life cycle.

ssDNA-binding protein A small protein that binds to single-stranded DNA during replication; it protects against attack by nucleases.

stable RNA rRNA and tRNA, which have long half-lives owing to slow rates of degradation.

stabilizing selection Natural selection favoring phenotypes near the population mean.

standard normal score An individual's normalized phenotypic value, $z = (x_i - \mu)/\sigma$.

start codon A triplet of nucleotides in mRNA, near the 5′ end, which codes for the first amino acid in a polypeptide; the sequence may be AUG or GUG.

statistic An empirically determined number that estimates a population parameter.

static equilibrium The tendency of allele frequencies to remain constant in a large population in which allele frequencies are the same in females and males, no forces act to change allele frequencies, and gametes unite independently of their haplotypes.

stem cell Any cell that gives rise to many differentiated cell types.

sticky ends Single-stranded ends of a piece of DNA produced when a restriction enzyme makes staggered cuts in double-stranded DNA.

stop codon A triplet of nucleotides in mRNA, which does not encode an amino acid and therefore signals the termination of translation. With few exceptions, stop codons are UAG, UGA, and UAA.

stringent control Regulation of the synthesis of stable RNAs in response to amino acid starvation.

strong promoter A promoter to which RNA polymerase readily binds; consequently, the gene is transcribed at a high rate.

structural gene A gene encoding anything except a gene-regulating protein.

sequence-tagged site (STS) A position in the genome that is identified by a unique DNA marker.

subclone (v.) To clone a DNA fragment cut from a larger cloned piece.

submetacentric chromosome A eukaryal chromosome, one of whose arms is somewhat longer than the other.

substitution Mutation in which one nucleotide is replaced by another.

substitution rate During evolution, the speed at which one nucleotide or amino acid is replaced by another.

sugar-phosphate backbone The part of a nucleic acid polymer exclusive of bases.

suicide enzyme An enzyme that can act only once, e.g., Ada methyltransferase in *Escherichia coli*.

supercoiling Overwinding or underwinding of the DNA double helix.

suppressor 1. A protein that prevents the binding step or the initiation step of transcription. 2. A gene whose product downregulates another gene.

synapsis The process of chromosome pairing, in which homologous chromosomes pair point for point. Synapsis occurs in zygonema of meiosis.

synaptonemal complex A protein structure that holds together synapsed chromosomes in zygonema.

syncitium A multinucleate cell, e.g., the *Drosophila* blastoderm.

syngamy During fertilization in sexual reproduction, the union of two gametes and the fusion of their nuclei, which produces a diploid zygote.

synonymous codon or substitution A nucleotide substitution within a codon, a mutation that does not result in an amino acid substitution.

systematic sequencing A method of determining the DNA sequence of an entire genome by sequencing overlapping segments within every cloned piece of DNA.

t A statistical distribution, useful in constructing confidence intervals for means and in testing hypotheses about mean differences.

tandem duplication A mutation in which two exactly repeated sequences are adjacent.

targeted gene replacement A method of making transgenic organisms in which the introduced gene replaces the genomic version by recombination.

TATA-associated factor (TAF) A kind of transcription factor for RNA polymerase II; TAFs bind to TATA-binding protein and other transcription factors.

TATA binding protein (TBP) A transcription factor that binds to the TATA box.

TATA box A short segment of DNA in the promoter of a class II eukaryal gene, located at about -25; the consensus sequence is $5'TATAAA3'$.

taxon (pl. taxa) A group of organisms representing a single evolutionary lineage.

T-DNA A 23-kb transposon within the Ti plasmid of *Agrobacterium tumefaciens* ; vectors for plant transfection are constructed from T-DNA.

telomerase An enzyme that lengthens eukaryal telomeres; without telomerase, linear chromosomes would erode at both telomeres.

telomere A structure at the end of a linear (noncircular) chromosome.

telophase The part of mitosis or meiosis that follows anaphase. During telophase, chromosomes are pulled to the poles by the spindle.

temperate phage A bacteriophage capable of forming a lysogen by integrating the phage chromosome into the host chromosome.

temperature-sensitive mutation A mutation whose phenotype is abnormal when the temperature is in a certain range but otherwise is normal.

template strand A strand of nucleic acid that is copied by RNA polymerase during transcription.

terminase A phage protein that cuts chromosomes and helps to package them.

termination step The final step of transcription or translation, stopping the process.

terminator A sequence at the 3' end of a gene, signaling termination of transcription.

tertiary structure In a protein, the folding pattern of secondary shapes.

testcross A genetic experiment in which a heterozygote is back-crossed to the recessive homozygote.

testis-determining factor (Tdf) A protein that signals the gonad of mammals to develop into a testis; the *Tdf* gene is Y-linked.

testosterone A male sex hormone of vertebrates; testosterone induces genes that promote male traits.

tetrad The four strands making up a pair of homologous chromosomes in prophase and metaphase of meiosis I.

theta replication Bidirectional synthesis of a circular dsDNA chromosome, beginning at the origin and terminating 180° from the origin.

3′ end The end of a strand of RNA or DNA whose final nucleotide terminates in a hydroxyl group on the 3′ carbon.

3′ untranslated region The segment of mRNA containing the 3′ end and in the 3′ direction from the stop codon(s).

transposition 1. The movement of a transposon from one locus to another. 2. The translocation of a chromosome segment into the interior of another, nonhomologous chromosome.

thymine (T) A pyrimidine base commonly found in DNA.

Ti plasmid A 200-kb plasmid of *Agrobacterium tumefaciens*.

totipotency The ability of a cell to develop into a reproductively competent multicellular organism.

trans-acting Of regulatory genes whose protein products diffuse in the cell and can bind to DNA of other chromosomes.

trans (repulsion) configuration The arrangement of genotypes on two copies of a chromosome, in which each chromosome has a mutation: $a\ b^+//a^+\ b$.

transcriptase A viral RNA-dependent RNA polymerase.

transcription RNA synthesis from a DNA template, or in some RNA viruses, from an RNA template. The newly synthesized RNA molecule is a **transcript**.

transcription factor A protein that stimulates transcription by binding to DNA or to RNA polymerase.

transduction The transfer of genomic DNA from one cell to another via a virus.

transfection The alteration of an organism's genotype by foreign DNA introduced via a vector.

transformation 1. The alteration of a cell's genome by acquiring exogenous DNA directly from the surrounding medium. 2. In mammalian cells, the process of becoming cancerous.

transgene A gene that has been modified *in vitro* and introduced into an organism's genome; often, the donor and recipient are of different taxa. The recipient is a therefore transgenic.

transient gene expression Short-term expression of transgenes introduced into somatic cells, often cultured somatic cells.

transition A mutation in which a purine substitutes for the other purine, or a pyrimidine for the other pyrimidine.

translation Synthesis of a polypeptide in the ribosome.

translocation A chromosome mutation in which DNA moves from one chromosome to a nonhomologous chromosome.

transposable element Transposon.

transposase An enzyme that catalyzes the recombination reactions in the movement of a transposon from one genomic locus to another.

transposon A segment of DNA, capable of moving locations within a genome.

transposon tagging Mutagenesis of a gene by insertion of a transposon into (or near) that gene; the transposon carries a genetic marker.

transversion A substitution mutation, either a purine for a pyrimidine, or vice versa.

trisomy Aneuploidy in which the normally diploid organism has three copies of one of its chromosomes $(2N + 1)$.

trivalent A grouping of three chromosomes formed during zygonema in a triploid organism.

transfer RNA (tRNA) A kind of RNA that carries an amino acid into the ribosome.

true-breeding Homozygous.

t test A statistical test comparing two means.

tubulin A family of proteins that makes up microtubules.

tumor A clone of abnormal cells growing in an uncontrolled manner within a multicellular organism.

tumor suppressor A gene that encodes a component of cell-cycle control or of the apoptosis machinery. Homozygosity for a loss-of-function mutation in a tumor suppressor contributes to tumor development.

twist The normal helical turning of relaxed double-stranded DNA.

two-hybrid screen A method to find genes encoding proteins that bind to a protein of interest, most commonly using genetically engineered strains of yeast.

tyrosine kinase An enzyme that phosphorylates certain tyrosine residues of proteins, thereby changing their activity.

ubiquitin A small, highly conserved protein of eukarya; ubiquitin attaches covalently to other proteins, to regulate their activity or to tag them for degradation.

underwound Referring to supercoiling of DNA in the direction opposite to its twist.

unequal crossing-over An asymmetric meiotic chromosomal exchange in which one recombinant homolog has a duplication and the other has a deficiency.

unique sequence A DNA sequence found once per haploid genome.

univalent An unpaired chromosome in meiosis – for example, produced in an aneuploid organism.

unlinked Referring to two genes the assort independently.

unsegmented viral genome A viral genome packed into one chromosome.

unstable RNA Any quickly degraded RNA, mainly mRNA.

upstream In a strand of RNA or DNA, in the 5′ direction.

upstream activating sequence (UAS) A regulatory DNA sequence upstream (in the 5′ direction) of a promoter.

uracil (U) A pyrimidine base commonly in RNA.

uRNA A class of small nuclear RNAs that help in spliceosomic splicing of mRNA.

UVA, UVB, and UVC Wavelength bands of ultraviolet light; UVA = 315 to 400 nm, UVB = 280 to 315 nm, and UVC = 100 to 280 nm; in nature, UVB makes most pyrimidine dimers.

USS (uptake signal sequence) A short DNA sequence scattered throughout the genomes of certain bacteria and required for transformation in those species.

variable expressivity Phenotypic variation among individuals of one genotype.

variable region A domain of an immunoglobin polypeptide that shows huge sequence diversity among cells; it binds to antigens.

variance A number that measures variation in a population for some random variable; it is the average of squared differences between variates and the mean.

variegation A random patchwork of phenotypes within a plant; e.g., a leaf may have patches of green and white, depending on chloroplast functioning.

variety In plants, a strain or genetic type within a species, usually artificially produced or selected.

vector A piece of DNA used to carry a transgene into a cell.

vertical gene transfer The transfer of DNA from an organism to its progeny, for example, from mother cell to daughter cell.

virion A complete virus particle.

viroid An acellular, parasitic life form consisting of small circular RNA molecules that do not encode protein; viroids infect plants.

virulent phage A bacteriophage that kills the host and never makes a prophage.

virus An obligate intracellular parasite consisting of one or more chromosomes and a protein capsid; some viruses also have a membrane borrowed from the host.

virus envelope A membrane surrounding the capsid of some viruses, as in HIV.

virusoid An acellular, parasitic life form consisting of a circular RNA chromosome and lacking a protein coat; virusoids infect plants.

W chromosome In species whose males are homogametic (e.g., chickens), the sex chromosome found in females: male = ZZ, female = ZW.

weak allele A recessive allele whose phenotype is weaker than wild type; also called a hypomorph.

weak promoter A promoter for which RNA polymerase has a weak binding affinity; consequently, the gene is transcribed at a low rate.

wild type The most common phenotype or genotype in a species for any gene that is not polymorphic.

wobble Alternative base-pairing between codon and anticodon, at position 3 of the codon.

writhe An integer that measures supercoiling in a piece of DNA.

X chromosome In species whose females are homogametic, the sex chromosome that is present in both sexes; female = XX, male = XY.

X-gal A colorless artificial substrate for β-galactosidase, which converts X-gal to a blue product.

X inactivation Turning off transcription of genes on an X chromosome.

X-linked Located on the X chromosome.

YAC (yeast artificial chromosome) A cloning vector of yeast, capable of mitotic inheritance and carrying a very large insert, $\sim 10^3$ kb.

Y chromosome In species whose females are homogametic, the sex chromosome that is present in males: female = XX, male = XY.

Z chromosome In species whose males are homogametic, the sex chromosome that is present in both sexes; male = ZZ, female = ZW.

Z form A rare, left-handed configuration of a DNA double helix, containing a zigzag arrangement.

Z ring A ring of FtsZ protein that forms at the equator of a dividing bacterium.

zygonema (adj., zygotene) The second stage of prophase I of meiosis, when homologs synapse.

zygotic gene A gene that is expressed in the zygote as opposed to the oocyte.

zygote In sexually reproducing eukarya, a diploid cell resulting from syngamy of two haploid gametes.

Index